APPLICATIONS OF DERIVATIVES MADE EASY

For School Students of Various Boards
Classes—XI and XII

OTHER BOOKS IN MADE EASY SERIES BY THE SAME AUTHOR

1. Limits and Derivatives Made Easy
2. Definite Integral Made Easy
3. Indefinite Integral Made Easy
4. Co-ordinate Geometry Made Easy
5. Complex Numbers Made Easy

APPLICATIONS OF DERIVATIVES MADE EASY

For School Students of Various Boards
Classes—XI and XII

By
DEEPAK BHARDWAJ
M.Sc., M.Phil. (Mathematics)
Head of Deptt. Mathematics
Parth Public Sr. Sec. School, Gharaunda
Karnal (Haryana)

FIREWALL MEDIA
(An Imprint of Laxmi Publications Pvt. Ltd.)
An ISO 9001:2015 Company
BENGALURU • CHENNAI • GUWAHATI • HYDERABAD
KOCHI • KOLKATA • LUCKNOW • RANCHI
NEW DELHI

Printed and bound in India

First Edition : 2007, Reprint : 2009, New Edition : 2025

ISBN : 978-81-318-0652-4

Branches

Bengaluru	080-26 75 69 30
Chennai	044-24 34 47 26
Guwahati	0361-254 36 69
Hyderabad	040-27 55 53 83
Kochi	0484-405 13 03
Kolkata	033-40 04 77 79
Lucknow	0522-430 36 13
Ranchi	0651-224 24 64

PUBLISHED IN INDIA BY

FIREWALL MEDIA
(An Imprint of Laxmi Publications Pvt. Ltd.)

An ISO 9001:2015 Company
113, GOLDEN HOUSE, GURUDWARA ROAD, DARYAGANJ,
NEW DELHI - 110002, INDIA

Corporate Office : B-33, Sector-5, Noida,
Gautambuddha Nagar (U.P.) - 201 301, INDIA
Telephone : +91-120-44 46 900
www.laxmipublications.com
info@laxmipublications.com

PREFACE

This book on Applications of Derivatives is designed for students of schools and colleges. The aim of the book is to present the subject in such a way that even an average student can feel no difficulty in understanding it.

There are, of course, a number of books available on the subject, which are too concise to be intelligible for the students. There is hardly any single book which covers the entire syllabus of Applications of Derivatives prescribed in various state boards.

Every effort has been made to present the subject matter in the most lucid style. The following are some of the salient features of this book :

(*i*) The book serves the purpose of a text as well as a help book.

(*ii*) A large number of typical and important solved examples have been provided to enable the student to have a clear grasp of the subject and to equip him for attempting problems in the board examination without any difficulty.

(*iii*) Care has been taken to grade the problems in such a way that students move from basic intricate problems with ease.

(*iv*) The 'Remarks and Notes' have been added quite often in the book so that they should help in understanding the ideas better.

(*v*) At the end of each chapter, a 'Revision Exercise' has been incorporated. The purpose of this exercise is to give an opportunity to students to revise the entire chapter in the minimum possible time.

It is hoped with all the above characteristics the book will be found really useful by teachers and students alike.

A serious effort has been made to keep the book free from errors, but even then some errors might have crept in.

I am grateful to M/s. Laxmi Publications (P) Ltd., New Delhi, for the keen interest and active co-operation.

Suggestions from the teachers as well as the students for further improvement will be warmly welcomed.

—Author

CONTENTS

SYMBOLS

Greek Alphabets

A	α	Alpha	I	ι	Iota	P	ρ	Rho
B	β	Beta	K	κ	Kappa	Σ	σ	Sigma
Γ	γ	Gamma	Λ	λ	Lambda	T	τ	Tau
D	δ	Delta	M	μ	Mu	Y	υ	Upsilon
E	ε	Epsilon	N	ν	Nu	Φ	φ	Phi
Z	ζ	Zeta	Ξ	ξ	Xi	X	χ	Chi
H	η	Eta	O	o	Omicron	Ψ	ψ	Psi
Θ	θ	Theta	Π	π	Pi	Ω	ω	Omega
	∃	there exists		∀	for all			

Metric Weights and Measures

LENGTH

10 millimetres	= 1 centimetre
10 centimetres	= 1 decimetre
10 decimetres	= 1 metre
10 metres	= 1 decametre
10 dekametres	= 1 hectometre
10 hectometres	= 1 kilometre

CAPACITY

10 millilitres	= 1 centilitre
10 centilitres	= 1 decilitre
10 decilitres	= 1 litre
10 litre	= 1 decalitre
10 dekalitres	= 1 hectolitre
10 hectolitres	= 1 kilolitre

VOLUME

1000 cubic centimetres	= 1 centigram
1000 cubic decimetres	= 1 cubic metre

AREA

100 square metres	= 1 are
100 ares	= 1 hectare
100 hectares	= 1 square kilometre

WEIGHT

10 milligrams	= 1 centigram
10 centigrams	= 1 decigram
10 decigrams	= 1 gram
10 grams	= 1 decagram
10 decagrams	= 1 hectogram
10 hectograms	= 1 kilogram
100 kilograms	= 1 quintal
10 quintals	= 1 metric ton (tonne)

ABBREVIATIONS

kilometre	km	tonne	t
metre	m	quintal	q
centimetre	cm	kilogram	kg
millimetre	mm	gram	g
kilolitre	kl	are	a
litre	l	hectare	ha
millilitre	ml	centiare	ca

1

Rate of Change of Quantities

1.1 INTRODUCTION

We have already discussed the methods of evaluating derivatives of various types of functions. The theory of derivatives is very useful. It has various dimensions. Derivative has many applications in engineering, science, social and behaviour sciences, information science and many other fields. In the present chapter, we learn how the derivative can be used to determine the rates of change of quantities.

1.2 RATE OF CHANGE OF QUANTITIES

Let x and y be any variables and y be a function of x.

i.e.,
$$y = f(x)$$

Then, $\frac{dy}{dx}$ denotes the rate of change of y with respect to x.

If x and y are differentiable functions of parameter t, then we have

$$\frac{dy}{dt} = \frac{dy}{dx} \cdot \frac{dx}{dt}$$

Thus, we see that, if the time rate of change of one of the variables x and y is given, then the equation $\frac{dy}{dt} = \frac{dy}{dx} \cdot \frac{dx}{dt}$ gives the time rate of change of the other variable.

Remark. *(i)* The value of $\frac{dy}{dx}$ at $x = x_0$ *i.e.,* $\left(\frac{dy}{dx}\right)_{x=x_0}$ represents the rate of change of y w.r.t. x at $x = x_0$.

(ii) Throughout this chapter, by rate of change we mean the instantaneous rate of change irrespective of the fact that the change is with respect to time or not.

(iii) If the rate of change of a variable is positive, then the value of the variable increases with the increase in the value of the independent variable.

(iv) If the rate of change of a variable is negative, then the value of the variable decreases with the increase in the value of the independent variable.

1.3 SOME USEFUL RESULTS

The following formulae of mensuration will be useful in solving the applied problems in the present chapter :

(*i*) Area of an equilateral triangle = $\frac{\sqrt{3}}{4}a^2$

where $a \rightarrow$ side of triangle.

(*ii*) For a rectangle of side x and y :

Area = xy, Perimeter = $2(x + y)$.

(*iii*) For a square of side x :

Area = x^2, Perimeter = $4x$.

(*iv*) For a circle of radius r :

Area = πr^2, circumference = $2\pi r$.

(*v*) The surface area (S) and volume (V) of a sphere of radius r is given by :

$$S = 4\pi r^2, \qquad V = \frac{4}{3}\pi r^3.$$

(*vi*) For a right circular cylinder of radius r and height h :

Volume = $\pi r^2 h$, Curved surface area = $2\pi rh$.

Total surface area = $2\pi r(h + r)$ [For closed cylinder]

Total surface area = $\pi r(2h + r)$ [For open cylinder]

(*vii*) For a right circular cone of radius r and height h :

$$\text{Volume} = \frac{1}{3}\pi r^2 h$$

Curved surface area = πrl.

where $l = \sqrt{r^2 + h^2}$ is the slant height of cone.

Total surface area = $\pi r(l + r)$.

(*viii*) For a cube of edge of length : a

Volume = a^3, Surface area = $6a^2$.

(*ix*) For a cuboid of edges of lengths a, b and c.

Volume = abc

Surface area = $2(ab + bc + ca)$.

SOME SOLVED EXAMPLES

Example 1. (*i*) *Find the rate of change of area of a circle of radius r, when the radius varies.*

(*ii*) *Find the rate of change of the area of a circle w.r.t. its radius r, when r = 4 cm.*

(*iii*) *The radius of a circle is increasing at the rate of 0.7 cm/sec. What the rate of increase of its circumference.*

(*iv*) *The radius of the circle is increasing uniformly at the rate of 3 cm/sec. At what rate is the area increasing when the radius is 10 cm.*

(*v*) *The radius of a circle is increasing uniformly at the rate of 4 cm/sec. Find the rate at which the area of the circle is increasing when the radius is 8 cm.*

Solution. (*i*) Let A be the area of a circle of radius r.

$$\therefore \qquad A = \pi r^2 \qquad \text{...(1)}$$

Differentiating both sides w.r.t. r, we get

$$\frac{dA}{dr} = 2\pi r$$

$\therefore$ The rate of change of the area of the circle $= 2\pi r$.

(*ii*) Proceed as in part (*i*) :

$$\frac{dA}{dr} = 2\pi r$$

When $r = 4$ cm

$$\therefore \quad \left(\frac{dA}{dr}\right)_{r=4} = 2\pi(4) = 8\pi \text{ cm}.$$

$\therefore$ The rate of change of the area of the circle at $r = 4$ cm is 8π cm.

(*iii*) Let r be the radius of the circle and C be the circumference at any time t.

$$\therefore \quad C = 2\pi r \quad ...(1)$$

$\therefore$ Rate of change of radius w.r.t. $t = \frac{dr}{dt} = 0.7$ cm/sec. ...(2) [Given]

Differentiating (1) w.r.t. t, we get

$$\frac{dC}{dt} = \frac{d}{dt}(2\pi r) = 2\pi \, . \frac{dr}{dt}$$

$$= 2\pi(0.7) = 1.4\pi \text{ cm/sec.} \quad [\because \text{ By using (2)}]$$

$\therefore$ Rate of change of circumference is 1.4π cm/sec.

(*iv*) Let r be the radius of the circle and A be the area at any time t.

$$\therefore \quad A = \pi r^2 \quad ...(1)$$

$\therefore$ Rate of change of radius w.r.t. $t = \left(\frac{dr}{dt}\right) = 3$ cm/sec. ...(2) [Given]

Differentiating (1) w.r.t. t, we get

$$\frac{dA}{dt} = \frac{d}{dt}(\pi r^2) = 2\pi r \, . \frac{dr}{dt}$$

$$= 2\pi r(3) = 6\pi r \text{ cm}^2\text{/sec.} \quad [\because \text{ By using (2)}]$$

When $r = 10$ cm

$$\therefore \quad \left(\frac{dA}{dt}\right)_{r=10} = 6\pi(10) = 60\pi \text{ cm}^2\text{/sec.}$$

$\therefore$ Rate of change of area is 60π cm^2/sec.

(*v*) Please try yourself.

[**Hint.** Proceed as in part (*iv*).] [**Ans.** 64π cm^2/sec.]

Example 2. (*i*) *An edge of a variable cube is increasing at the rate of 10 cm / sec. How fast the volume of the cube is increasing when the edge is 5 cm long.*

(*ii*) *The side of a square is increasing at the rate of 0.2 cm / sec. Find the rate of increase of the perimeter of the square.*

(*iii*) *An edge of a variable cube is increasing at the rate of 3 cm / sec. How fast is the volume of the cube increasing when the edge is 10 cm long.*

(*iv*) *The side of a square sheet of metal is increasing at 3 cm/sec. At what rate is the area increasing when the side is 10 cm long.*

(*v*) *The side of a square is increasing at 4 cm/min. At what rate is the area increasing when the side is 8 cm long.*

Solution. (*i*) Let x be the length of the edge of the cube and V be the volume at any time t.

$$\therefore \quad V = x^3 \quad ...(1)$$

$\therefore$ Rate of change of edge of the cube w.r.t. $t = \left(\frac{dx}{dt}\right) = 10$ cm/sec ...(2) [Given]

Differentiating (1) w.r.t. t, we get

$$\frac{dV}{dt} = \frac{d}{dt}(x^3) = 3x^2 \cdot \frac{dx}{dt}$$

$$= 3x^2 (10) = 30x^2 \text{ cm}^3\text{/sec.} \quad [\because \text{ By using (2)}]$$

When $x = 5$ cm

$$\therefore \quad \left(\frac{dV}{dt}\right)_{\text{at } x=5} = 30(5)^2 = 30 \times 25 = 750 \text{ cm}^3\text{/sec.}$$

$\therefore$ The volume is increasing at the rate of 750 cm^3/sec.

(*ii*) Let x be side of the square and P be the perimeter at any time t.

$$\therefore \quad P = 4x \quad ...(1)$$

$\therefore$ Rate of change of side w.r.t. $t = \frac{dx}{dt} = 0.2$ cm/sec. ...(2) [Given]

Differentiating (1) w.r.t. t, we get

$$\frac{dP}{dt} = \frac{d}{dt}(4x) = 4\frac{dx}{dt}$$

$$= 4(0.2) = 0.8 \text{ cm/sec.} \quad [\because \text{ By using (2)}]$$

$\therefore$ Perimeter is increasing at the rate of 0.8 cm/sec.

(*iii*) Please try yourself.

[**Hint.** Proceed as in part (*i*) of the same example.] [**Ans.** 900 cm^3/sec.]

(*iv*) Let x be the side of the square sheet of metal and A be the area at any time t.

$$\therefore \quad A = x^2 \quad ...(1)$$

$\therefore$ Rate of change of side w.r.t. $t = \frac{dx}{dt} = 3$ cm/sec. ...(2) [Given]

Differentiating (1) w.r.t. t, we get

$$\frac{dA}{dt} = \frac{d}{dt}(x^2) = 2x \cdot \frac{dx}{dt}$$

$$= 2x(3) = 6x \text{ cm}^2\text{/sec.} \quad [\because \text{ By using (2)}]$$

When $x = 10$ cm

$$\therefore \quad \left(\frac{dA}{dt}\right)_{\text{at } x=10} = 6(10) = 60 \text{ cm}^2\text{/sec.}$$

$\therefore$ Area is increasing at the rate of 60 cm^2/sec.

(*v*) Please try yourself.

[**Hint.** Proceed as in part (*iv*) of the same example.] [**Ans.** 64 cm^2/min.]

Example 3. (*i*) *The radius of a spherical balloon is decreasing at the rate of 10 cm/sec. At what rate is the surface area of the balloon decreasing when its radius is 15 cm.*

(*ii*) *A spherical soap bubble is expanding so that its radius is increasing at the rate of 0.02 cm/sec. At what rate is the surface area increasing when its radius is 4 cm.*

(*iii*) *The volume of a spherical balloon is increasing at the rate of 20 cm^3/sec. Find the rate of change of its surface area at the instant when its radius is 8 cm.*

(*iv*) *The volume of a spherical balloon is increasing at the rate of 25 cm^3/sec. Find the rate of change of its surface area at the instant when its radius is 5 cm.*

(*v*) *A stone is dropped into a quiet lake and waves move in a circle at a speed of 3.5 cm/sec. At the instant when the radius of the circular wave is 7.5 cm, how fast is the enclosed area increasing.*

Solution. (*i*) Let *r* be the radius of the spherical balloon and S be the surface area at any time *t*.

$$\therefore \qquad S = 4\pi r^2 \qquad ...(1)$$

$\therefore$ Rate of change (decrease) of radius w.r.t. *t*

$$= \left(\frac{dr}{dt}\right) = -\ 10 \text{ cm/sec.} \qquad ...(2) \quad [\text{Given}]$$

Differentiating (1) w.r.t. *t*, we get

$$\frac{dS}{dt} = \frac{d}{dt}(4\pi r^2) = 8\pi r \cdot \frac{dr}{dt}$$

$$= 8\pi r\,(-\ 10) = -\ 80\pi r \text{ cm}^2\text{/sec.} \qquad [\because \text{ By using (2)}]$$

When $r = 15$ cm

$$\therefore \qquad \left(\frac{dS}{dt}\right)_{\text{at } r=15} = -\ 80\pi\,(15) = -\ 1200\pi \text{ cm}^2\text{/sec.}$$

$\therefore$ The surface area of spherical balloon decreases at the rate of 1200π cm^2/sec.

(*ii*) Please try yourself.

[**Hint.** Proceed as in part (*i*)]. [**Ans.** 0.64π cm^2/sec.]

(*iii*) Let *r* be the radius of the balloon, V be its volume and S be the surface area at any time *t*.

$$\therefore \qquad V = \frac{4}{3}\pi r^3 \qquad ...(1)$$

$$\therefore \quad \text{Rate of change of volume w.r.t. } t = \frac{dV}{dt} = 20 \text{ cm}^3\text{/sec.} \qquad ...(2) \quad [\text{Given}]$$

Differentiating (1) w.r.t. *t*, we get

$$\frac{dV}{dt} = \frac{d}{dt}\left(\frac{4}{3}\pi r^3\right) = 3 \cdot \frac{4}{3}\pi r^2 \cdot \frac{dr}{dt} = 4\pi r^2\,\frac{dr}{dt} \qquad ...(3)$$

$\therefore$ From equations (2) and (3), we get

$$20 = 4\pi r^2 \cdot \frac{dr}{dt} \quad \Rightarrow \quad \frac{dr}{dt} = \frac{5}{\pi r^2} \qquad ...(4)$$

$$\therefore \qquad S = 4\pi r^2$$

Differentiating w.r.t. t, we get

$$\frac{dS}{dt} = \frac{d}{dt}(4\pi r^2) = 8\pi r \cdot \frac{dr}{dt}$$

$$= 8\pi r \cdot \left(\frac{5}{\pi r^2}\right) = \frac{40}{r}$$ [$\because$ By using (4)]

When $r = 8$ cm

$\therefore$ $$\left(\frac{dS}{dt}\right)_{\text{at } r=8} = \left(\frac{40}{8}\right) = 5 \text{ cm}^2/\text{sec.}$$

$\therefore$ Rate of change of surface area = 5 cm^2/sec.

(*iv*) Please try yourself.

[**Hint.** Proceed as in part (*iii*) of the same example.]

(*v*) Let r be the radius of the circular wave and A be the area at any time t.

$\therefore$ $$A = \pi r^2 \quad \text{...(1)}$$

$\therefore$ Rate of change of radius w.r.t. $t = \left(\frac{dr}{dt}\right) = 3.5$ cm/sec. ...(2) [Given]

Differentiating (1) w.r.t. t, we get

$$\frac{dA}{dt} = \frac{d}{dt}(\pi r^2) = 2\pi r \cdot \frac{dr}{dt}$$

$$= 2\pi r\ (3.5) = 7\pi r \text{ cm}^2/\text{sec.}$$ [$\because$ By using (2)]

When $r = 7.5$ cm

$\therefore$ $$\left(\frac{dA}{dt}\right)_{\text{at } r=7.5} = 7\pi\ (7.5) = 52.5\pi \text{ cm}^2/\text{sec.}$$

$\therefore$ Rate of increase of area of circular wave is 52.5π cm^2/sec.

Example 4. (*i*) *The volume of a cube is increasing at a constant rate. Prove that the increase in surface area varies inversely as the length of the edge of the cube.*

(*ii*) *The area of an expanding rectangle is increasing at the rate of 48 cm^2/sec. The length of the rectangle is always equal to the square of the breadth. At what rate the length is increasing at the instant when the breadth is 4.5 cm.*

(*iii*) *The volume of a cube is increasing at a rate of 7 cm^3/sec. How fast is the surface area increasing when the length of an edge is 12 cm.*

Solution. (*i*) Let x be the length of each edge of the cube, V be its volume and S be the surface area at any time t.

$\therefore$ $$V = x^3 \quad \text{...(1)}$$

$\therefore$ Rate of change of volume w.r.t. $t = \frac{dV}{dt} = k$ (constant) ...(2) [Given]

Differentiating (1) w.r.t. t, we get

$$\frac{dV}{dt} = \frac{d}{dt}(x^3) = 3x^2 \cdot \frac{dx}{dt} \quad \text{...(3)}$$

$\Rightarrow$ $$k = 3x^2 \frac{dx}{dt}$$

$\Rightarrow \quad \dfrac{dx}{dt} = \dfrac{k}{3x^2}$ ...(4) [$\because$ By using (2)]

Now, $\quad S = 6x^2$

$\Rightarrow \quad \dfrac{dS}{dt} = 12x \cdot \dfrac{dx}{dt}$

$= 12x \cdot \left(\dfrac{k}{3x^2}\right) = \dfrac{4k}{x} \Rightarrow \dfrac{dS}{dt} \propto \dfrac{1}{x}.$ [$\because$ By using (4)]

Thus, the rate of increase of surface area varies inversely as the length of the edge of the cube.

(*ii*) Let x be the length and y be the breadth of the rectangle at any time t.

Let A be the area of the rectangle at any time t.

$\therefore \quad A = xy$ ...(1)

Also, It is given that $\quad x = y^2$ ...(2)

$\therefore \quad A = (y^2)\, y = y^3$

Differentiating w.r.t. t, we get

$$\frac{dA}{dt} = \frac{d}{dt}(y^3) = 3y^2 \cdot \frac{dy}{dt} \quad ...(3)$$

$\therefore$ Rate of change of area w.r.t. $t = \left(\dfrac{dA}{dt}\right) = 48$ cm^2/sec. ...(4) [Given]

$\therefore$ From (3) and (4), we have

$$3y^2 \frac{dy}{dt} = 48 \Rightarrow \frac{dy}{dt} = \frac{48}{3y^2} = \frac{16}{y^2} \quad ...(5)$$

$\therefore$ Rate of change of length w.r.t. $t = \left(\dfrac{dx}{dt}\right) = \dfrac{d}{dt}(y^2)$ [$\because$ By using (2)]

$= 2y \cdot \dfrac{dy}{dt} = 2y\left(\dfrac{16}{y^2}\right) = \dfrac{32}{y}$ cm/sec. [$\because$ By using (5)]

When $y = 4.5$ cm.

$$\therefore \quad \left(\frac{dx}{dt}\right)_{\text{at } y=4.5} = \frac{32}{4.5} = 7.11 \text{ cm/sec.}$$

$\therefore$ The rate of increase of length of rectangle is 7.11 cm/sec.

(*iii*) Please try yourself.

[**Hint.** Proceed as in part (*i*) of the same example. Put $k = 7$ cm^3/sec.]

$\left[\textbf{Ans. } \dfrac{7}{3} \text{ cm}^2\text{/sec}\right]$

Example 5. *(i) A particle moves along the curve $6y = x^3 + 2$. Find the points on the curve at which the y-coordinate is changing 8 times as fast as the x-coordinate.*

(ii) A particle moves along the curve $y = \frac{2}{3}x^3 + 1$. Find the points on the curve at which the y-coordinate is changing twice as fast as the x-coordinate.

(iii) A balloon which always remains spherical has a variable diameter $\frac{3}{2}(2x + 3)$. Find the rate of change of volume with respect to x.

Solution. (*i*) We have, $6y = x^3 + 2$ ...(1)

It is given that ;

The y-coordinate is changing 8 times as fast as the x-coordinate.

$$\therefore \quad \frac{dy}{dt} = 8 . \frac{dx}{dt} \quad ...(2)$$

Differentiating (1) w.r.t. t, we get

$$6 . \frac{dy}{dt} = \frac{d}{dt}(x^3 + 2)$$

$$\Rightarrow \quad 6\frac{dy}{dt} = 3x^2 \frac{dx}{dt} + 0$$

$$\Rightarrow \quad 6\left(8\frac{dx}{dt}\right) = 3x^2 \frac{dx}{dt} \quad [\because \text{ By using (2)}]$$

$$\Rightarrow \quad 48\frac{dx}{dt} = 3x^2 \frac{dx}{dt} \Rightarrow 16 = x^2 \Rightarrow x = \pm 4.$$

$\therefore$ When $x = + 4$

$\therefore$ From (1) $\quad 6y = (4)^3 + 2 = 64 + 2 = 66$

$$\Rightarrow \quad y = \frac{66}{6} = 11.$$

When $x = - 4$.

$\therefore$ From (1), $\quad 6y = (- 4)^3 + 2 = - 64 + 2 = - 62$

$$\Rightarrow \quad 6y = - 62 \Rightarrow y = -\frac{62}{6} = -\frac{31}{3}$$

$\therefore$ The required points are (4, 11) and $\left(-4, -\frac{31}{3}\right)$.

(*ii*) We have $\quad y = \frac{2}{3}x^3 + 1$ ...(1)

It is given that :

The y-coordinate is changing 2 times as fast as the x-coordinate.

$$\therefore \quad \frac{dy}{dt} = 2\frac{dx}{dt} \quad ...(2)$$

Differentiating (1) w.r.t. t, we get

$$\frac{dy}{dt} = \frac{d}{dt}\left(\frac{2}{3}x^3 + 1\right) = \frac{2}{3} . 3x^2 \frac{dx}{dt} + 0$$

$\Rightarrow$ $$\frac{dy}{dt} = 2x^2 \frac{dx}{dt} \Rightarrow 2\frac{dx}{dt} = 2x^2 \frac{dx}{dt}$$ [$\because$ By using (2)]

$\Rightarrow$ $$x^2 = 1 \Rightarrow x = \pm 1$$

$\therefore$ When $x = + 1$,

$\therefore$ From (1), $$y = \frac{2}{3}(1)^3 + 1 = \frac{2}{3} + 1 = \frac{5}{3}$$

When $x = - 1$,

$\therefore$ From (1), $$y = \frac{2}{3}(-1)^3 + 1 = -\frac{2}{3} + 1 = \frac{1}{3}$$

$\therefore$ The required points are $\left(1, \frac{5}{3}\right)$ and $\left(-1, \frac{1}{3}\right)$.

(*iii*) It is given that,

Diameter of the sphere $= \frac{3}{2}(2x + 3)$

$\therefore$ Radius of the sphere $= \frac{1}{2}\left(\frac{3}{2}(2x+3)\right) = \frac{3}{4}(2x + 3)$.

$\therefore$ Volume of the sphere $V = \frac{4}{3}\pi r^3$

$$= \frac{4}{3}\pi\left[\frac{3}{4}(2x+3)\right]^3 = \frac{4}{3}\pi \times \frac{27}{64}(2x + 3)^3$$

$$= \frac{9\pi}{16}(2x + 3)^3.$$

$\therefore$ Rate of change of volume w.r.t. $x = \left(\frac{dV}{dx}\right) = \frac{d}{dx}\left[\frac{9\pi}{16}(2x+3)^3\right]$

$$= \frac{9\pi}{16} . 3(2x + 3)^2 . \frac{d}{dx}(2x + 3)$$

$$= \frac{27\pi}{16}(2x + 3)^2 . (2 + 0)$$

$$= \frac{27\pi}{8}(2x + 3)^2.$$

Example 6. (*i*) *The radius of spherical air bubble is increasing at the rate of $\frac{1}{2}$ cm/sec. At what rate is the volume of the bubble increasing, when the radius is 1 cm.*

(*ii*) *Find the rate of change of volume of the sphere w.r.t. diameter.*

(*iii*) *A man 2 metres high, walks at a uniform speed of 6 metres per minute away from a lamp post, 5 metres high. Find the rate at which the length of his shadow increases.*

(*iv*) *A man 2 m high walks at a uniform speed of 5 km/hr. away from a lamp post 6 m high. Find the rate at which the length of his shadow increases.*

Solution. (*i*) Let r be the radius of the spherical bubble and V be the volume at any time t.

$$\therefore \qquad V = \frac{4}{3}\pi r^3 \qquad \text{...(1)}$$

$\therefore$ Rate of change of radius w.r.t. $t = \left(\frac{dr}{dt}\right) = \frac{1}{2}$ cm/sec. ...(2) [Given]

Differentiating (1) w.r.t. t, we get

$$\frac{dV}{dt} = \frac{d}{dt}\left(\frac{4}{3}\pi r^3\right) = \frac{4}{3}\pi \cdot 3r^2 \cdot \frac{dr}{dt}$$

$$= 4\pi r^2 \frac{dr}{dt}$$

$$= 4\pi r^2 \left(\frac{1}{2}\right) = 2\pi r^2 \text{ cm}^3\text{/sec.} \qquad [\because \text{ By using (2)}]$$

When $r = 1$ cm.

$$\therefore \qquad \left(\frac{dV}{dt}\right)_{\text{at } r=1} = 2\pi(1)^2 = 2\pi \text{ cm}^3\text{/sec.}$$

$\therefore$ Rate of increase of volume of the bubble is 2π cm^3/sec.

(*ii*) Let x be the diameter and V be the volume of the sphere.

$$\therefore \qquad V = \frac{4}{3}\pi r^3 \qquad \left[\Rightarrow \text{ radius} = \frac{\text{diameter}}{2} \quad \therefore \ r = \frac{x}{2}\right]$$

$$\Rightarrow \qquad V = \frac{4}{3}\pi\left(\frac{x}{2}\right)^3 = \frac{4}{3}\pi\left(\frac{x^3}{8}\right) = \frac{\pi}{6}x^3.$$

Differentiating w.r.t. x, we get

$$\frac{dV}{dx} = \frac{d}{dx}\left(\frac{\pi}{6}x^3\right) = \frac{\pi}{6}(3x^2) = \frac{\pi}{2}x^2.$$

(*iii*) Let AB be the lamp post.

Let at any time t, the man CD is at a distance x m from the lamp post and y m. be the length of his shadow CE.

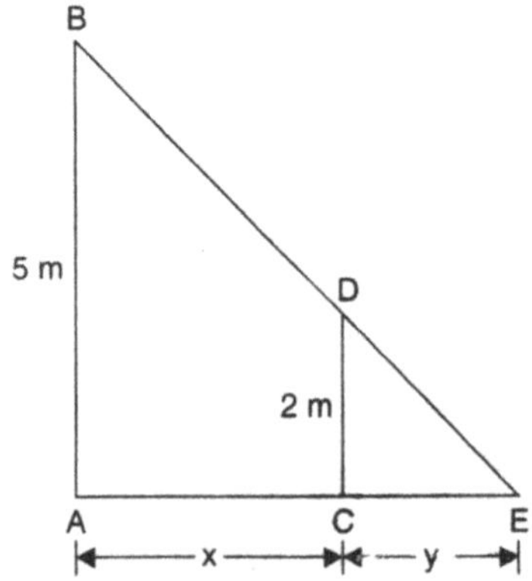

$\therefore$ Speed of man $= \left(\dfrac{dx}{dt}\right) = 6$ m/min. ...(1)

[Given]

In similar triangles ABE and CDE, we have

$$\frac{AB}{CD} = \frac{AE}{CE}$$

$$\Rightarrow \quad \frac{5}{2} = \frac{x+y}{y}$$

$$\Rightarrow \quad 5y = 2(x+y) \Rightarrow 5y = 2x + 2y$$

$$\Rightarrow \quad 3y = 2x \qquad ...(2)$$

Differentiating w.r.t. t, we get

$$3\frac{dy}{dt} = 2\frac{dx}{dt}$$

$$\frac{dy}{dt} = \frac{2}{3}(6) = 4 \text{ m/min.} \qquad [\because \text{ By using (1)}]$$

$\therefore$ The shadow of the man increases at the rate of 4 m/min.

(*iv*) Please try yourself.

[**Hint.** Proceed as in part (*iii*) of the same example.] **[Ans.** 2.5 km/hr.]

Example 7. (*i*) *A man 160 cm tall, walks away from a source of light situated at the top of a pole 6 m high, at the rate of 1.1 m/sec. How fast is the length of his shadow increasing when he is 1 m away from the pole.*

(*ii*) *At what point of the ellipse $16x^2 + 9y^2 = 400$, does the ordinate decrease at the same rate at which the abscissa increases.*

(*iii*) *A ladder 5 m long is leaning against a wall. The bottom of the ladder is pulled along the ground away from the wall, at the rate of 2 m/sec. How fast is its height on the wall decreasing when the foot of ladder is 4 m away from the wall.*

(*iv*) *A ladder 16 cm long is leaning against a wall. The bottom of the ladder is pulled along the ground, away from the wall, at the rate of 2 cm/sec. How fast is its height on the wall decreasing when the foot of the ladder is 4 cm away from wall.*

Solution. (*i*) Let AB be the pole, B being the source of light.

Let at any time t, the man CD is at a distance x m from the pole and y m be the length of his shadow CE.

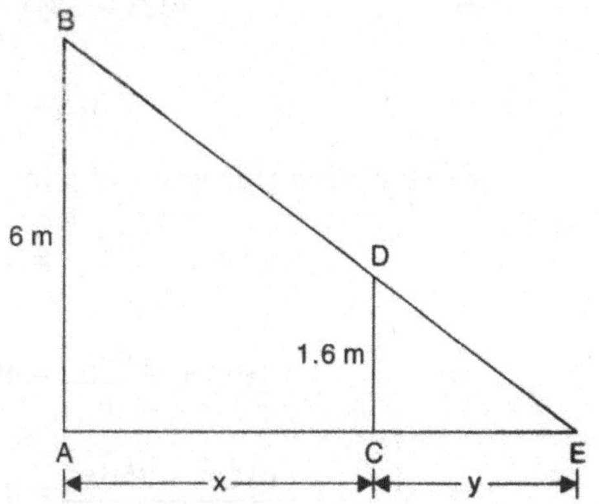

$\therefore$ Speed of man $= \left(\dfrac{dx}{dt}\right) = 1.1$ m/sec. ...(1)

[Given]

In similar triangles ABE and CDE, we have

$$\frac{AB}{CD} = \frac{AE}{CE}$$

$$\Rightarrow \quad \frac{6}{1.6} = \frac{x+y}{y}$$

$\Rightarrow \quad 6y = 1.6\,(x + y)$

$\Rightarrow \quad 6y = 1.6x + 1.6y$

$\Rightarrow \quad 4.4y = 1.6x$ ...(2)

Differentiating w.r.t. t, we get

$$4.4\,\frac{dy}{dt} = 1.6\,\frac{dx}{dt}$$

$$\Rightarrow \quad \frac{dy}{dt} = \frac{1.6}{4.4}\,(1.1) \qquad [\because \text{ By using (1)}]$$

$$= \frac{16 \times 11}{44 \times 10} = \frac{4}{10} = 0.4 \text{ m/sec.}$$

Since, $\frac{dy}{dt}$ is independent of x.

$$\therefore \quad \left.\frac{dy}{dt}\right]_{\text{at } x=1} = 0.4 \text{ m/sec.}$$

$\therefore$ The length of shadow is increasing at the rate of 0.4 m/sec.

(*ii*) We have, $16x^2 + 9y^2 = 400$ ...(1)

It is given that :

The ordinate decreases at the same rate at which the abscissa increases.

$$\therefore \quad -\frac{dy}{dt} = +\frac{dx}{dt}$$

$$\Rightarrow \quad \frac{dy}{dt} = -\frac{dx}{dt} \qquad \text{...(2)}$$

Differentiating (1), w.r.t. t, we get

$$32x\,.\,\frac{dx}{dt} + 18y\,.\,\frac{dy}{dt} = 0$$

$$\Rightarrow \quad 32x\,\frac{dx}{dt} + 18y\left(-\frac{dx}{dt}\right) = 0 \qquad [\because \text{ By using (2)}]$$

$$\Rightarrow \quad (32x - 18y)\,\frac{dx}{dt} = 0$$

$$\Rightarrow \quad 32x - 18y = 0 \;\Rightarrow\; y = \frac{32}{18}\,x \;\Rightarrow\; y = \frac{16x}{9} \qquad \text{...(3)}$$

Substituting this value of y in (1), we get

$$16x^2 + 9\left(\frac{16x}{9}\right)^2 = 400$$

$$\Rightarrow \quad 16x^2 + \frac{256x^2}{9} = 400$$

$$\Rightarrow \quad \frac{144x^2 + 256x^2}{9} = 400$$

$\Rightarrow \quad \dfrac{400x^2}{9} = 400 \Rightarrow x^2 = 9 \Rightarrow x = \pm 3$

When $x = +3$

$\therefore$ From (3) $\quad y = \dfrac{16(3)}{9} = \dfrac{16}{3}$

When $x = -3$

$\therefore$ From (3) $\quad y = \dfrac{16(-3)}{9} = -\dfrac{16}{3}$

$\therefore$ The required points are $\left(3, \dfrac{16}{3}\right)$ and $\left(-3, -\dfrac{16}{3}\right)$.

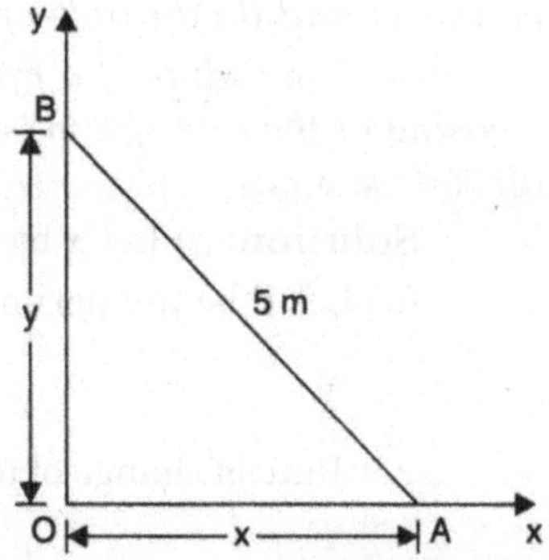

(*iii*) Let AB be the ladder of length 5 m.

At any time t, let OA = x m and OB = y m.

$\therefore$ In right triangle OAB :

$$OA^2 + OB^2 = AB^2$$

$$\Rightarrow \quad x^2 + y^2 = 5^2$$

$$\Rightarrow \quad x^2 + y^2 = 25 \quad \ldots(1)$$

Differentiating (1) w.r.t. t, we get

$$\Rightarrow \quad 2x \cdot \frac{dx}{dt} + 2y\frac{dy}{dt} = 0$$

$$\Rightarrow \quad x\frac{dx}{dt} + y\frac{dy}{dt} = 0 \quad \ldots(2)$$

It is given that the bottom A of the ladder is pulled away from the wall at the rate of 2 m/sec.

$\therefore$ Rate of change of x w.r.t. t

$$= \left(\frac{dx}{dt}\right) = 2 \text{ m/sec.} \quad \ldots(3)$$

$\therefore$ From (2) and (3), we have

$$x(2) + y\frac{dy}{dt} = 0$$

$$\Rightarrow \quad y\frac{dy}{dt} = -2x \Rightarrow \frac{dy}{dt} = -\frac{2x}{y} \quad \ldots(4)$$

When $x = 4$.

$\therefore$ From (1), $\quad (4)^2 + y^2 = 25 \Rightarrow 16 + y^2 = 25 \Rightarrow y^2 = 25 - 16$

$\Rightarrow \quad y^2 = 9 \Rightarrow y = 3$

$\therefore$ From equation (4), we have

$$\frac{dy}{dt} = -\frac{2(4)}{3} = -\frac{8}{3} \text{ m/sec.}$$

$\therefore$ The rate of decrease in the height of the ladder on the wall is $\dfrac{8}{3}$ m/sec.

(*iv*) Please try yourself.

[**Hint.** Proceed as in part (*iii*) of the same example.] $\left[\textbf{Ans. } \dfrac{2}{\sqrt{15}} \text{ cm/sec.}\right]$

Example 8. (*i*) *The length x of a rectangle is decreasing at the rate of 3 cm/min. and the width y is increasing at the rate of 2 cm/min. when x = 10 cm and y = 6 cm, find the rates of change of (a) the perimeter and (b) the area of the rectangle.*

(*ii*) *The length x of a rectangle is decreasing at the rate of 2 cm/sec and the width y is increasing at the rate of 2 cm/sec. when x = 12 cm and y = 5 cm, find the rate of change of (a) the perimeter and (b) the area of the rectangle.*

(*iii*) *The radius of a cylinder is increasing at the rate of 2 m/sec and its altitude is decreasing at the rate of 3 m/sec. Find the rate of change of volume when radius is 3 m and the altitude is 5 m.*

Solution. (*i*) Let x be the length and y be the breadth of the rectangle.

(*a*) Let P be the perimeter of the rectangle at any time t.

$$\therefore \quad P = 2(x + y) \quad ...(1)$$

$\therefore$ Rate of change of (decrease) length w.r.t. $t = \left(\dfrac{dx}{dt}\right) = -3$ cm/min. ...(2) [Given]

Rate of change of (increase) breadth w.r.t. $t = \left(\dfrac{dy}{dt}\right) = 2$ cm/min. ...(3) [Given]

Differentiating (1) w.r.t. t, we get

$$\frac{dP}{dt} = \frac{d}{dt}(2x + 2y)$$

$$= 2\frac{dx}{dt} + 2\frac{dy}{dt} \quad ...(4)$$

$$\Rightarrow \quad \frac{dP}{dt} = 2(-3 + 2) = -2 \text{ cm/min.} \quad [\because \text{ By using (2) and (3)}]$$

$\therefore$ Rate of change of perimeter of rectangle = – 2 cm/min.

(*b*) Let A be the area of the rectangle at any time t.

$$\therefore \quad A = xy \quad ...(5)$$

Differentiating w.r.t. t we get

$$\frac{dA}{dt} = \frac{d}{dt}(xy)$$

$$= x \cdot \frac{dy}{dt} + y \cdot \frac{dx}{dt}$$

$$= x(2) + y(-3) \quad [\because \text{ By using (2) and (3)}]$$

$$\frac{dA}{dt} = 2x - 3y$$

When $x = 10$ cm and $y = 6$ cm.

$$\left.\frac{dA}{dt}\right]_{\substack{\text{at } x=10 \\ \text{at } y=6}} = 2(10) - 3(6) = 20 - 18 = 2 \text{ cm}^2/\text{min.}$$

of change of area of rectangle = 2 cm^2/min.

(*ii*) Let x be the length and y be the breadth of the rectangle.

(*a*) Let P be the perimeter of the rectangle at any time t.

$\therefore$ $$P = 2(x + y) \quad ...(1)$$

$\therefore$ Rate of change of (decrease) length w.r.t. $t = \frac{dx}{dt} = -2$ cm/sec. ...(2) [Given]

$\therefore$ Rate of change of (increase) breadth w.r.t. $t = \frac{dy}{dt} = 2$ cm/sec. ...(3) [Given]

Differentiating (1) w.r.t. t, we get

$$\frac{dP}{dt} = \frac{d}{dt}(2x + 2y)$$

$$\Rightarrow \frac{dP}{dt} = 2\frac{dx}{dt} + 2\frac{dy}{dt} \quad ...(4)$$

$$= 2(-2) + 2(2) \quad [\because \text{ By using (2) and (3)}]$$

$$\Rightarrow \frac{dP}{dt} = -4 + 4 = 0 \text{ cm/sec.}$$

$\therefore$ The perimeter remains constant.

(*b*) Let A be the area of the rectangle at any time t.

$\therefore$ $$A = xy \quad ...(5)$$

Differentiating w.r.t. t, we get

$$\frac{dA}{dt} = \frac{d}{dt}(xy)$$

$$= x \cdot \frac{dy}{dt} + y \cdot \frac{dx}{dt}$$

$$= x \cdot (2) + y \cdot (-2) \quad [\because \text{ By using (2) and (3)}]$$

$$\Rightarrow \frac{dA}{dt} = 2x - 2y$$

When $x = 12$ cm and $y = 5$ cm.

$\therefore$ $$\left.\frac{dA}{dt}\right]_{\substack{\text{at } x=12 \\ \text{at } y=5}} = 2(12) - 2(5) = 24 - 10 = 14 \text{ cm}^2\text{/sec.}$$

$\therefore$ Rate of change of area of rectangle = 14 cm^2/sec.

(*iii*) Let at any time t, x m be the radius and y m be the altitude of the cylinder and V be the volume.

$\therefore$ $$V = \pi r^2 h = \pi x^2 y \quad ...(1)$$

$\therefore$ Rate of change of radius w.r.t. $t = \frac{dx}{dt} = 2$ m/sec. ...(2) [Given]

$\therefore$ Rate of change of altitude w.r.t. $t = \frac{dy}{dt} = -3$ m/sec. ...(3) [Given]

Differentiating (1) w.r.t. t, we get

$$\frac{dV}{dt} = \frac{d}{dt}(\pi x^2 y) = \pi x^2 \cdot \frac{dy}{dt} + \pi y \cdot \frac{d}{dt}(x^2)$$

$$= \pi x^2 \frac{dy}{dt} + 2\pi xy \frac{dx}{dt}$$

$$= \pi x^2(-3) + 2\pi xy(2) \qquad [\because \text{ By using (2) and (3)}]$$

$$\frac{dV}{dt} = -3\pi x^2 + 4\pi xy \qquad ...(4)$$

When $x = 3$ m and $y = 5$ m.

$$\therefore \quad \left.\frac{dV}{dt}\right]_{\substack{x=3\\y=5}} = -3\pi(3)^2 + 4\pi(3)(5) = -3\pi(9) + 4\pi(15)$$

$$= -27\pi + 60\pi = 33\pi \text{ m}^3\text{/sec.}$$

$\therefore$ Rate of change of volume = 33π m^3/sec.

Example 9. (*i*) *A man is walking at the rate of 6.5 km/h towards the foot of a tower 120 m high. At what rate is he approaching the top of the tower when he is 50 m away from the tower.*

(*ii*) *A hemisphere is constructed on a circular base. If the radius of the base is increasing at the rate of 0.5 cm/sec. Find the rate at which the volume of the hemisphere is increasing when the radius is 10 cm.*

Solution. (*i*) Let AB be the tower. Let at any time t, the distances of man from the foot of tower and the top of tower are x metre and y metre respectively.

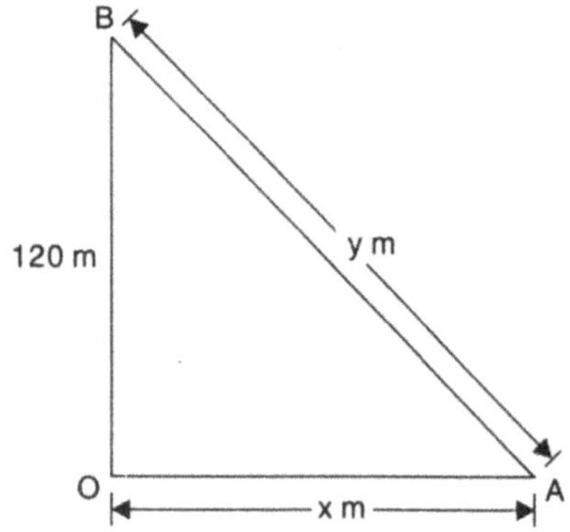

$\therefore$ In right triangle OAB :

$$AB^2 = OA^2 + OB^2$$

$$\Rightarrow \quad y^2 = x^2 + (120)^2 \qquad ...(1)$$

Differentiating w.r.t. t, we get

$$\Rightarrow \quad 2y\frac{dy}{dt} = 2x \cdot \frac{dx}{dt} + 0$$

$$\Rightarrow \quad \frac{dy}{dt} = \frac{x}{y} \cdot \frac{dx}{dt} \qquad ...(2)$$

$\therefore$ Rate of change of x (decrease) w.r.t. $t = \frac{dx}{dt} = -6.5$ km/h. ...(3) [Given]

$\therefore$ From equations (2) and (3), we have

$$\frac{dy}{dt} = \frac{x}{y}(-6.5) = -\frac{6.5x}{y} \qquad ...(4)$$

When $x = 50$ m

$\therefore$ From (1), $y^2 = (50)^2 + (120)^2$

$$\Rightarrow \quad y = \sqrt{2500 + 14400} = \sqrt{16900}$$

$$\Rightarrow \quad y = 130$$

$\therefore$ From equation (4), we have

$$\frac{dy}{dt} = -\frac{6.5(50)}{130} = -2.5 \text{ km/hr.}$$

$\therefore$ The man is approaching to the top of the tower at the rate of 2.5 km/hr.

(*ii*) Let r be the radius and V be the volume of the hemisphere at any time t.

$$\therefore \quad V = \frac{2}{3}\pi r^3 \quad ...(1)$$

Rate of change of radius w.r.t. $t = \frac{dr}{dt} = 0.5$ cm/sec. ...(2) [Given]

Differentiating (1) w.r.t. t, we get :

$$\frac{dV}{dt} = \frac{2}{3}\pi 3r^2 \cdot \frac{dr}{dt} = 2\pi r^2 \frac{dr}{dt}$$

$$= 2\pi r^2 (0.5) = \pi r^2 \quad ...(3) \; [\because \text{ By using (2)}]$$

When $r = 10$ cm

$$\therefore \quad \left.\frac{dV}{dt}\right]_{at\, r=10} = \pi(10)^2 = 100\pi \text{ cm}^3\text{/sec.}$$

$\therefore$ Rate of change of volume is $= 100\pi$ cm^3/sec.

Example 10. (*i*) *A man moving away from a tower 41.6 m high at the rate of 2 m/sec. Find the rate at which the angle of elevation of the top of the tower is changing, when he is at a distance of 30 m from the foot of the tower. Assume that the eye level of the man is 1.6 m from the ground.*

(*ii*) *A ladder 13 m long leans against a wall. The foot of the ladder is pulled along the ground away from the wall at the rate of 1.5 m/sec. How fast is the angle θ between the ladder and the ground is changing when the foot of the ladder is 12 m away from the wall.*

(*iii*) *A man is moving away from a tower 85 feet high at a speed of 4 feet per second. Find the rate at which his angle of elevation of the top of the tower is changing, when he is at a distance of 60 feet from the foot of the tower. Assume that the eye-level of man is 5 feet from the ground.*

Solution. (*i*) Let AB be the tower.

Let at any time t, the man be at a distance of x metres from the tower AB and let θ be the angle of elevation at that time.

$\therefore$ In right triangle BCD :

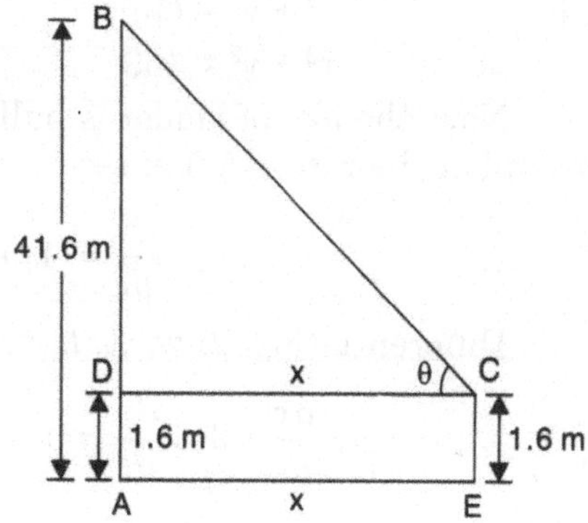

$$\tan\theta = \frac{BD}{CD} \quad ...(1)$$

Now, $\quad BD = AB - AD$

$$= 41.6 - 1.6 = 40 \text{ m.}$$

$\therefore$ From (1), we have

$$\tan\theta = \frac{40}{x}$$

$$\Rightarrow \quad \theta = \tan^{-1}\left(\frac{40}{x}\right) \quad \left[\because \frac{d}{dx}(\tan^{-1}x) = \frac{1}{1+x^2}\right]$$

Differentiating w.r.t. t, we get

$$\frac{d\theta}{dt} = \frac{1}{1+\left(\frac{40}{x}\right)^2} \cdot \frac{d}{dt}\left(\frac{40}{x}\right)$$

$$= \frac{1}{1+\frac{1600}{x^2}}\left(-\frac{40}{x^2}\right).\frac{dx}{dt}$$

$$= \frac{1}{\frac{x^2+1600}{x^2}}.\left(-\frac{40}{x^2}\right).\frac{dx}{dt} = -\frac{x^2}{x^2+1600}\times\frac{40}{x^2}.\frac{dx}{dt}$$

$$= -\frac{40}{x^2+1600}.\frac{dx}{dt} \qquad ...(2)$$

Now, the man is moving away from the tower at the rate of 2 m/sec.

$\therefore$ $\frac{dx}{dt} = 2$ m/sec. ...(3) [Given]

$\therefore$ From (2) and (3), we have

$$\frac{d\theta}{dt} = \frac{-40}{x^2+1600}(2) = \frac{-80}{x^2+1600}$$

When $x = 30$ m

$$\therefore \quad \left.\frac{d\theta}{dt}\right]_{\text{at } x=30} = \frac{-80}{(30)^2+1600} = \frac{-80}{900+1600} = \frac{-80}{2500} = \frac{-4}{125} \text{ rad./sec.}$$

$\therefore$ The angle of elevation of the top of tower is changing at the rate of $\frac{4}{125}$ rad./sec.

(*ii*) Let BC be the ladder leaning against the wall AC. Let at any time t, angle of elevation of the ladder to the wall is θ and the distance of foot of ladder to the wall is x metre.

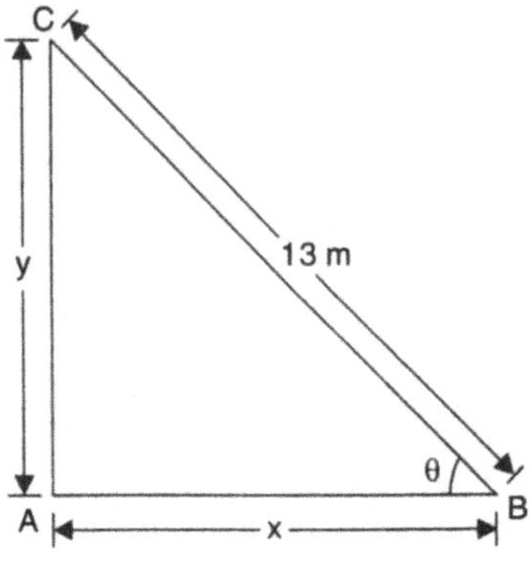

Let y be the height of the wall.

$\therefore$ In right triangle ABC :

$$\tan\theta = \frac{AC}{AB} = \frac{y}{x} \qquad ...(1)$$

and $x^2 + y^2 = (13)^2$

$\Rightarrow$ $x^2 + y^2 = 169$ ...(2)

Now, the foot of ladder is pulled along the ground from the wall at the rate of 1.5 m/sec.

$\therefore$ $\frac{dx}{dt} = (1.5) = \frac{3}{2}$ m/sec. ...(3) [Given]

Differentiating (2) w.r.t. t, we get

$$2x\frac{dx}{dt} + 2y.\frac{dy}{dt} = 0$$

$\Rightarrow$ $y\frac{dy}{dt} = -x\frac{dx}{dt}$ [$\because$ By using (3)]

$$\frac{dy}{dt} = -\frac{x}{y}\left(\frac{3}{2}\right) = -\frac{3x}{2y} \qquad ...(4)$$

Differentiating (1) w.r.t. t, we get

$$\sec^2\theta \cdot \frac{d\theta}{dt} = \frac{x \cdot \frac{dy}{dt} - y \cdot \frac{dx}{dt}}{x^2}$$

$$= \frac{x\left(-\frac{3x}{2y}\right) - y\left(\frac{3}{2}\right)}{x^2} \qquad [\because \text{ By using (3) and (4)}]$$

$$= \frac{-\frac{3x^2}{2y} - \frac{3y}{2}}{x^2} = \left(\frac{-3x^2 - 3y^2}{2x^2 y}\right)$$

$$\Rightarrow \qquad \frac{d\theta}{dt} = \frac{1}{\sec^2\theta}(-3)\left[\frac{x^2 + y^2}{2x^2 y}\right]$$

$$= \frac{-3(x^2 + y^2)}{(1 + \tan^2\theta)(2x^2 y)} = \frac{-3(x^2 + y^2)}{\left(1 + \frac{y^2}{x^2}\right)(2x^2 y)} \qquad \left[\because \tan\theta = \frac{y}{x}\right]$$

$$= \frac{-3(x^2 + y^2)\,x^2}{(x^2 + y^2)(2x^2 y)} = -\frac{3}{2y} \qquad \ldots(5)$$

When $x = 12$ m, then, from (2)

$$(12)^2 + y^2 = 169$$

$$\Rightarrow \qquad 144 + y^2 = 169 \quad \Rightarrow \quad y^2 = 169 - 144 = 25$$

$$\Rightarrow \qquad y = 5 \text{ m.}$$

$\therefore$ When $y = 5$ metre.

$$\left.\frac{d\theta}{dt}\right]_{\text{at } y=5} = \frac{-3}{2(5)} = -\frac{3}{10} \text{ rad./sec.}$$

$\therefore$ Rate of change of angle between the ladder and the ground is $\left(\frac{3}{10}\right)$ rad./sec.

(*iii*) Please try yourself.

[Hint. Proceed as in part (*i*) of the same example.] **[Ans.** (0.032) rad./sec.]

Example 11. *An airforce plane is ascending vertically at the rate of 100 km/h. If the radius of the earth is r km. How fast is the area of the earth, visible from the plane increasing at 3 minutes after it started ascending. It is given that if h is the height of the plane above the earth, then visible area is equal to* $\frac{2\pi r^2 h}{r + h}$.

Solution. Let h be the height of the plane above the earth and A be visible area from the plane at time t.

$\therefore \quad A = \dfrac{2\pi r^2 h}{r + h}$ ...(1)

Now, the height of plane is increasing at the rate of 100 km/h.

$\therefore \quad \dfrac{dh}{dt} = 100$ km/h. ...(2) [Given]

Differentiating (1) w.r.t. t, we get

$$\frac{dA}{dt} = 2\pi r^2 \frac{d}{dt}\left(\frac{h}{r+h}\right) = 2\pi r^2 \cdot \left[\frac{(r+h).1 - h(0+1)}{(r+h)^2}\right] \cdot \frac{dh}{dt}$$

$$= 2\pi r^2 \frac{(r+h-h)}{(r+h)^2} \times 100 \qquad [\because \text{ By using (2)}]$$

$\Rightarrow \quad \dfrac{dA}{dt} = \dfrac{200\pi r^3}{(r+h)^2}$ ...(3)

Now, height covered by plane is 60 min. = 100 km.

Height covered by plane in 3 min. $= \dfrac{100}{60} \times 3$

$\therefore \quad h = 5$ km.

$\therefore$ When $h = 5$ km

$$\left.\frac{dA}{dt}\right]_{h=5} = \frac{200\pi r^3}{(r+5)^2} \text{ km}^2\text{/hr.}$$

Example 12. *A kite is 120 m high and 130 m of string is out. If the kite is moving away horizontally at the rate of 52 m/sec. Find the rate at which the string is being paid out.*

Solution. Let at any time t, C be the position of kite and BC be the string.

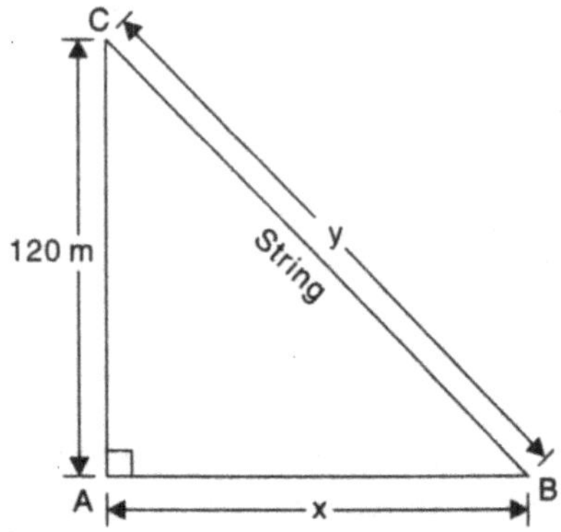

From the figure, we have

$$(BC)^2 = (AB)^2 + (AC)^2$$

$$y^2 = x^2 + (120)^2$$

$\Rightarrow \quad y^2 = x^2 + 14400$ ...(1)

When $y = 130$ m

$\therefore \quad (130)^2 = x^2 + 14400$

$\Rightarrow \quad 16900 - 14400 = x^2 \Rightarrow x^2 = 2500 \Rightarrow x = 50$ m.

Now, the kite is moving away horizontally at the rate of 52 m/sec.

$\therefore \quad \dfrac{dx}{dt} = 52$ m/sec. ...(2) [Given]

Differentiating (1) w.r.t. t, we get

$$2y\frac{dy}{dt} = 2x\frac{dx}{dt} + 0$$

$\Rightarrow \quad \dfrac{dy}{dt} = \left(\dfrac{x}{y}\right) \cdot \left(\dfrac{dx}{dt}\right)$

$$\Rightarrow \qquad \frac{dy}{dt} = \left(\frac{x}{y}\right)(52) \qquad \text{...(3)} \quad [\because \text{ By using (2)}]$$

When $y = 130$ m and $x = 50$ m.

$$\therefore \qquad \left.\frac{dy}{dt}\right]_{\substack{\text{at } y=130 \\ \text{at } x=50}} = \frac{50}{130} \times 52 = 20 \text{ m/sec.}$$

$\therefore$ The string is paid out at the rate of 20 m/sec.

Example 13. *Sand is pouring from a pipe at the rate of 12 cm³/sec. The falling sand forms a cone on the ground in such a way that the height of the cone is always one-sixth of the radius of the base. How fast is the height of the sand-cone is increasing when the height is 4 cm.*

Solution. Let r be the radius and h be the height of the sand cone at any time t.

$$\therefore \qquad h = \frac{r}{6} \qquad \text{...(1)} \quad [\text{Given}]$$

Let V be the volume of the cone at any time t.

$$\therefore \qquad V = \frac{1}{3}\pi r^2 h \qquad \text{...(2)}$$

$$\Rightarrow \qquad V = \frac{1}{3}\pi(6h)^2 h = \frac{1}{3}\pi(36h^2)h = 12\pi h^3 \qquad \text{...(3)}$$

Now, sand is pouring from a pipe at the rate of 12 cm³/sec.

$$\therefore \qquad \frac{dV}{dt} = 12 \text{ cm}^3/\text{sec.} \qquad \text{...(4)} \quad [\text{Given}]$$

Differentiating (3) w.r.t. t, we get

$$\frac{dV}{dt} = 36\pi h^2 \cdot \frac{dh}{dt} \qquad \text{...(5)}$$

$\therefore$ From (4) and (5), we have

$$36\pi h^2 \frac{dh}{dt} = 12 \quad \Rightarrow \quad \frac{dh}{dt} = \frac{12}{36\pi h^2} = \frac{1}{3\pi h^2} \text{ cm/sec.}$$

When $h = 4$ cm

$$\therefore \qquad \left.\frac{dh}{dt}\right]_{\text{at } h=4} = \frac{1}{3\pi(4)^2} = \frac{1}{3\pi(16)} = \frac{1}{48\pi} \text{ cm/sec.}$$

$\therefore$ The height of cone is increasing at the rate of $\frac{1}{48\pi}$ cm/sec.

Example 14. *A balloon which remains spherical, is being inflated by pumping at 900 cm³/sec. Find the rate at which radius of the balloon is increasing, when the radius is 15 cm.*

Solution. Let r be the radius and V be the volume of the spherical balloon at any time t.

$$\therefore \qquad V = \frac{4}{3}\pi r^3 \qquad \text{...(1)}$$

Now, rate of change of volume w.r.t. $t = \frac{dV}{dt} = 900$ cm³/sec. ...(2) [Given]

Differentiating (1) w.r.t. t, we get

$$\frac{dV}{dt} = \frac{4}{3}\pi.\left(3r^2 \frac{dr}{dt}\right)$$

$$\frac{dV}{dt} = 4\pi r^2 \frac{dr}{dt} \qquad ...(3)$$

$\therefore$ From (2) and (3), we have

$$4\pi r^2 . \frac{dr}{dt} = 900$$

$$\Rightarrow \qquad \frac{dr}{dt} = \frac{900}{4\pi r^2} \qquad ...(4)$$

When $r = 15$ cm.

$$\therefore \qquad \left.\frac{dr}{dt}\right]_{\text{at } r=15} = \frac{900}{4\pi(15)^2} = \frac{900}{4\pi(225)} = \frac{900}{900\pi} = \frac{1}{\pi} \text{ cm/sec.}$$

$\therefore$ Radius of balloon is increasing at the rate of $\frac{1}{\pi}$ cm/sec.

Example 15. *Two parallel sides of a rectangle are being lengthened at the rate of 2 cm/sec. while the other two sides are shortened in such a way that the figure remains a rectangle with constant area 50 cm². What is the rate of change of the perimeter P when the length of the increasing side is (i) 5 cm (ii) 10 cm.*

Solution. Let x cm be the length of the sides which are being lengthened and y cm be the length of the other sides at any time t.

$$\therefore \qquad \text{Area} = xy \text{ cm}^2$$

$$\Rightarrow \qquad xy = 50 \qquad [\because \text{ Area} = 50 \text{ cm}^2 \text{ is constant}]$$

$$\Rightarrow \qquad y = \frac{50}{x} \qquad ...(1)$$

Also, perimeter, $\qquad P = 2(x + y) = 2\left(x + \frac{50}{x}\right)$

$$P = 2x + \frac{100}{x} \qquad ...(2)$$

Differentiating w.r.t. t, we get

$$\frac{dP}{dt} = \left[2 + \left(-\frac{100}{x^2}\right)\right].\frac{dx}{dt} = \left[2 - \frac{100}{x^2}\right].\frac{dx}{dt} \qquad ...(3)$$

Now, rate of change of sides w.r.t. $t = \frac{dx}{dt} = 2$ cm/sec. ...(4) [Given]

$$\Rightarrow \qquad \frac{dP}{dt} = \left(2 - \frac{100}{x^2}\right)(2) = 4 - \frac{200}{x^2}$$

(*i*) When $x = 5$ cm.

$$\therefore \qquad \left.\frac{dP}{dt}\right]_{\text{at } x=5} = 4 - \frac{200}{(5)^2} = 4 - \frac{200}{25} = 4 - 8 = -4 \text{ cm/sec.}$$

$\therefore$ Perimeter is decreasing at the rate of 4 cm/sec.

(*ii*) When $x = 10$ cm.

$$\therefore \quad \left.\frac{dP}{dt}\right]_{\text{at } x=10} = 4 - \frac{200}{(10)^2} = 4 - \frac{200}{100} = 4 - 2 = 2 \text{ cm/sec.}$$

$\therefore$ Perimeter is increasing at the rate of 2 cm/sec.

Example 16. *Water is running out of a conical funnel at the rate of 5 cm³/sec. If the radius of the base of the funnel is 10 cm and the altitude is 20 cm. Find the rate at which the water level is dropping when it is 5 cm from the top.*

Solution. Let r be the radius and h be the height of the surface of water at any time t.

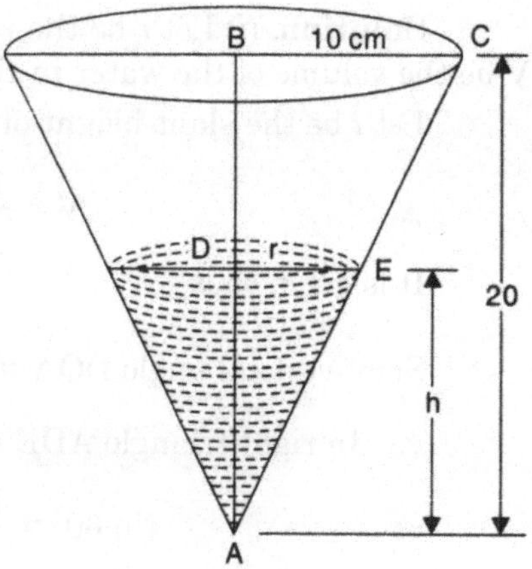

Let V be the volume of water in the funnel.

$$\therefore \quad V = \frac{1}{3}\pi r^2 h \qquad ...(1)$$

From the figure :

In ΔABC

BC = 10 cm

BA = 20 cm

In ΔADE

DE = r and AD = h

Now, ΔABC and ΔADE are similar.

$$\therefore \quad \frac{BC}{AB} = \frac{DE}{DA}$$

$$\Rightarrow \quad \frac{10}{20} = \frac{r}{h} \Rightarrow \frac{1}{2} = \frac{r}{h} \Rightarrow r = \frac{h}{2} \qquad ...(2)$$

$\therefore$ From (1) and (2), we have

$$V = \frac{1}{3}\pi\left(\frac{h}{2}\right)^2 . h = \frac{1}{3}\pi\left(\frac{h^2}{4}\right)h = \frac{1}{12}\pi h^3.$$

Differentiating w.r.t. t, we get

$$\frac{dV}{dt} = \frac{1}{12}\pi\left(3h^2 . \frac{dh}{dt}\right) = \frac{1}{4}\pi h^2 . \frac{dh}{dt} \qquad ...(3)$$

Since, water is running out of the funnel at the rate of 5 cm³/sec.

$$\therefore \quad \frac{dV}{dt} = -5 \text{ cm}^3\text{/sec.} \qquad ...(4) \quad \text{[Given]}$$

$\therefore$ From (3) and (4), we have

$$\frac{1}{4}\pi h^2 . \frac{dh}{dt} = -5 \Rightarrow \frac{dh}{dt} = \frac{-20}{\pi h^2}$$

when water level is 5 cm from the top, then $h = 20 - 5 = 15$ cm.

$$\therefore \quad \left.\frac{dh}{dt}\right]_{\text{at } h=15} = \frac{-20}{\pi(15)^2} = \frac{-20}{225\pi} = \frac{-4}{45\pi} \text{ cm/sec.}$$

$\therefore$ Rate of dropping of the water level is $\frac{4}{45\pi}$ cm/sec.

Example 17. (*i*) *Water is dripping out from a conical funnel at the uniform rate of 2 cm³/sec through a tiny hole at the vertex at the bottom. When the slant height of the water is 4 cm, find the rate of decrease of the slant height of the water, given that the vertical angle of the funnel is 120.*

(*ii*) *Water is dripping out from a conical funnel at a uniform rate of 4 cm³/sec through a tiny hole at the vertex at the bottom. When the slant height of the water is 3 cm. Find the rate of decrease of the slant height of the water. Given that the vertical angle of the funnel is 120.*

Solution. (*i*) Let r be the radius, h be the height and V be the volume of the water in the funnel at any time t.

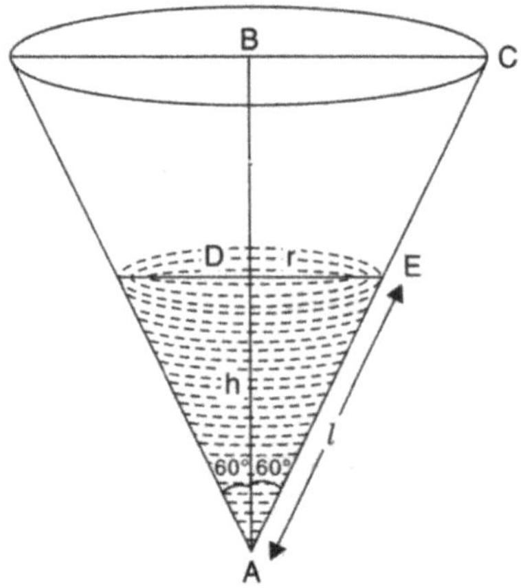

Let l be the slant-height of water-cone.

$$\therefore \qquad V = \frac{1}{3}\pi r^2 h \qquad ...(1)$$

It is given that,

Semi-vertical angle DOA = $\frac{120°}{2} = 60°$

$\therefore$ In right triangle ADE :

$$\Rightarrow \qquad \sin 60° = \frac{DE}{AE}$$

and
$$\cos 60° = \frac{AD}{AE}$$

$$\Rightarrow \qquad \frac{\sqrt{3}}{2} = \frac{r}{l} \quad \text{and} \quad \frac{1}{2} = \frac{h}{l}$$

$$\Rightarrow \qquad r = \frac{\sqrt{3}}{2}l \quad \text{and} \quad h = \frac{l}{2}.$$

Putting these values of r and h in equation (1), we have

$$V = \frac{1}{3}\pi\left(\frac{\sqrt{3}}{2}l\right)^2\left(\frac{l}{2}\right) = \frac{1}{3}\pi\left(\frac{3l^2}{4}\right)\left(\frac{l}{2}\right) = \frac{\pi}{8}l^3 \qquad ...(2)$$

Differentiating w.r.t. t, we get

$$\frac{dV}{dt} = \frac{\pi}{8}\left(3l^2 \cdot \frac{dl}{dt}\right) = \frac{3}{8}\pi l^2 \cdot \frac{dl}{dt} \qquad ...(3)$$

Now, rate of change (decrease) of volume of water w.r.t. $t = \frac{dV}{dt} = -2$ cm³/sec.

...(4) [Given]

$\therefore$ From (4) and (5), we have

$$\Rightarrow \qquad \frac{3}{8}\pi l^2 \cdot \frac{dl}{dt} = -2$$

$$\Rightarrow \qquad \frac{dl}{dt} = \frac{-16}{3\pi l^2} \qquad ...(5)$$

When $l = 4$ cm

$$\therefore \quad \left.\frac{dl}{dt}\right]_{\text{at } l=4} = \frac{-16}{3\pi(4)^2} = \frac{-16}{3\pi(16)} = \frac{-1}{3\pi} \text{ cm/sec.}$$

$\therefore$ Rate of decrease of the slant height of water is $\frac{1}{3\pi}$ cm/sec.

(*ii*) Please try yourself.

[**Hint.** Proceed as in part (*i*)] $\left[\textbf{Ans.} -\frac{32}{27\pi} \text{ cm/sec.}\right]$

Example 18. *From a cylindrical drum containing oil and kept vertical, the oil is leaking at the rate of 10 cm³/sec. If the radius of the drum is 10 cm and height is 50 cm. Find the rate at which level of oil is changing when oil level is 20 cm.*

Solution. Let r be the radius, h be the height and V be the volume of the cylindrical drum at any time t.

$$\therefore \quad V = \pi r^2 h \quad \quad ...(1)$$

When radius $r = 10$ cm

$$\Rightarrow \quad V = \pi(10)^2 h = 100\pi h \quad \quad ...(2)$$

Differentiating w.r.t. t, we get

$$\frac{dV}{dt} = 100\pi \frac{dh}{dt} \quad \quad ...(3)$$

Now, rate of change (decrease) of volume of oil w.r.t. t

$$= \left(\frac{dV}{dt}\right) = -10 \text{ cm}^3\text{/sec.} \quad \quad ...(4) \quad [\text{Given}]$$

$\therefore$ From (3) and (4), we have

$$100\pi \frac{dh}{dt} = -10$$

$$\Rightarrow \quad \frac{dh}{dt} = \frac{-10}{100\pi} = \frac{-1}{10\pi} \text{ cm/sec.}$$

$\therefore$ Rate of decrease of oil level is $\frac{1}{10\pi}$ cm/sec.

Example 19. *The two equal sides of an isosceles triangle with fixed base b are decreasing at the rate of 3 cm/sec. How fast is the area decreasing when the two equal sides are equal to the base.*

Solution. Let ABC be an isosceles triangle with AB and AC are two equal sides.

Let $\quad AB = AC = x$ (say)

Let at any time t, A be the area of ΔABC.

$$\therefore \quad A = \frac{1}{2} (\text{base} \times \text{height})$$

$$A = \frac{1}{2} \times BC \times AD \quad \quad ...(1)$$

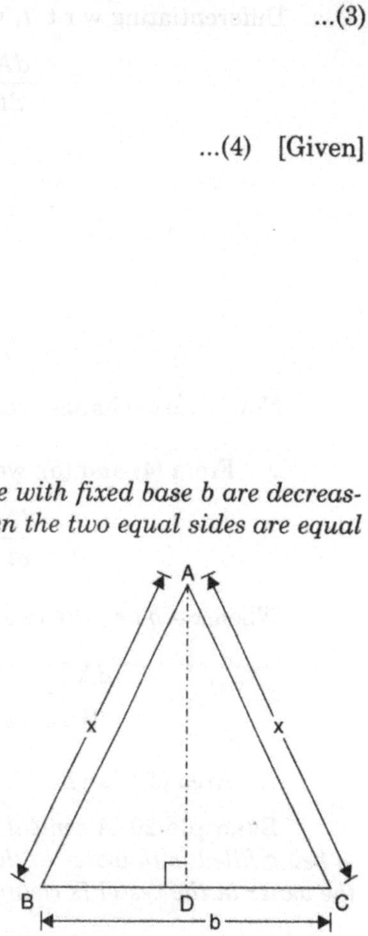

Since, ABC be an isosceles triangle, therefore height AD bisects BC at point D.

i.e., $$BD = DC = \frac{1}{2} DC = \frac{b}{2}$$

∴ In right triangle ADC :

$$\Rightarrow \quad (AC)^2 = AD^2 + DC^2 \quad \Rightarrow \quad x^2 = AD^2 + \left(\frac{b}{2}\right)^2$$

$$\Rightarrow \quad AD^2 = x^2 - \frac{b^2}{4} \quad \Rightarrow \quad AD = \sqrt{\frac{4x^2 - b^2}{4}}$$

$$\Rightarrow \quad AD = \frac{1}{2}\sqrt{4x^2 - b^2} \qquad ...(2)$$

∴ Equation (1) becomes :

$$A = \frac{1}{2} \times b \times \frac{1}{2}\sqrt{4x^2 - b^2}$$

$$= \frac{1}{4} b \sqrt{4x^2 - b^2} \qquad ...(3)$$

Differentiating w.r.t. t, we get

$$\frac{dA}{dt} = \frac{1}{4} b \frac{1}{2\sqrt{4x^2 - b^2}} \cdot \frac{d}{dt}(4x^2 - b^2)$$

$$= \frac{b}{8\sqrt{4x^2 - b^2}} (8x - 0) \cdot \frac{dx}{dt}$$

$$= \frac{bx}{\sqrt{4x^2 - b^2}} \cdot \frac{dx}{dt} \qquad ...(4)$$

Now, rate of change (decrease) of equal sides w.r.t. $t = \left(\frac{dx}{dt}\right) = -3$ cm/sec. ...(5) [Given]

∴ From (4) and (5), we have

$$\frac{dA}{dt} = \frac{bx}{\sqrt{4x^2 - b^2}} \cdot (-3) = \frac{-3bx}{\sqrt{4x^2 - b^2}} \text{ cm}^2\text{/sec.}$$

When $x = b$ *i.e.*, the two equal sides are equal to the base.

$$\therefore \quad \left.\frac{dA}{dt}\right]_{\text{at } x=b} = \frac{-3b.(b)}{\sqrt{4b^2 - b^2}} = \frac{-3b^2}{\sqrt{3b^2}} = -\sqrt{3}b \text{ cm}^2\text{/sec.}$$

∴ Area of triangle ABC is decreasing at the rate of $\sqrt{3}b$ cm²/sec.

Example 20. *A conical vessel whose height is 10 m and the radius of whose base is 5 m is being filled with water at the uniform rate of 1.5 m³/sec. Find the rate at which the level of the water in the vessel is rising, when the depth is 4 m.*

Solution. Let ABC be the cone having radius 5 m and height 10 m.

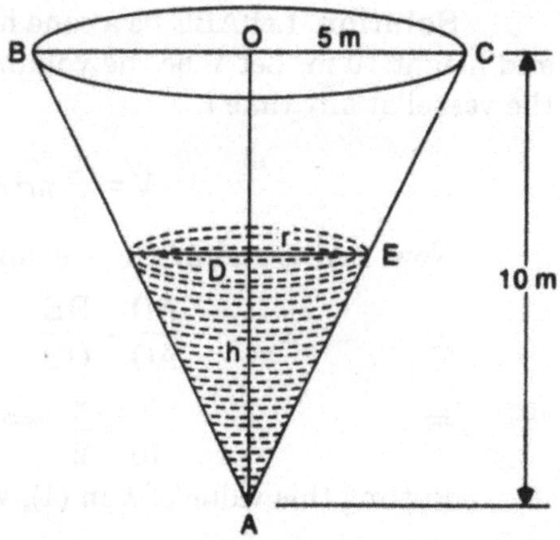

Let V be the volume of the water in the vessel at any time t.

$$\therefore \quad V = \frac{1}{3}\pi r^2 h \quad ...(1)$$

Now, ΔAOC and ΔADF are similar.

$$\therefore \quad \frac{DE}{OC} = \frac{DA}{OA}$$

$$\Rightarrow \quad \frac{r}{5} = \frac{h}{10}$$

$$\Rightarrow \quad r = \frac{5h}{10} \Rightarrow r = \frac{h}{2}$$

Putting this value of r in (1), we get

$$V = \frac{1}{3}\pi\left(\frac{h}{2}\right)^2 . h = \frac{1}{3}\pi\frac{h^2}{4} \times h = \frac{1}{12}\pi h^3$$

$$\Rightarrow \quad V = \frac{1}{12}\pi h^3 \quad ...(2)$$

Differentiating (2) w.r.t. t, we get

$$\frac{dV}{dt} = \frac{1}{12}\pi\, 3h^2 . \frac{dh}{dt} = \frac{1}{4}\pi h^2 \frac{dh}{dt} \quad ...(3)$$

Now, rate of change of volume of water w.r.t. $t = \frac{dV}{dt} = 1.5\ \text{m}^3/\text{sec.}$ ...(4) [Given]

$$= \frac{3}{2}\ \text{m}^3/\text{sec.}$$

$\therefore$ From (3) and (4), we get

$$\frac{3}{2} = \frac{1}{4}\pi h^2 . \frac{dh}{dt}$$

$$\Rightarrow \quad \pi h^2 \frac{dh}{dt} = 6 \Rightarrow \frac{dh}{dt} = \frac{6}{\pi h^2} \quad ...(5)$$

When $h = 4$ m

$$\therefore \quad \left.\frac{dh}{dt}\right]_{\text{at } h=4} = \frac{6}{\pi(4)^2} = \frac{6}{16\pi} = \frac{3}{8\pi}\ \text{m/sec.}$$

Example 21. *Height of a tank in the form of an inverted cone is 10 m and radius of its circular base is 2 m. The tank contains water and it is leaking through a hole at its vertex at the rate of 0.02 m^3/sec. Find the rate :*

(i) At which the water level changes.

(ii) At which the radius of water surface changes.

when the height of water level is 5 m.

Solution. Let ABC be a cone having radius 2 m and height 10 m. Let V be the volume of the water in the vessel at any time t.

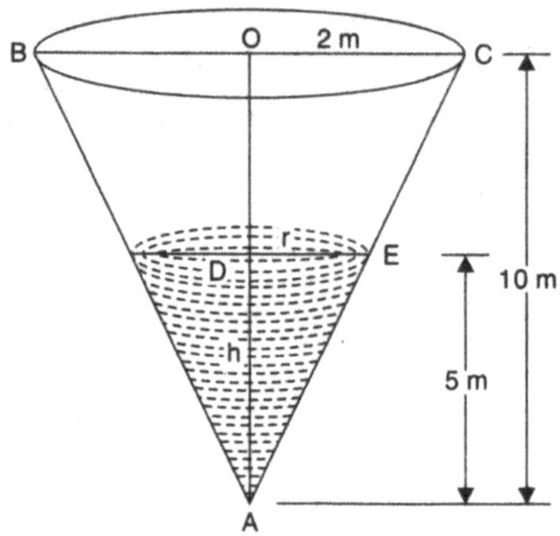

$$\therefore \quad V = \frac{1}{3}\pi r^2 h \quad ...(1)$$

Now, ΔADE and ΔAOC are similar.

$$\therefore \quad \frac{AD}{AO} = \frac{DE}{OC}$$

$$\Rightarrow \quad \frac{h}{10} = \frac{r}{2} \Rightarrow h = 5r$$

Putting this value of h in (1), we get

$$V = \frac{1}{3}\pi r^2 (5r) = \frac{5}{3}\pi r^3 \quad ...(2)$$

Differentiating w.r.t. t, we get

$$\frac{dV}{dt} = \frac{5}{3}\pi 3r^2 . \frac{dr}{dt} \Rightarrow \frac{dV}{dt} = 5\pi r^2 \frac{dr}{dt} \quad ...(3)$$

$\therefore$ Rate of change (decrease) of volume of water w.r.t. t

$$\frac{dV}{dt} = -0.02 \text{ m}^3/\text{sec.} \quad ...(4) \quad \text{[Given]}$$

$\therefore$ From (3) and (4), we have

$$-0.02 = 5\pi r^2 . \frac{dr}{dt} \Rightarrow \frac{dr}{dt} = -\frac{0.02}{5\pi r^2} \text{ m/sec.}$$

$\because h = 5r$
if $h = 5 \Rightarrow 5 = 5r$
$\Rightarrow r = 1$

When $r = 1$,

$$\therefore \quad \frac{dr}{dt} = \frac{-0.02}{5\pi(1)^2} = \frac{-0.02}{5\pi} = \frac{-0.004}{\pi} \text{ m/sec.} \quad ...(5)$$

$\therefore$ Rate of change of radius of water surface is $\frac{0.004}{\pi}$ m/sec.

Also, $h = 5r$

Differentiating w.r.t. t, we get

$$\frac{dh}{dt} = 5 . \frac{dr}{dt} \quad [\because \text{ By using (5)}]$$

$$= 5 . \left(\frac{-0.004}{\pi}\right) = \frac{-0.02}{\pi} \text{ m/sec.}$$

$\therefore$ Rate of change of water level w.r.t. t is $\left(\frac{0.02}{\pi}\right)$ m/sec.

Example 22. *Water is running into a conical vessel 15 cm deep and 5 cm in radius at the rate of 0.1 cm^3/sec when the water is 6 cm deep, find at what rate is :*

(i) The water level rising.
(ii) The water surface area increasing.
(iii) The water surface of the vessel increasing.

Solution. Let ABC be a cone having radius 5 cm and height 15 cm.

Let V be the volume of water in the vessel at any time t.

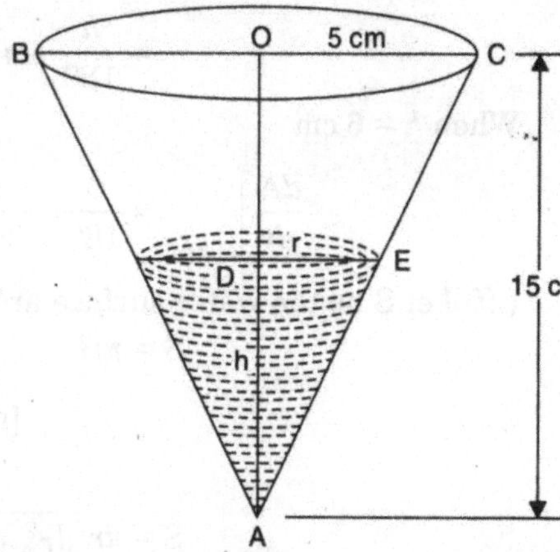

$\therefore \qquad V = \frac{1}{3} \pi r^2 h \qquad ...(1)$

Now, ΔADE and ΔAOC are similar

$$\therefore \qquad \frac{AD}{AO} = \frac{DE}{OC}$$

$$\frac{h}{15} = \frac{r}{5}$$

$$\Rightarrow \qquad h = 3r \quad \Rightarrow \quad r = \frac{h}{3}$$

Putting this value of r in (1), we get

$$(i) \qquad V = \frac{1}{3}\pi\left(\frac{h}{3}\right)^2 . h = \frac{1}{3}\pi \frac{h^2}{9} . h = \frac{1}{27}\pi h^3 \qquad ...(2)$$

Differentiating w.r.t. t, we get

$$\frac{dV}{dt} = \frac{1}{27}\pi 3h^2 . \frac{dh}{dt}$$

$$\frac{dV}{dt} = \frac{1}{9}\pi h^2 \frac{dh}{dt} \qquad ...(3)$$

$\therefore$ Rate of change of volume of water.w.r.t. $t = \frac{dV}{dt} = 0.1$ cm³/sec. ...(4) [Given]

$\therefore$ From (3) and (4), we have

$$\frac{1}{10} = \frac{1}{9}\pi h^2 \frac{dh}{dt} \quad \Rightarrow \quad \frac{dh}{dt} = \frac{9}{10\pi h^2} \text{ cm/sec.} \qquad ...(5)$$

When $h = 6$ cm

$$\therefore \qquad \left.\frac{dh}{dt}\right]_{\text{at } h=6} = \frac{9}{10\pi(6)^2} = \frac{9}{10\pi(30)} = \frac{1}{40\pi} \text{ cm/sec.} \qquad ...(6)$$

(*ii*) Let A be the water surface area at any time t.

$$\therefore \qquad A = \pi r^2 \qquad ...(7) \qquad \left[\because \ r = \frac{h}{3}\right]$$

$$\Rightarrow \qquad A = \pi\left(\frac{h}{3}\right)^2 = \frac{1}{9}\pi h^2$$

Differentiating w.r.t. t, we get

$$\frac{dA}{dt} = \frac{2}{9}\pi h . \frac{dh}{dt} = \frac{2}{9}\pi h\left(\frac{1}{40\pi}\right) \qquad [\because \text{ By using (6)}]$$

$$= \frac{h}{180} \text{ cm}^2/\text{sec.}$$

When $h = 6$ cm

$$\therefore \quad \left.\frac{dA}{dt}\right]_{\text{at } h=6} = \frac{6}{180} = \frac{1}{30} \text{ cm}^2/\text{sec.}$$

(*iii*) Let S be the water surface area of the vessel at any time t.

$$\therefore \quad S = \pi r l \qquad \ldots(8)$$

[where $l = \sqrt{r^2 + h^2}$ is the slant height of water-cone.]

$$\therefore \quad S = \pi r \sqrt{r^2 + h^2} = \frac{\pi h}{3}\sqrt{\left(\frac{h}{3}\right)^2 + h^2} \qquad \left[\because \ r = \frac{h}{3}\right]$$

$$= \frac{\pi h}{3}\sqrt{\frac{h^2}{9} + h^2} = \frac{\pi h}{3}\sqrt{\frac{h^2 + 9h^2}{9}} = \frac{\pi h}{3}\sqrt{\frac{10h^2}{9}} = \frac{\pi h}{3}\left(\frac{\sqrt{10}h}{3}\right)$$

$$\Rightarrow \quad S = \frac{\pi\sqrt{10}}{9} h^2 \qquad \ldots(9)$$

Differentiating w.r.t. t, we get

$$\frac{dS}{dt} = \frac{\pi\sqrt{10}}{9} . 2h . \frac{dh}{dt}$$

$$= \frac{2\pi\sqrt{10}h}{9}\left(\frac{1}{40\pi}\right) \qquad [\because \text{ By using (6)}]$$

$$= \frac{\sqrt{10}}{180} h \text{ cm}^2/\text{sec.}$$

When $h = 6$ cm

$$\therefore \quad \left.\frac{dS}{dt}\right]_{\text{at } h=6} = \frac{\sqrt{10}}{180}(6) = \frac{\sqrt{10}}{30} \text{ cm}^2/\text{sec.}$$

EXERCISE FOR PRACTICE

1. The radius of a spherical soap bubble is increasing at the rate of 0.2 cm/sec. Find the rate of increase of its surface area, when the radius is 7 cm.
2. A man 160 cm tall, walks away from a source of light situated at the top of a pole 6 m high at the rate of 1.1 m/sec. How fast is the length of his shadow increasing when he is 1 m away from the pole.
3. A particle moves along the curve $y = x^2 + 2x$. At what point(s) on the curve are the x and y coordinates of the particle changing at the same rate.
4. A balloon is in the form of a right circular cone surmounted by a hemisphere, having a diameter equal to the height of the cone, is being inflated. How fast is its volume changing with respect to its total height h, when $h = 9$ cm.

5. **The surface area of a spherical bubble is increasing at the rate of 2 cm^2/sec. When the radius of the bubble is 6 cm, at what rate is the volume of the bubble increasing.**
6. **A kite is 120 m high and 130 m string is out. If the kite is moving away horizontally at the rate of 52 m/sec. Find rate at which the string is being paid out.**
7. **An edge of a variable cube is increasing at the rate of 10 cm/sec. How fast is the volume of the cube increasing when the edge is 10 cm long.**
8. **If the side of an equilateral triangle increases uniformly at the rate of 3 cm/sec. At what rate per second is the area increasing, when the side is 10 cm.**

Answers

1. 11.2 π cm^2/sec	**2.** 0.4 m/sec	**3.** $\left(-\frac{1}{2}, -\frac{3}{4}\right)$ km/h	**4.** 12 π cm^3/sec
5. 6 cm^3/sec	**6.** 20 m/sec	**7.** 3000 cm^3/sec	**8.** $15\sqrt{3}$ cm^2/sec.

2

Tangents and Normals

2.1 INTRODUCTION

In this section, we shall use differentiation to find the equations of the tangent and normal to a curve at a given point. For a given point on a straight line and its slope, we can find the equation of straight line. As we have already learnt that the equation of a straight line passing through a given point $P(x_0, y_0)$ having slope m is given by : $y - y_0 = m(x - x_0)$.

By using the derivative of the equation of a curve, we can find the slope of any tangent of the curve, which helps us in finding the equations of tangent and normal to a given curve at any point on the curve.

2.2 SLOPE (GRADIENT) OF A LINE

Definition. (*i*) Slope of a line is defined as the trigonometric tangent of the angle made by the line with positive x-axis in anticlockwise direction and is denoted by m, *i.e.*, $m = \tan \theta$.

where θ is the angle which the line makes with positive x-axis in the anticlockwise direction. *e.g.*, slope of a line making an angle 45° with positive x-axis is $\tan 45° = 1$.

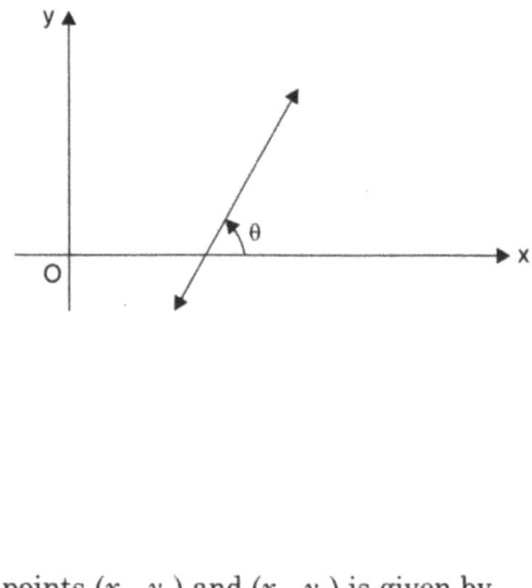

Definition. (*ii*) Slope of a line $ax + by + c = 0$ is given by

$$m = -\frac{\text{co-eff. of } x}{\text{co-eff. of } y} = -\frac{a}{b}$$

e.g., slope of line $5x + 3y + 4 = 0$ is $-\frac{5}{3}$.

Definition. (*iii*) Slope of a line joining the two points (x_1, y_1) and (x_2, y_2) is given by

$$m = \frac{y_2 - y_1}{x_2 - x_1}.$$

e.g., slope of the line joining the points (1, 2) and (3, 4) is $\frac{4-2}{3-1} = \frac{2}{2} = 1$.

2.3 EQUATION OF TANGENT AT ANY POINT OF THE CURVE y = f(x)

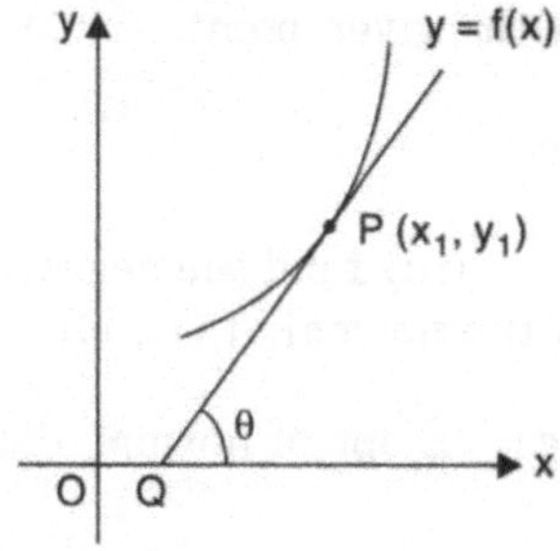

Let $P(x_1, y_1)$ be any point on the curve $y = f(x)$.

∴ Slope of the tangent at $P(x_1, y_1)$

$$m = \left(\frac{dy}{dx}\right)_{\text{at }(x_1, y_1)}.$$

Now, the tangent at $P(x_1, y_1)$ is a line through $P(x_1, y_1)$ having slope m.

∴ Equation of the tangent at $P(x_1, y_1)$ is given by

$$y - y_1 = m(x - x_1).$$

2.3.1. Rule to Find the Equation of the Tangent at a Point

(*i*) Find $\left(\frac{dy}{dx}\right)$ from the given equation of the curve. This gives the slope of the tangent at any point $P(x, y)$.

(*ii*) Find the value of $\frac{dy}{dx}$ at the given point (x_1, y_1). This gives the slope of the tangent at (x_1, y_1).

i.e.,

$$m = \left(\frac{dy}{dx}\right)_{\text{at }(x_1, y_1)}.$$

(*iii*) Now, the equation of the tangent is given by

$$(y - y_1) = m(x - x_1)$$

where m is the slope found in step (*ii*).

2.4 EQUATION OF THE NORMAL AT ANY POINT OF THE CURVE y = f(x)

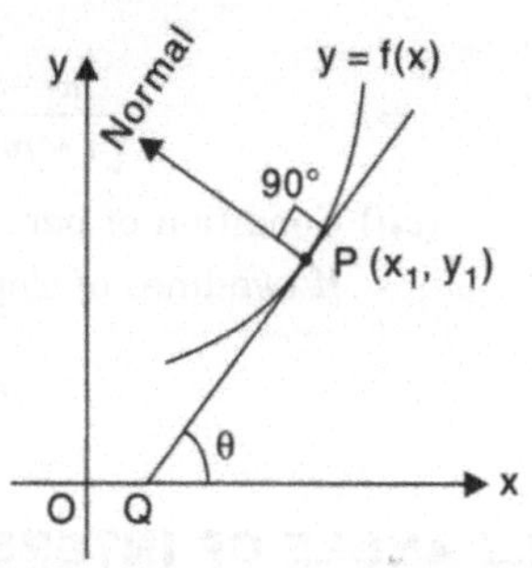

The normal to a curve at any point $P(x_1, y_1)$ is the straight line through the point perpendicular to the tangent to the curve at that point.

Let the slope of the tangent to the curve at $P(x_1, y_1)$ is m.

∴ Slope of normal to the curve at

$$P(x_1, y_1) = -\frac{1}{m}$$

= the negative reciprocal of the slope of tangent.

i.e.,

$$= -\frac{1}{m} = \frac{-1}{\left(\frac{dy}{dx}\right)_{\text{at }(x_1, y_1)}}.$$

∴ Equation of the normal to the curve at $P(x_1, y_1)$ is given by

$$(y - y_1) = m(x - x_1).$$

2.4.1 Rule to Find the Equation of the Normal at a Point

(*i*) Find $\left(\frac{dy}{dx}\right)$ from the equation of the curve. This gives the slope of the tangent at any point $P(x, y)$.

(*ii*) Find the value of $\left(\frac{dy}{dx}\right)$ at the given point (x_1, y_1). This gives the slope of the tangent at the given point.

i.e., $$m = \left(\frac{dy}{dx}\right)_{\text{at }(x_1, y_1)}.$$

(*iii*) Find the negative reciprocal of the slope of the tangent at (x_1, y_1). This gives the slope of the normal at (x_1, y_1).

i.e., Slope of normal $$= -\frac{1}{m} = -\frac{1}{\left(\frac{dy}{dx}\right)_{\text{at }(x_1, y_1)}}.$$

(*iv*) Now, the equation of the normal at (x_1, y_1) is given by

$$(y - y_1) = m(x - x_1)$$

where m is the slope found in step (*iii*).

Remember. (*i*) Slope of a line parallel to x-axis $= m = \tan 0° = 0$.

(*ii*) Slope of a line parallel to y-axis $= m = \tan 90° = \infty$.

$\therefore$ Slope of a vertical line is undefined.

(*iii*) Slope of a line equally inclined to the axes :

$$m = \tan 45° \text{ or } \tan 135° = +1 \text{ or } -1 = \pm 1.$$

(*iv*) Slope of a line making equal intercepts on the axes $m = \pm 1$.

(*v*) Angle between the two lines.

If θ is the acute angle between the two lines having slopes m_1 and m_2. Then ;

$$\tan \theta = \left|\frac{m_1 - m_2}{1 + m_1 m_2}\right|.$$

(*vi*) Condition of parallelism :

If the two lines are parallel, then $\theta = 0$

$\therefore$ $$\tan \theta = \tan 0° = 0.$$

$\Rightarrow$ $$\frac{m_1 - m_2}{1 + m_1 m_2} = 0 \Rightarrow m_1 - m_2 = 0 \Rightarrow m_1 = m_2.$$

(*vii*) Condition of perpendicularity :

If two lines of slopes m_1 and m_2 are perpendicular, then ;

$$m_1 m_2 = -1 \quad \text{or} \quad m_2 = -\frac{1}{m_1}.$$

2.5 ANGLE OF INTERSECTION OF TWO CURVES

The angle of intersection of two curves is defined as the angle between the tangents to the curves at their point of intersection.

Let us consider that the two curves $y = f(x)$ and $y = \phi(x)$ intersecting at point $P(x_1, y_1)$.

Let m_1 and m_2 be the slopes of the tangents at P to the curves. If θ be the angle between the tangents at P, then ;

$$\tan \theta = \left|\frac{m_1 - m_2}{1 + m_1 m_2}\right|.$$

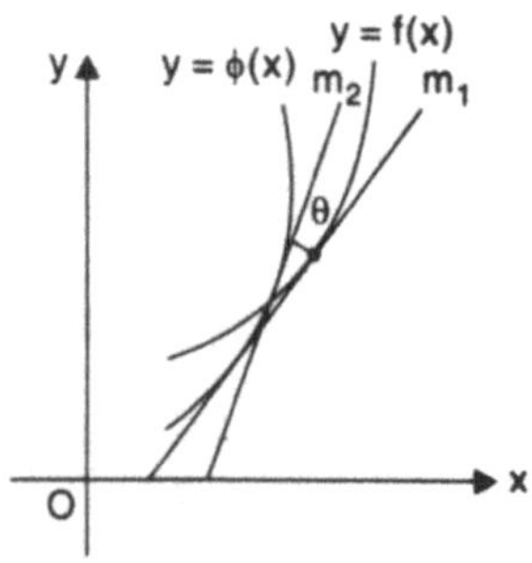

Remark. (*i*) If $m_1 m_2 = -1$, the curves cut orthogonally and if $m_1 = m_2$, the curves touches each other.

(*ii*) **Orthogonal Curves.** If the angle of intersection of two curves is a right angle, then the two curves are said to intersect orthogonally and the curves are called orthogonal curves.

SOME SOLVED EXAMPLES

Example 1. *Find the slopes of tangents and normals to the following curves :*

(*i*) $y = 8x^2 - 3$ *at* $x = \frac{1}{4}$ (*ii*) $x^2 + 3y + y^2 = 5$ *at* $(1, 1)$

(*iii*) $y = \cos x$ *at* $x = \frac{\pi}{4}$ (*iv*) $y = \sin x$ *at* $x = \frac{\pi}{4}$

(*v*) $x = a(\theta - \sin\theta),\ y = a(1 - \cos\theta)$ *at* $\theta = \frac{\pi}{2}$

(*vi*) $x^3 + y^3 = 3axy$ *at* $\left(\frac{3a}{2}, \frac{3a}{2}\right)$ (*vii*) $4x^2 + 9y^2 - 36 = 0$ *at* $\left(\frac{3\sqrt{2}}{2}, \sqrt{2}\right)$

(*viii*) $x = (1 - a\sin\theta),\ y = b\cos^2\theta$ *at* $\theta = \frac{\pi}{2}$

(*ix*) $x = \theta + \sin\theta,\ y = 1 + \cos\theta$ *at* $\theta = \frac{\pi}{2}$ (*x*) $x = a\cos^3\theta,\ y = a\sin^3\theta$ *at* $\theta = \frac{\pi}{4}$.

Solution. (*i*) We have, $y = 8x^2 - 3$ ...(1)

Differentiating w.r.t. x, we get

$$\frac{dy}{dx} = 16x$$

$\therefore$ Slope of tangent at $x = \frac{1}{4}$, *i.e.*, $\left(\frac{dy}{dx}\right)_{\text{at } x=\frac{1}{4}} = 16\left(\frac{1}{4}\right) = 4.$

$\therefore$ Slope of normal at $x = \frac{1}{4}$, *i.e.*, $\frac{-1}{\left(\frac{dy}{dx}\right)_{\text{at } x=\frac{1}{4}}} = \frac{-1}{4}$.

(*ii*) We have, $x^2 + 3y + y^2 = 5$ ...(1)

Differentiating w.r.t. x, we get

$$2x + 3\frac{dy}{dx} + 2y.\frac{dy}{dx} = 0$$

$$\Rightarrow \quad (3 + 2y)\frac{dy}{dx} = -2x \Rightarrow \left(\frac{dy}{dx}\right) = \frac{-2x}{3 + 2y}.$$

$\therefore$ Slope of tangent at (1, 1), *i.e.*, $\left(\frac{dy}{dx}\right)_{(1,1)} = \frac{-2(1)}{3 + 2(1)} = \frac{-2}{5}$

Slope of normal at (1, 1), *i.e.*, $\dfrac{-1}{\left(\dfrac{dy}{dx}\right)_{(1,1)}} = \dfrac{-1}{-2/5} = \dfrac{5}{2}$.

(*iii*) We have, $\quad y = \cos x \quad$ ...(1)

Differentiating w.r.t. x, we get

$$\frac{dy}{dx} = -\sin x$$

$\therefore$ Slope of tangent at $x = \dfrac{\pi}{4}$, *i.e.*, $\left(\dfrac{dy}{dx}\right)_{x=\frac{\pi}{4}} = -\sin\dfrac{\pi}{4} = \dfrac{-1}{\sqrt{2}}$

Slope of normal at $x = \dfrac{\pi}{4}$, *i.e.*, $\dfrac{-1}{\left(\dfrac{dy}{dx}\right)_{x=\frac{\pi}{4}}} = \dfrac{-1}{-1/\sqrt{2}} = \sqrt{2}$.

(*iv*) We have, $\quad y = \sin x \quad$ ...(1)

Differentiating w.r.t. x, we get

$$\frac{dy}{dx} = \cos x$$

$\therefore$ Slope of tangent at $x = \dfrac{\pi}{4}$, *i.e.*, $\left(\dfrac{dy}{dx}\right)_{x=\frac{\pi}{4}} = \cos\dfrac{\pi}{4} = \dfrac{1}{\sqrt{2}}$

$\therefore$ Slope of normal at $x = \dfrac{\pi}{4}$, *i.e.*, $\dfrac{-1}{\left(\dfrac{dy}{dx}\right)_{x=\frac{\pi}{4}}} = \dfrac{-1}{1/\sqrt{2}} = -\sqrt{2}$.

(*v*) We have, $\quad x = a(\theta - \sin\theta)$ and $y = a(1 - \cos\theta)$

$\Rightarrow \quad \dfrac{dx}{d\theta} = a(1 - \cos\theta)$ and $\dfrac{dy}{d\theta} = a\,[0 - (-\sin\theta)]$

$$\frac{dy}{d\theta} = a\sin\theta.$$

$$\left[\begin{array}{ll} \because & \sin 2A = 2\sin A\cos A \\ \Rightarrow & \sin A = 2\sin\dfrac{A}{2}\cos\dfrac{A}{2} \\ & 1-\cos 2A = 2\sin^2 A \\ \Rightarrow & 1-\cos A = 2\sin^2\dfrac{A}{2} \end{array}\right]$$

$\therefore$
$$\frac{dy}{dx} = \frac{dy/d\theta}{dx/d\theta} = \frac{a\sin\theta}{a(1-\cos\theta)}$$
$$= \frac{2\sin\theta/2\cos\theta/2}{2\sin^2\theta/2}$$
$$= \frac{\cos\theta/2}{\sin\theta/2} = \cot\frac{\theta}{2}.$$

$\therefore$ Slope of tangent at $\theta = \dfrac{\pi}{2}$, *i.e.*, $\left(\dfrac{dy}{dx}\right)_{\theta=\frac{\pi}{2}} = \cot\left(\dfrac{\pi/2}{2}\right) = \cot\dfrac{\pi}{4} = 1$.

$\therefore$ Slope of normal at $\theta = \frac{\pi}{2}$, *i.e.*, $\frac{-1}{\left(\frac{dy}{dx}\right)_{\theta=\frac{\pi}{2}}} = \frac{-1}{1} = -1.$

(*vi*) We have, $x^3 + y^3 = 3axy$ ...(1)

Differentiating w.r.t. x, we get

$$3x^2 + 3y^2 \cdot \frac{dy}{dx} = 3a\left(x \cdot \frac{dy}{dx} + y.1\right)$$

$$\Rightarrow \quad 3x^2 + 3y^2 \frac{dy}{dx} = 3ax \frac{dy}{dx} + 3ay$$

$$\Rightarrow \quad 3(y^2 - ax) \frac{dy}{dx} = 3(ay - x^2)$$

$$\Rightarrow \quad \frac{dy}{dx} = \left(\frac{ay - x^2}{y^2 - ax}\right).$$

$\therefore$ Slope of tangent at $\left(\frac{3a}{2}, \frac{3a}{2}\right)$,

i.e., $$\left(\frac{dy}{dx}\right)_{\left(\frac{3a}{2}, \frac{3a}{2}\right)} = \frac{a\left(\frac{3a}{2}\right) - \left(\frac{3a}{2}\right)^2}{\left(\frac{3a}{2}\right)^2 - a\left(\frac{3a}{2}\right)} = \frac{\frac{3a}{2}\left[a - \frac{3a}{2}\right]}{\frac{3a}{2}\left[\frac{3a}{2} - a\right]} = \frac{-\frac{a}{2}}{+\frac{a}{2}} = -1.$$

$\therefore$ Slope of normal at $\left(\frac{3a}{2}, \frac{3a}{2}\right)$, *i.e.*, $\frac{-1}{\left(\frac{dy}{dx}\right)_{\left(\frac{3a}{2}, \frac{3a}{2}\right)}} = \frac{-1}{-1} = 1.$

(*vii*) We have, $4x^2 + 9y^2 - 36 = 0$ ...(1)

Differentiating (1) w.r.t. x, we get

$$8x + 18y \frac{dy}{dx} - 0 = 0 \quad \Rightarrow \quad 18y \frac{dy}{dx} = -8x$$

$$\Rightarrow \quad \frac{dy}{dx} = \frac{-8x}{18y} = \frac{-4x}{9y}$$

$\therefore$ Slope of tangent at $\left(\frac{3\sqrt{2}}{2}, \sqrt{2}\right)$, *i.e.*, $\left(\frac{dy}{dx}\right)_{\left(\frac{3\sqrt{2}}{2}, \sqrt{2}\right)} = \frac{-4\left(\frac{3\sqrt{2}}{2}\right)}{9\left(\sqrt{2}\right)} = \frac{-2}{3}$

$\therefore$ Slope of normal at $\left(\frac{3\sqrt{2}}{2}, \sqrt{2}\right)$, *i.e.*, $\frac{-1}{\left(\frac{dy}{dx}\right)_{\left(\frac{3\sqrt{2}}{2}, \sqrt{2}\right)}} = \frac{-1}{-\frac{2}{3}} = \frac{3}{2}.$

(*viii*) We have, $x = 1 - a\sin\theta$ and $y = b\cos^2\theta$

$$\Rightarrow \quad \frac{dx}{d\theta} = 0 - a\cos\theta \quad \text{and} \quad \frac{dy}{d\theta} = 2b\cos\theta \cdot \frac{d}{d\theta}(\cos\theta)$$

$$\Rightarrow \quad \frac{dx}{d\theta} = -a\cos\theta \quad \text{and} \quad \frac{dy}{d\theta} = 2b\cos\theta(-\sin\theta)$$

$$\frac{dy}{d\theta} = -2b\cos\theta\sin\theta$$

$$\therefore \quad \frac{dy}{dx} = \frac{dy/d\theta}{dx/d\theta} = \frac{-2b\cos\theta\sin\theta}{-a\cos\theta} = \frac{2b}{a}\sin\theta$$

$\therefore$ Slope of tangent at $\theta = \frac{\pi}{2}$, *i.e.*, $\left(\frac{dy}{dx}\right)_{\theta=\frac{\pi}{2}} = \frac{2b}{a}\sin\frac{\pi}{2} = \frac{2b}{a}\cdot 1 = \frac{2b}{a}$.

$\therefore$ Slope of normal at $\theta = \frac{\pi}{2}$, *i.e.*, $\frac{-1}{\left(\frac{dy}{dx}\right)_{\theta=\frac{\pi}{2}}} = \frac{-1}{\frac{2b}{a}} = \frac{-a}{2b}$.

(*ix*) We have, $x = \theta + \sin\theta$ and $y = 1 + \cos\theta$

$$\Rightarrow \quad \frac{dx}{d\theta} = 1 + \cos\theta \quad \text{and} \quad \frac{dy}{d\theta} = 0 - \sin\theta = -\sin\theta.$$

$$\therefore \quad \frac{dy}{dx} = \frac{dy/d\theta}{dx/d\theta} = \frac{-\sin\theta}{1+\cos\theta}$$

$$\left[\begin{array}{ll} \because & \sin 2A = 2\sin A\cos A \\ \Rightarrow & \sin A = 2\sin\frac{A}{2}\cos\frac{A}{2} \\ & 1+\cos 2A = 2\cos^2 A \\ \Rightarrow & 1+\cos A = 2\cos^2\frac{A}{2} \end{array}\right]$$

$$= \frac{-2\sin\frac{\theta}{2}\cos\frac{\theta}{2}}{2\cos^2\frac{\theta}{2}} = -\frac{\sin\frac{\theta}{2}}{\cos\frac{\theta}{2}} = -\tan\frac{\theta}{2}$$

$\therefore$ Slope of tangent at $\theta = \frac{\pi}{2}$, *i.e.*, $\left(\frac{dy}{dx}\right)_{\theta=\frac{\pi}{2}} = -\tan\left(\frac{\pi/2}{2}\right) = -\tan\frac{\pi}{4} = -1$

$\therefore$ Slope of normal at $\theta = \frac{\pi}{2}$, *i.e.*, $\frac{-1}{\left(\frac{dy}{dx}\right)_{\theta=\frac{\pi}{2}}} = \frac{-1}{-1} = 1$.

(*x*) We have, $x = a\cos^3\theta$ and $y = a\sin^3\theta$

$$\Rightarrow \quad \frac{dx}{d\theta} = 3a\cos^2\theta \cdot \frac{d}{d\theta}(\cos\theta) \quad \text{and} \quad \frac{dy}{d\theta} = 3a\sin^2\theta \cdot \frac{d}{d\theta}(\sin\theta)$$

$$\Rightarrow \quad \frac{dx}{d\theta} = 3a\cos^2\theta\,(-\sin\theta) \quad \text{and} \quad \frac{dy}{d\theta} = 3a\sin^2\theta\,(\cos\theta)$$

$$\Rightarrow \quad \frac{dx}{d\theta} = -3a\cos^2\theta\sin\theta \quad \text{and} \quad \frac{dy}{d\theta} = 3a\sin^2\theta\cos\theta$$

$$\therefore \quad \frac{dy}{dx} = \frac{dy/d\theta}{dx/d\theta} = \frac{3a\sin^2\theta\cos\theta}{-3a\cos^2\theta\sin\theta} = -\frac{\sin\theta}{\cos\theta} = -\tan\theta.$$

$\therefore$ Slope of tangent at $\theta = \frac{\pi}{4}$, *i.e.*, $\left(\frac{dy}{dx}\right)_{\theta=\frac{\pi}{4}} = -\tan\frac{\pi}{4} = -1$

$\therefore$ Slope of normal at $\theta = \frac{\pi}{4}$, *i.e.*, $\frac{-1}{\left(\frac{dy}{dx}\right)_{\theta=\frac{\pi}{4}}} = \frac{-1}{-1} = 1.$

Example 2. (*i*) *Find the points on the curve $y = x^3 - 3x^2 - 9x + 7$ at which the tangents are parallel to the x-axis.*

(*ii*) *Find the points on the curve $3b^2y = x^3 - 3ax^2$ at which the tangent is parallel to the x-axis.*

Solution. (*i*) We have, $y = x^3 - 3x^2 - 9x + 7$ ...(1)

Differentiating w.r.t. x, we get

$$\frac{dy}{dx} = 3x^2 - 6x - 9 \quad \text{...(2)}$$

Since, the tangent is parallel to the x-axis.

$\therefore$ Slope of tangent *i.e.*, $\left(\frac{dy}{dx}\right) = 0$

$\Rightarrow \quad 3x^2 - 6x - 9 = 0$ [$\because$ By using (2)]

$\Rightarrow \quad x^2 - 2x - 3 = 0$

$\Rightarrow \quad (x - 3)(x + 1) = 0$

$\Rightarrow \quad x = 3, -1$

$\therefore$ When $x = 3$,

$\therefore$ From (1), we have, $y = (3)^3 - 3(3)^2 - 9(3) + 7$

$= 27 - 27 - 27 + 7 = -20$

When $x = -1$,

$\therefore$ From (1), we have, $y = (-1)^3 - 3(-1)^2 - 9(-1) + 7$

$= -1 - 3 + 9 + 7 = 12$

$\therefore$ The required points are $(3, -20)$ and $(-1, 12)$.

(*ii*) We have, $3b^2y = x^3 - 3ax^2$ ...(1)

Differentiating w.r.t. x, we get

$$\Rightarrow \quad 3b^2\frac{dy}{dx} = 3x^2 - 6ax$$

$$\Rightarrow \qquad \frac{dy}{dx} = \frac{3x(x-2a)}{3b^2} = \frac{x(x-2a)}{b^2} \qquad ...(2)$$

Since, the tangent is parallel to the x-axis.

$\therefore$ Slope of tangent *i.e.*, $\left(\frac{dy}{dx}\right) = 0 \Rightarrow \frac{x(x-2a)}{b^2} = 0$ [$\because$ By using (2)]

$\Rightarrow \quad x(x-2a) = 0$

$\Rightarrow \quad x = 0, 2a$

$\therefore$ When $x = 0$,

$\therefore$ From (1), we have $3b^2y = 0 - 3a \cdot 0 = 0 \Rightarrow y = 0$

When $x = 2a$,

$\therefore$ From (1), we have $3b^2y = (2a)^3 - 3a(2a)^2$

$= 8a^3 - 3a(4a^2) = 8a^3 - 12a^3 = -4a^3$

$\Rightarrow \quad y = \frac{-4a^3}{3b^2}.$

$\therefore$ The required points are (0, 0) and $\left(2a, \frac{-4a^3}{3b^2}\right)$.

Example 3. (*i*) *At what points on the curve $y = \sin x$ is the tangent parallel to the x-axis.*

(*ii*) *Find the points on the curve $x^2 + y^2 - 2x - 3 = 0$ at which the tangents are parallel to the x-axis.*

(*iii*) *Find the point on the curve $y = 2x^2 - 6x - 4$ at which the tangent is parallel to the x-axis.*

Solution. (*i*) We have, $y = \sin x$ ...(1)

Differentiating w.r.t. x, we get

$$\frac{dy}{dx} = \cos x \qquad ...(2)$$

Since, the tangent is parallel to the axis.

$\therefore$ Slope of tangent *i.e.*, $\left(\frac{dy}{dx}\right) = 0$

$\Rightarrow \quad \cos x = 0$ [$\because$ By using (2)]

$\Rightarrow \quad \cos x = 0 = \cos (2n+1)\frac{\pi}{2}; \quad n \in Z$

$\Rightarrow \quad x = (2n+1)\frac{\pi}{2}; \quad n \in Z.$

$\therefore$ When $x = (2n+1)\frac{\pi}{2}$

$\therefore$ From (1), we have $y = \sin (2n+1)\frac{\pi}{2} = \sin\left(n\pi + \frac{\pi}{2}\right)$

$= (-1)^n \cdot \sin \frac{\pi}{2}$ [$\because$ $\sin (n\pi + \theta) = (-1)^n \sin \theta$]

$= (-1)^n$

$\therefore$ The required points are $\left((2n+1)\frac{\pi}{2}, (-1)^n\right)$; $n \in Z$.

(*ii*) We have, $x^2 + y^2 - 2x - 3 = 0$ ...(1)

Differentiating w.r.t. x, we have

$$2x + 2y \cdot \frac{dy}{dx} - 2 - 0 = 0 \Rightarrow 2y\frac{dy}{dx} = 2 - 2x$$

$$\Rightarrow \quad \frac{dy}{dx} = \frac{2(1-x)}{2y} = \frac{1-x}{y} \quad ...(2)$$

Since, the tangent is parallel to the x-axis.

$\therefore$ Slope of tangent *i.e.*, $\left(\frac{dy}{dx}\right) = 0 \Rightarrow \frac{1-x}{y} = 0$ [$\because$ By using (2)]

$$\Rightarrow \quad 1 - x = 0 \Rightarrow x = 1$$

$\therefore$ When $x = 1$

From (1), we have

$$(1)^2 + y^2 - 2(1) - 3 = 0$$

$$\Rightarrow \quad y^2 = 4 \Rightarrow y = \pm 2$$

$\therefore$ The required points are (1, 2) and (1, – 2).

(*iii*) Please try yourself.

$$\left[\textbf{Ans.} \left(\frac{3}{2}, \frac{-17}{2}\right)\right]$$

Example 4. (*i*) *At what points on the circle* $x^2 + y^2 - 2x - 4y + 1 = 0$, *the tangent is (a) parallel to x-axis. (b) parallel to y-axis.*

(*ii*) *Find the points on the curve* $\frac{x^2}{9} - \frac{y^2}{16} = 1$ *at which the tangents are parallel to the : (a) x-axis (b) y-axis.*

(*iii*) *Find the points on the curve* $\frac{x^2}{4} + \frac{y^2}{25} = 1$ *at which the tangents are parallel to the (a) x-axis (b) y-axis.*

Solution. (*i*) We have

$$x^2 + y^2 - 2x - 4y + 1 = 0 \quad ...(1)$$

Differentiating w.r.t. x, we get

$$2x + 2y\frac{dy}{dx} - 2 - 4\frac{dy}{dx} = 0$$

$$\Rightarrow \quad (2y - 4)\frac{dy}{dx} = 2 - 2x$$

$$\Rightarrow \quad \frac{dy}{dx} = \frac{2(1-x)}{2(y-2)} = \frac{1-x}{y-2} \quad ...(2)$$

(*a*) If the tangent is parallel to x-axis

$$\therefore \quad \frac{dy}{dx} = 0 \Rightarrow \frac{1-x}{y-2} = 0 \quad [\because \text{ By using (2)}]$$

$$\Rightarrow \quad 1 - x = 0 \Rightarrow x = 1.$$

Putting this value of x in (1), we get

$$\Rightarrow \quad (1)^2 + y^2 - 2(1) - 4y + 1 = 0$$

$$\Rightarrow \quad y^2 - 4y = 0 \Rightarrow y(y-4) = 0 \Rightarrow y = 0, 4.$$

$\therefore$ The required points are (1, 0) and (1, 4).

(*b*) If the tangent is parallel to y-axis

$$\therefore \quad \frac{dx}{dy} = 0 \Rightarrow \frac{y-2}{1-x} = 0 \qquad \left[\because \text{ From (2), } \frac{dy}{dx} = \frac{1-x}{y-2} \Rightarrow \frac{dx}{dy} = \frac{1}{\left(\frac{1-x}{y-2}\right)} = \frac{y-2}{1-x}\right]$$

$$\Rightarrow \quad y - 2 = 0 \Rightarrow y = 2$$

$\therefore$ Putting this value of y in (1), we get

$$x^2 + (2)^2 - 2x - 4(2) + 1 = 0$$

$$\Rightarrow \quad x^2 + 4 - 2x - 8 + 1 = 0$$

$$\Rightarrow \quad x^2 - 2x - 3 = 0$$

$$\Rightarrow \quad (x-3)(x+1) = 0$$

$$\Rightarrow \quad x = 3, -1.$$

$\therefore$ The required points are (3, 2) and (– 1, 2).

(*ii*) We have, $\dfrac{x^2}{9} - \dfrac{y^2}{16} = 1$ ...(1)

Differentiating w.r.t. x, we get

$$\frac{2x}{9} - \frac{2y}{16} \cdot \frac{dy}{dx} = 0$$

$$\Rightarrow \quad -\frac{1}{8} y \cdot \frac{dy}{dx} = -\frac{2x}{9}$$

$$\Rightarrow \quad \frac{dy}{dx} = \frac{2x}{9} \times \frac{8}{y} = \frac{16x}{9y} \qquad \text{...(2)}$$

(*a*) If the tangent is parallel to the x-axis :

$$\therefore \quad \frac{dy}{dx} = 0 \Rightarrow \frac{16x}{9y} = 0 \qquad [\because \text{ By using (2)}]$$

$$\Rightarrow \quad 16x = 0 \Rightarrow x = 0.$$

Putting this value of x in (1), we get

$$\Rightarrow \quad \frac{0}{9} - \frac{y^2}{16} = 1$$

$$\Rightarrow \quad -y^2 = 16 \Rightarrow y^2 = -16 \text{ which is not possible.}$$

$\therefore$ There is no point on the curve where tangent is parallel to x-axis.

(*b*) If the tangent is parallel to y-axis.

$$\therefore \quad \frac{dx}{dy} = 0 \qquad \left[\because \text{ From (2), } \frac{dy}{dx} = \frac{16x}{9y} \Rightarrow \frac{dx}{dy} = \frac{9y}{16x}\right]$$

$$\Rightarrow \quad \frac{9y}{16x} = 0 \Rightarrow 9y = 0 \Rightarrow y = 0$$

Putting this value of y in (1), we get

$$\frac{x^2}{9} - \frac{0}{16} = 1 \Rightarrow x^2 = 9 \Rightarrow x = \pm 3.$$

$\therefore$ The required points are (3, 0) and (– 3, 0).

(*iii*) Please try yourself.

[**Hint.** Proceed as part (*ii*)]

[**Ans.** (*a*) If the tangent is parallel to x-axis (0, 5) and (0, – 5).
(*b*) If the tangent is parallel to y-axis (2, 0) and (– 2, 0).]

Example 5. (*i*) *Show that the tangents to the curve $y = x^2 - 5x + 6$ at the point (2, 0) and (3, 0) are at right angle.*

(*ii*) *Show that the tangents to the curve $y = 2x^3 - 3$ at the points where $x = 2$ and $x = -2$ are parallel.*

Solution. (*i*) We have, $\quad y = x^2 - 5x + 6 \quad$ ...(1)

Differentiating w.r.t. x, we get

$$\frac{dy}{dx} = 2x - 5.$$

$\therefore$ Slope of tangent at (2, 0) = m_1, *i.e.*, $\left(\frac{dy}{dx}\right)_{(2,0)} = 2(2) - 5 = 4 - 5 = -1$

Slope of tangent at (3, 0) = m_2, *i.e.*, $\left(\frac{dy}{dx}\right)_{(3,0)} = 2(3) - 5 = 6 - 5 = 1.$

Now, $\quad m_1 \cdot m_2 = (-1)(1) = -1.$

Since, the tangents are perpendicular if $m_1 m_2 = -1$.

$\therefore$ Tangents at (2, 0) and (3, 0) are perpendicular to each other.

(*ii*) We have, $\quad y = 2x^3 - 3 \quad$ ...(1)

Differentiating w.r.t. x, we get

$$\frac{dy}{dx} = 6x^2$$

$\therefore$ Slope of tangent at $x = 2 = m_1$, *i.e.*, $\left(\frac{dy}{dx}\right)_{x=2} = 6(2)^2 = 6(4) = 24.$

$\therefore$ Slope of tangent at $x = -2 = m_2$, *i.e.*, $\left(\frac{dy}{dx}\right)_{x=-2} = 6(-2)^2 = 6(4) = 24$

$\therefore \quad m_1 = m_2 = 24.$

Since the tangents are parallel if $m_1 = m_2$.

$\therefore$ The tangents at $x = 2$ and $x = -2$ are parallel to each other.

Example 6. (*i*) *Find the point at which the tangent to the curve* $y = \sqrt{4x-3} - 1$ *has slope* $\frac{2}{3}$.

(*ii*) *Find the point on the curve* $y = 3x^2 + 4$ *at which the tangent is perpendicular to a line whose slope is* $\frac{-1}{6}$.

(*iii*) *Determine the point at which the normal to the curve* $y = x^2 - 4x + 3$ *is parallel to the line whose slope is* $\frac{1}{2}$.

Solution. (*i*) We have $\qquad y = \sqrt{4x-3} - 1 \qquad$ ...(1)

Differentiating w.r.t. x, we get

$$\frac{dy}{dx} = \frac{1}{2\sqrt{4x-3}} \cdot \frac{d}{dx}(4x-3) - 0$$

$$= \frac{1}{2\sqrt{4x-3}}(4-0)$$

$\therefore$ Slope of tangent, *i.e.*, $\frac{dy}{dx} = \frac{2}{\sqrt{4x-3}} \qquad$ ...(2)

Now, the slope of tangent $\left(\frac{dy}{dx}\right) = \frac{2}{3} \qquad$ ...(3) [Given]

$\therefore$ We have, $\frac{2}{\sqrt{4x-3}} = \frac{2}{3}$

$\Rightarrow \quad \sqrt{4x-3} = 3$

$\Rightarrow \quad 4x - 3 = 9 \qquad$ (Squaring both sides)

$\Rightarrow \quad 4x = 9 + 3 \Rightarrow 4x = 12 \Rightarrow x = 3$

When $x = 3$.

$\therefore$ From (1), we have, $y = \sqrt{4(3)-3} - 1 = \sqrt{12-3} - 1 = \sqrt{9} - 1 = 3 - 1 = 2.$

$\therefore$ The required point is (3, 2).

(*ii*) We have, $\qquad y = 3x^2 + 4 \qquad$ ...(1)

Differentiating w.r.t. x, we get

$\therefore$ Slope of tangent $= \left(\frac{dy}{dx}\right) = 6x + 0 = 6x \qquad$ ...(2)

Now, the slope of the line perpendicular to the tangent $= \frac{-1}{\left(\frac{dy}{dx}\right)} = \frac{-1}{6x} \qquad$ ...(3)

Also, the slope the line perpendicular to the tangent $= \frac{-1}{6} \qquad$ ...(4) [Given]

$\therefore$ We have, $\frac{-1}{6x}=\frac{-1}{6} \Rightarrow x=1$

When $x = 1$, from (1), we have

$$y = 3(1)^2 + 4 = 3 + 4 = 7$$

$\therefore$ The required point is (1, 7).

(*iii*) We have, $y = x^2 - 4x + 3$ ...(1)

Differentiating w.r.t. x, we get

$\therefore$ Slope of tangent $=\left(\frac{dy}{dx}\right) = 2x - 4$ ...(2)

$\therefore$ Slope of the normal $=\frac{-1}{\left(\frac{dy}{dx}\right)}=\frac{-1}{(2x-4)}$ ...(3)

Also, the slope of the normal to the curve is parallel to the line

$$=\frac{1}{2} \quad ...(4) \quad \text{[Given]}$$

$\therefore$ We have, $\frac{-1}{2x-4}=\frac{1}{2} \Rightarrow 2x-4=-2$

$\Rightarrow$ $2x = -2 + 4 \Rightarrow 2x = 2$

$\Rightarrow$ $x = 1$

When $x = 1$,

$\therefore$ From (1), we have

$$y = (1)^2 - 4(1) + 3$$
$$y = 1 - 4 + 3 = 0.$$

$\therefore$ The required point is (1, 0).

Example 7. (*i*) *Find a point on the curve $y = (x - 2)^2$ where the tangent is parallel to the chord joining (2, 0) and (4, 4).*

(*ii*) *Find the point on the curve $y = (x - 3)^2$ where the tangent is parallel to the line joining (3, 0) and (4, 1).*

Solution. (*i*) We have, $y = (x - 2)^2$ ...(1)

Differentiating w.r.t. x, we get

$\therefore$ Slope of tangent $=\frac{dy}{dx} = 2(x-2)$ ...(2)

Now, the slope of chord joining the points (2, 0) and (4, 4) is given by

$$m=\frac{4-0}{4-2}=\frac{4}{2}=2 \quad ...(3) \quad \left[\because m=\frac{y_2-y_1}{x_2-x_1}\right]$$

If the tangent is parallel to the chord joining the points (2, 0) and (4, 4), then, their slopes are equal.

$\therefore$ From (2) and (3), we have

$$2(x-2) = 2 \Rightarrow x-2=1 \Rightarrow x=3$$

When $x = 1$,

$\therefore$ From (1), we have

$$y = (3 - 2)^2 = (1)^2 = 1$$

$\therefore$ The required point is (3, 1).

(*ii*) Please try yourself.

[**Hint.** Proceed as in Part (*i*)] $\left[\textbf{Ans.} \left(\frac{7}{2}, \frac{1}{4}\right)\right]$

Example 8. (*i*) *Find the points on the curve $4x^2 + 9y^2 = 1$, where tangents are perpendicular to the line $2y + x = 0$.*

(*ii*) *Find the point on the curve $y = 3x^2 - 2x - 4$ at which tangent is perpendicular to the line $x + 10y - 7 = 0$.*

Solution. (*i*) We have,

$$4x^2 + 9y^2 = 1 \qquad ...(1)$$

Differentiating w.r.t. x, we get

$$8x + 18y \cdot \frac{dy}{dx} = 0$$

$$\Rightarrow \quad 18y \frac{dy}{dx} = -8x \Rightarrow \frac{dy}{dx} = \frac{-8x}{18y} = \frac{-4x}{9y}$$

$$\therefore \text{ Slope of tangent } (m_1) = \frac{dy}{dx} = \frac{-4x}{9y} \qquad ...(2)$$

Now, the slope of line $2y + x = 0$ is given by

$$m_2 = \frac{-1}{2} \qquad ...(3) \qquad \left[\because \text{ Slope of line } ax + by + c = 0 \quad m = -\frac{\text{co-eff. of } x}{\text{co-eff. of } y} = -\frac{a}{b}\right]$$

If the tangent and the line are perpendicular to each other, then product of their slopes is -1 *i.e.* $m_1 \cdot m_2 = -1$.

$$\Rightarrow \quad \left(\frac{-4x}{9y}\right) \times \left(\frac{-1}{2}\right) = -1$$

$$\Rightarrow \quad 2x = -9y$$

$$\Rightarrow \quad y = -\frac{2x}{9} \qquad ...(4)$$

Putting this value of y in (1), we get

$$4x^2 + 9\left(-\frac{2x}{9}\right)^2 = 1$$

$$\Rightarrow \quad 4x^2 + 9\left(\frac{4x^2}{81}\right) = 1 \Rightarrow 4x^2 + \frac{4x^2}{9} = 1$$

$$\Rightarrow \quad \frac{36x^2 + 4x^2}{9} = 1 \Rightarrow 40x^2 = 9$$

$$\Rightarrow \qquad x^2 = \frac{9}{40}$$

$$\Rightarrow \qquad x = \pm \frac{3}{\sqrt{40}} = \pm \frac{3}{2\sqrt{10}}$$

When $x = \frac{3}{2\sqrt{10}}$,

$\therefore$ From (4), we have,

$$y = -\frac{2x}{9} = -\frac{2}{9}\left(\frac{3}{2\sqrt{10}}\right) = -\frac{1}{3\sqrt{10}}$$

When $x = \frac{-3}{2\sqrt{10}}$,

$\therefore$ From (4), we have,

$$y = \frac{-2x}{9} = \frac{-2}{9}\left(\frac{-3}{2\sqrt{10}}\right) = \frac{1}{3\sqrt{10}}$$

$\therefore$ The required points are $\left(\frac{3}{2\sqrt{10}}, \frac{-1}{3\sqrt{10}}\right)$ and $\left(\frac{-3}{2\sqrt{10}}, \frac{1}{3\sqrt{10}}\right)$.

(*ii*) We have, $\qquad y = 3x^2 - 2x - 4 \qquad$ ...(1)

Differentiating w.r.t. x, we get

$$\frac{dy}{dx} = 6x - 2$$

$\therefore$ Slope of tangent $= m_1 = \left(\frac{dy}{dx}\right) = 6x - 2 \qquad$ ...(2)

Slope of line $x + 10y - 7 = 0$ is given by

$$m_2 = \frac{-1}{10} \qquad ...(3)$$

$\left[\because \text{ Slope of line } ax + by + c = 0 \quad m = -\frac{\text{co-eff. of } x}{\text{co-eff. of } y} = -\frac{a}{b}\right]$

If the tangent and the line are perpendicular to each other, then the product of their slopes is -1 *i.e.*, $m_1 \cdot m_2 = -1$.

$$\Rightarrow \qquad (6x - 2)\left(\frac{-1}{10}\right) = -1 \quad \Rightarrow \quad 6x - 2 = 10$$

$$\Rightarrow \qquad 6x = 12 \quad \Rightarrow \quad x = \frac{12}{6} = 2.$$

Putting this value of x in (1), we have

$$y = 3(2)^2 - 2(2) - 4 = 3(4) - 4 - 4$$
$$= 12 - 8 = 4$$

$\therefore$ The required point is (2, 4).

Example 9. (*i*) *Find the point on the curve $y = x^3 - 11x + 5$ at which the tangent has the equation $y = x - 11$.*

(*ii*) *Find the points on the circle $x^2 + y^2 = 13$, at which the tangent is parallel to the line $2x + 3y = 7$.*

(*iii*) *Determine the points on the curve $2y = (3 - x^2)$ at which the tangent is parallel to the line $x + y = 0$.*

Solution. (*i*) We have, $y = x^3 - 11x + 5$ ...(1)

Differentiating w.r.t. x, we get

$$\frac{dy}{dx} = 3x^2 - 11 + 0$$

$\therefore$ Slope of tangent $= m_1 = \frac{dy}{dx} = 3x^2 - 11$ ...(2)

Now, the slope of line $y = x - 11$ is given by

$m_2 = 1$ ...(3) [$\because$ Slope of line $y = mx + c$ is m]

$\therefore$ From (2) and (3), we have

$$3x^2 - 11 = 1$$

$\Rightarrow$ $3x^2 = 1 + 11 = 12$

$\Rightarrow$ $x^2 = 4 \Rightarrow x = \pm 2$

When $x = 2$,

$\therefore$ From (1), we have

$$y = (2)^3 - 11(2) + 5 = 8 - 22 + 5 = -9$$

When $x = -2$,

$\therefore$ From (1), we have

$$y = (-2)^3 - 11(-2) + 5 = -8 + 22 + 5 = 19.$$

$\therefore$ The points are $(2, -9)$ and $(-2, 19)$.

Now, if the tangent has the equation $y = x - 11$, then these points must lies on this line.

i.e., For $(2, -9)$, we have, $y = x - 11$

$\Rightarrow$ $-9 = 2 - 11 = -9$

which is true.

For $(-2, 19)$, we have, $y = x - 11$

$19 = -2 - 11 = -13$

which is not true.

$\therefore$ The required point is $(2, -9)$.

(*ii*) We have, $x^2 + y^2 = 13$ ...(1)

Differentiating w.r.t. x, we get

$$2x + 2y \cdot \frac{dy}{dx} = 0 \Rightarrow 2y\frac{dy}{dx} = -2x$$

$$\frac{dy}{dx} = -\frac{x}{y}$$

$\therefore$ Slope of tangent $= m_1 = \left(\frac{dy}{dx}\right) = -\frac{x}{y}$ ...(2)

Now, the slope of the line $2x + 3y - 7 = 0$ is given by

$$m_2 = \frac{-2}{3} \quad ...(3) \quad \left[\because \text{ Slope of line } ax + by + c = 0 \quad m = -\frac{\text{co-eff. of } x}{\text{co-eff. of } y} = -\frac{a}{b}\right]$$

If the tangent is parallel to the line $2x + 3y - 7 = 0$, then their slopes are equal.

i.e., $$m_1 = m_2$$

$\therefore$ From (2) and (3), we have

$$\frac{-x}{y} = \frac{-2}{3} \Rightarrow 3x = 2y$$

$$\Rightarrow \quad x = \left(\frac{2y}{3}\right) \quad ...(4)$$

Putting this value of x in (1), we have

$$\left(\frac{2y}{3}\right)^2 + y^2 = 13$$

$$\Rightarrow \quad \frac{4}{9}y^2 + y^2 = 13$$

$$\Rightarrow \quad \frac{4y^2 + 9y^2}{9} = 13 \Rightarrow 13y^2 = 117 \Rightarrow y^2 = 9$$

$$\Rightarrow \quad y = \pm 3.$$

$\therefore$ When $y = +3$,

$\therefore$ From (4), we have $x = \frac{2}{3}(3) = 2$

When $x = -3$,

$\therefore$ From (4), we have $x = \frac{2}{3}(-3) = -2$

$\therefore$ The required points are (2, 3) and (– 2, – 3).

(*iii*) Please try yourself.

[**Hint.** Proceed as in part (*i*)] [**Ans.** (1, 1)]

Example 10. (*i*) *Find the points on the curve* $9y^2 = x^3$ *where normal to the curve makes equal intercepts with the axes.*

(*ii*) *Find the point on the curve* $y = x^3$ *at which the slope of the tangent is equal to the y-co-ordinate of the point.*

Solution. (*i*) We have, $9y^2 = x^3$ ...(1)

Differentiating w.r.t. x, we get

$$18y \cdot \frac{dy}{dx} = 3x^2 \Rightarrow \frac{dy}{dx} = \frac{3x^2}{18y} = \frac{x^2}{6y}$$

$\therefore$ Slope of tangent $= m_1 = \left(\dfrac{dy}{dx}\right) = \dfrac{x^2}{6y}$ ...(2)

Since the normal to the curve makes equal intercepts with the axes.

$\therefore$ The slope of normal $= \pm 1$.

$$\Rightarrow \quad \frac{-1}{\left(\dfrac{dy}{dx}\right)} = \pm 1 \Rightarrow \pm\left(\frac{dy}{dx}\right) = -1$$

$$\Rightarrow \quad \pm\frac{x^2}{6y} = -1 \Rightarrow \pm x^2 = -6y$$

$$\Rightarrow \quad x^4 = 36y^2 \quad \text{[On squaring]}$$

$$\Rightarrow \quad x^4 = 36\left(\frac{x^3}{9}\right) \quad \left[\because \text{ By (1), } 9y^2 = x^3 \Rightarrow y^2 = \frac{x^3}{9}\right]$$

$$\Rightarrow \quad x^4 = 4x^3$$

$$\Rightarrow \quad x^4 - 4x^3 = 0 \Rightarrow x^3(x-4) = 0$$

$$\Rightarrow \quad x = 0, 4.$$

When $x = 0$,

$\therefore$ From (1), we have

$$9y^2 = 0 \Rightarrow y = 0$$

When $x = 4$,

$\therefore$ From (1), we have

$$9y^2 = (4)^3$$

$$\Rightarrow \quad 9y^2 = 64 \Rightarrow y^2 = \frac{64}{9} \Rightarrow y = \pm\frac{8}{3}.$$

But, the line making equal intercepts with the axes cannot pass through the origin.

$\therefore$ The required points are $\left(4, \dfrac{8}{3}\right)$ and $\left(4, \dfrac{-8}{3}\right)$.

(*ii*) We have, $y = x^3$ ...(1)

Differentiating w.r.t. x, we get

$$\frac{dy}{dx} = 3x^2$$

$\therefore$ Slope of tangent $= m_1 = \left(\dfrac{dy}{dx}\right) = 3x^2$ ...(2)

Since, the slope of the tangent is equal to the y-coordinate.

$\therefore$ $3x^2 = y$ ...(3)

$\therefore$ From (1) and (3), we have

$$x^3 = 3x^2 \Rightarrow x^3 - 3x^2 = 0$$

$$\Rightarrow \quad x^2(x-3) = 0 \Rightarrow x = 0, 3$$

When $x = 0$,

$\therefore$ From (1), we have

$$y = (0)^3 = 0$$

When $x = 3$,

$\therefore$ From (1), we have

$$y = (3)^3 = 27.$$

$\therefore$ The required points are (0, 0) and (3, 27).

Example 11. *Find the equation of tangent and normal to the following curves at the points indicated in each case :*

(*i*) $y = x^2 - 4x - 5$ at $x = -2$ (*ii*) $y = x^4 - 6x^3 + 13x^2 - 10x + 5$ at (1, 3)

(*iii*) $y^2 = 4ax$ at $(at^2, 2at)$ (*iv*) $y^2 = \dfrac{x^3}{4-x}$ at (2, – 2).

Solution. (*i*) We have, $y = x^2 - 4x - 5$ ...(1)

At $x = -2$,

$$y = (-2)^2 - 4(-2) - 5$$
$$y = 4 + 8 - 5 = 7$$

$\therefore$ The co-ordinates of point of contact are (– 2, 7).

Differentiating (1) w.r.t. x, we get

$$\frac{dy}{dx} = 2x - 4 \quad \text{...(2)}$$

$\therefore$ Slope of tangent at (– 2, 7)

$$= m_1 = \left(\frac{dy}{dx}\right)_{(-2,7)} = 2(-2) - 4$$

$\Rightarrow$ $m_1 = -4 - 4 = -8.$

$\therefore$ The equation of tangent at (– 2, 7) is given by

$$y - 7 = -8(x - (-2)) \quad [\because \text{ By using } (y - y_1) = m(x - x_1)]$$

$\Rightarrow$ $y - 7 = -8(x + 2)$

$\Rightarrow$ $y - 7 = -8x - 16$

$\Rightarrow$ $8x + y + 9 = 0$ ...(3)

Now, slope of normal at (– 2, 7) $= m_2 = \dfrac{-1}{\left(\dfrac{dy}{dx}\right)_{(-2,7)}} = \dfrac{-1}{-8} = \dfrac{1}{8}$.

$\therefore$ Equation of normal at (– 2, 7) is given by

$$y - 7 = \frac{1}{8}(x - (-2)) \quad [\because \text{ By using } (y - y_1) = m(x - x_1)]$$

$\Rightarrow$ $8(y - 7) = (x + 2)$

$\Rightarrow$ $8y - 56 = x + 2$

$\Rightarrow$ $x - 8y + 58 = 0$

(*ii*) We have, $y = x^4 - 6x^3 + 13x^2 - 10x + 5$ ...(1)

The co-ordinates of the point of contact are (1, 3).

Differentiating (1) w.r.t. x, we get

$$\frac{dy}{dx} = 4x^3 - 18x^2 + 26x - 10 \quad ...(2)$$

$\therefore$ Slope of tangent at (1, 3), *i.e.*, $m_1 = \left(\frac{dy}{dx}\right)_{(1,3)}$

$$= 4(1)^3 - 18(1)^2 + 26(1) - 10 = 4 - 18 + 26 - 10 = 2$$

$\therefore$ Equation of tangent at (1, 3) is given by

$$(y - 3) = 2(x - 1) \quad [\because \text{ By using } (y - y_1) = m(x - x_1)]$$

$\Rightarrow$ $y - 3 = 2x - 2$

$\Rightarrow$ $2x - y + 1 = 0 \quad ...(3)$

Now, slope of normal at (1, 3), *i.e.*, $m_2 = \frac{-1}{\left(\frac{dy}{dx}\right)_{(1,3)}} = \frac{-1}{2}$

$\therefore$ Equation of normal at (1, 3) is given by

$$(y - 3) = \frac{-1}{2}(x - 1) \quad [\because \text{ By using } (y - y_1) = m(x - x_1)]$$

$\Rightarrow$ $2(y - 3) = -x + 1$

$\Rightarrow$ $2y - 6 = -x + 1$

$\Rightarrow$ $x + 2y - 7 = 0.$

(*iii*) We have, $y^2 = 4ax \quad ...(1)$

The co-ordinates of point of contact are $(at^2, 2at)$.

Differentiating (1) w.r.t. x, we get

$$2y\frac{dy}{dx} = 4a$$

$\Rightarrow$ $\frac{dy}{dx} = \frac{4a}{2y} = \frac{2a}{y} \quad ...(2)$

$\therefore$ Slope of tangent at $(at^2, 2at)$, *i.e.*, $m_1 = \left(\frac{dy}{dx}\right)_{(at^2, 2at)}$

$$m_1 = \frac{2a}{2at} = \frac{1}{t}$$

$\therefore$ The equation of tangent at $(at^2, 2at)$ is given by

$$(y - 2at) = \frac{1}{t}(x - at^2) \quad [\because \text{ By using } (y - y_1) = m(x - x_1)]$$

$\Rightarrow$ $t(y - 2at) = x - at^2$

$\Rightarrow$ $ty - 2at^2 = x - at^2$

$\Rightarrow$ $x - ty + at^2 = 0. \quad ...(3)$

Now, the slope of normal at $(at^2, 2at)$

i.e., $$m_2 = \frac{-1}{\left(\frac{dy}{dx}\right)_{(at^2,2at)}} = \frac{-1}{1/t} = -t$$

$\therefore$ The equation of normal at $(at^2, 2at)$ is given by

$$(y - 2at) = -t(x - at^2) \qquad [\because \text{ By using } (y - y_1) = m(x - x_1)]$$

$$\Rightarrow \quad y - 2at = -tx + at^3$$

$$\Rightarrow \quad tx + y + at^3 - 2at = 0$$

(*iv*) We have, $$y^2 = \frac{x^3}{4-x} \Rightarrow y^2(4-x) = x^3 \qquad ...(1)$$

The co-ordinates of the point of contact are (2, – 2).

Differentiating (1) w.r.t. x, we get

$$y^2(0-1) + (4-x)\,2y \cdot \frac{dy}{dx} = 3x^2$$

$$\Rightarrow \quad -y^2 + (4-x)\,2y\,\frac{dy}{dx} = 3x^2$$

$$\Rightarrow \quad (4-x)\,2y\,\frac{dy}{dx} = 3x^2 + y^2$$

$$\Rightarrow \quad \frac{dy}{dx} = \frac{3x^2 + y^2}{2y(4-x)} \qquad ...(2)$$

$\therefore$ Slope of tangent at $(2, -2) = m_1 = \left(\frac{dy}{dx}\right)_{(2,-2)}$

$$= \frac{3(2)^2 + (-2)^2}{2(-2)(4-2)} = \frac{3(4)+4}{-4(2)} = \frac{12+4}{-8} = \frac{16}{-8} = -2.$$

$\therefore$ Equation of tangent at (2, – 2) is given by

$$(y - (-2)) = -2(x-2) \qquad [\because \text{ By using } (y - y_1) = m(x - x_1)]$$

$$\Rightarrow \quad y + 2 = -2x + 4$$

$$\Rightarrow \quad 2x + y - 2 = 0 \qquad ...(3)$$

Now, the slope of normal at (2, – 2),

i.e., $$m_2 = \frac{-1}{\left(\frac{dy}{dx}\right)_{(2,-2)}} = \frac{-1}{-2} = \frac{1}{2}$$

$\therefore$ The equation of normal at (2, – 2) is given by

$$(y - (-2)) = \frac{1}{2}(x-2) \qquad [\because \text{ By using } (y - y_1) = m(x - x_1)]$$

$$\Rightarrow \quad (y+2) = \frac{1}{2}(x-2) \Rightarrow 2(y+2) = x - 2$$

$$\Rightarrow \quad 2y + 4 = x - 2 \Rightarrow x - 2y - 6 = 0 \qquad ...(4)$$

Example 12. *Find the equation of the tangent and normal to the following curves at the points indicated in each case :*

(*i*) $y = 2x^2 + 3\sin x$ at $x = 0$ (*ii*) $y = x - \sin x \cos x$ at $x = \frac{\pi}{2}$

(*iii*) $y = \cot^2 x - 2\cot x + 2$ at $x = \frac{\pi}{4}$

(*iv*) $y = 2\sin x + \sin 2x$ at $x = \frac{\pi}{3}$.

Solution. (*i*) We have, $y = 2x^2 + 3\sin x$ ...(1)

When $x = 0$, $y = 2(0) + 3\sin 0° = 0 + 0$

$y = 0$

$\therefore$ The co-ordinates of point of contact are (0, 0).

Differentiating (1), w.r.t. x, we get

$$\frac{dy}{dx} = 4x + 3\cos x \quad ...(2)$$

$\therefore$ Slope of tangent at (0, 0) $= m_1 = \left(\frac{dy}{dx}\right)_{(0,0)}$

$$= 4(0) + 3\cos 0° = 0 + 3(1) = 3.$$

$\therefore$ The equation of tangent at (0, 0) is given by

$$(y - 0) = 3(x - 0) \quad [\because \text{ By using } (y - y_1) = m(x - x_1)]$$

$\Rightarrow$ $y = 3x \Rightarrow 3x - y = 0$ ...(3)

Now, the slope of normal at (0, 0) $= m_2 = \dfrac{-1}{\left(\frac{dy}{dx}\right)_{(0,0)}} = \dfrac{-1}{3}$

$\therefore$ The equation of normal at (0, 0) is given by

$$(y - 0) = \frac{-1}{3}(x - 0) \quad [\because \text{ By using } (y - y_1) = m(x - x_1)]$$

$\Rightarrow$ $3y = -x \Rightarrow x + 3y = 0$

(*ii*) We have, $y = x - \sin x \cos x$ ...(1)

At $x = \frac{\pi}{2}$, $y = \frac{\pi}{2} - \sin\frac{\pi}{2}\cos\frac{\pi}{2} = \frac{\pi}{2} - 0 = \frac{\pi}{2}$

$\therefore$ The co-ordinates of point of contact are $\left(\frac{\pi}{2}, \frac{\pi}{2}\right)$.

From (1), we have, $y = x - \frac{1}{2}.(2\sin x \cos x)$

$\Rightarrow$ $y = x - \frac{1}{2}\sin 2x$ ...(2) $[\because \sin 2A = 2\sin A\cos A]$

Differentiating w.r.t. x, we get

$$\frac{dy}{dx} = 1 - \frac{1}{2}(\cos 2x).\frac{d}{dx}(2x) = 1 - \frac{1}{2}\cos 2x\,(2)$$

$\Rightarrow$ $\dfrac{dy}{dx} = 1 - \cos 2x$ ...(3)

$\therefore$ Slope of tangent at $\left(\frac{\pi}{2}, \frac{\pi}{2}\right) = m_1 = \left(\frac{dy}{dx}\right)_{\left(\frac{\pi}{2}, \frac{\pi}{2}\right)}$

$$= 1 - \cos 2\left(\frac{\pi}{2}\right) = 1 - \cos 2\pi = 1 - (-1) = 2.$$

$\therefore$ Equation of tangent at $\left(\frac{\pi}{2}, \frac{\pi}{2}\right)$ is given by

$$\left(y - \frac{\pi}{2}\right) = 2\left(x - \frac{\pi}{2}\right) \qquad [\because \text{ By using } (y - y_1) = m(x - x_1)]$$

$$\Rightarrow \quad y - \frac{\pi}{2} = 2x - \pi \Rightarrow 2x - y - \frac{\pi}{2} = 0$$

$$\Rightarrow \quad 4x - 2y - \pi = 0 \qquad ...(4)$$

Now, the slope of normal at $\left(\frac{\pi}{2}, \frac{\pi}{2}\right) = m_2 = \frac{-1}{\left(\frac{dy}{dx}\right)_{\left(\frac{\pi}{2}, \frac{\pi}{2}\right)}} = \frac{-1}{2}$.

$\therefore$ Equation of normal at $\left(\frac{\pi}{2}, \frac{\pi}{2}\right)$ is given by

$$\left(y - \frac{\pi}{2}\right) = \frac{-1}{2}\left(x - \frac{\pi}{2}\right) \qquad [\because \text{ By using } (y - y_1) = m(x - x_1)]$$

$$\Rightarrow \quad 2\left(y - \frac{\pi}{2}\right) = -1\left(x - \frac{\pi}{2}\right) \Rightarrow 2y - \pi = -x + \frac{\pi}{2}$$

$$\Rightarrow \quad x + 2y - \frac{3\pi}{2} = 0 \qquad \Rightarrow 2x + 4y - 3\pi = 0$$

(*iii*) We have, $\quad y = \cot^2 x - 2\cot x + 2 \qquad ...(1)$

At $x = \frac{\pi}{4}$, $\quad y = \cot^2\left(\frac{\pi}{4}\right) - 2\cot\left(\frac{\pi}{4}\right) + 2 = (1)^2 - 2(1) + 2 = 1$

$\therefore$ The co-ordinates of point of contact are $\left(\frac{\pi}{4}, 1\right)$.

Differentiating (1) w.r.t. x, we get

$$\frac{dy}{dx} = 2\cot x \cdot \frac{d}{dx}(\cot x) - 2(-\operatorname{cosec}^2 x) + 0$$

$$= 2\cot x(-\operatorname{cosec}^2 x) + 2\operatorname{cosec}^2 x$$

$$\Rightarrow \quad \frac{dy}{dx} = 2\operatorname{cosec}^2 x(1 - \cot x) \qquad ...(2)$$

$\therefore$ Slope of tangent at $\left(\frac{\pi}{4}, 1\right) = m_1 = \left(\frac{dy}{dx}\right)_{\left(\frac{\pi}{4}, 1\right)}$

$$= 2\,\text{cosec}^2 \frac{\pi}{4}\left(1 - \cot\frac{\pi}{4}\right) = 2\left(\sqrt{2}\right)^2 (1-1) = 0.$$

$\therefore$ Equation of tangent at $\left(\frac{\pi}{4}, 1\right)$ is given by

$$(y-1) = 0\left(x - \frac{\pi}{4}\right) \qquad [\because \text{ By using } (y - y_1) = m(x - x_1)]$$

$\Rightarrow$ $y - 1 = 0$

Now, the slope of normal at $\left(\frac{\pi}{4}, 1\right) = m_2 = \dfrac{-1}{\left(\frac{dy}{dx}\right)_{\left(\frac{\pi}{4}, 1\right)}} = \dfrac{-1}{0}$.

$\therefore$ Equation of normal at $\left(\frac{\pi}{4}, 1\right)$ is given by

$$(y-1) = \frac{-1}{0}\left(x - \frac{\pi}{4}\right) \qquad [\because \text{ By using } (y - y_1) = m(x - x_1)]$$

$\Rightarrow$ $0 = -x + \frac{\pi}{4}$

$\Rightarrow$ $x - \frac{\pi}{4} = 0 \Rightarrow 4x - \pi = 0.$

(*iv*) We have, $y = 2\sin x + \sin 2x$ ...(1)

At $x = \frac{\pi}{3}$, $y = 2\sin\frac{\pi}{3} + \sin\frac{2\pi}{3} = 2 \cdot \frac{\sqrt{3}}{2} + \frac{\sqrt{3}}{2} = \sqrt{3} + \frac{\sqrt{3}}{2} = \frac{3\sqrt{3}}{2}$.

$\therefore$ The co-ordinates of point of contact are $\left(\frac{\pi}{3}, \frac{3\sqrt{3}}{2}\right)$.

Differentiating (1) w.r.t. x, we get

$$\frac{dy}{dx} = 2\cos x + 2\cos 2x \qquad \text{...(2)}$$

$\therefore$ The slope of tangent at $\left(\frac{\pi}{3}, \frac{3\sqrt{3}}{2}\right) = m_1 = \left(\frac{dy}{dx}\right)_{\left(\frac{\pi}{3}, \frac{3\sqrt{3}}{2}\right)}$

$$= 2\cos\frac{\pi}{3} + 2\cos\frac{2\pi}{3} = 2\left(\frac{1}{2}\right) + 2\left(-\frac{1}{2}\right) = 0$$

$\therefore$ The equation of tangent at $\left(\frac{\pi}{3}, \frac{3\sqrt{3}}{2}\right)$ is given by

$$\left(y - \frac{3\sqrt{3}}{2}\right) = 0\left(x - \frac{\pi}{3}\right) \qquad [\because \text{ By using } (y - y_1) = m(x - x_1)]$$

$$\Rightarrow \qquad \frac{2y - 3\sqrt{3}}{2} = 0 \Rightarrow 2y - 3\sqrt{3} = 0 \qquad ...(3)$$

Now, the slope of normal at $\left(\frac{\pi}{3}, \frac{3\sqrt{3}}{2}\right) = m_2 = \dfrac{-1}{\left(\frac{dy}{dx}\right)_{\left(\frac{\pi}{3}, \frac{3\sqrt{3}}{2}\right)}} = \dfrac{-1}{0}$

$\therefore$ Equation of normal at $\left(\frac{\pi}{3}, \frac{3\sqrt{3}}{2}\right)$ is given by

$$\left(y - \frac{3\sqrt{3}}{2}\right) = \frac{-1}{0}\left(x - \frac{\pi}{3}\right) \qquad [\because \text{ By using } (y - y_1) = m(x - x_1)]$$

$$\Rightarrow \qquad 0 = -1\left(x - \frac{\pi}{3}\right)$$

$$\Rightarrow \qquad x - \frac{\pi}{3} = 0 \Rightarrow 3x - \pi = 0 \qquad ...(4)$$

Example 13. (*i*) *Find the equation of the normal at the point (am^2, am^3) for the curve $ay^2 = x^3$.*

(*ii*) *Find the equation of the normal at the point (2, 18) for the curve $y = x^3 + 2x + 6$.*

Solution. (*i*) We have, $ay^2 = x^3$ ...(1)

The co-ordinates of point of contact are (am^2, am^3).

Differentiating (1) w.r.t. x, we get

$$\Rightarrow \qquad 2ay\frac{dy}{dx} = 3x^2 \Rightarrow \frac{dy}{dx} = \frac{3x^2}{2ay} \qquad ...(2)$$

$\therefore$ The slope of tangent at $(am^2, am^3) = m_1 = \left(\frac{dy}{dx}\right)_{(am^2, am^3)}$

$$= \frac{3(am^2)^2}{2a(am^3)} = \frac{3a^2m^4}{2a^2m^3} = \frac{3m}{2}$$

$\therefore$ Slope of normal at (am^2, am^3)

$$= \frac{-1}{m_1} = \frac{-1}{\frac{3m}{2}} = \frac{-2}{3m}$$

$\therefore$ Equation of normal at (am^2, am^3) is given by

$$(y - am^3) = \frac{-2}{3m}(x - am^2) \qquad [\because \text{ By using } (y - y_1) = m(x - x_1)]$$

$\Rightarrow \quad 3m(y - am^3) = -2(x - am^2)$

$\Rightarrow \quad 3my - 3am^4 = -2x + 2am^2$

$\Rightarrow \quad 2x + 3my - 3am^4 - 2am^2 = 0$

$\Rightarrow \quad 2x + 3my - am^2(2 + 3m^2) = 0$ ...(3)

(*ii*) We have, $y = x^3 + 2x + 6$ ...(1)

The co-ordinates of point of contact are (2, 18).

Differentiating (1) w.r.t. x, we get

$$\frac{dy}{dx} = 3x^2 + 2 \qquad ...(2)$$

$\therefore$ The slope of tangent at (2, 18) $= m_1 = \left(\frac{dy}{dx}\right)_{(2, 18)}$

$= 3(2)^2 + 2 = 3(4) + 2 = 12 + 2 = 14.$

$\therefore$ The slope of normal at (2, 18) $= m_2 = \frac{-1}{m_1} = \frac{-1}{14}$

$\therefore$ The equation of normal at (2, 18) is given by

$$(y - 18) = \frac{-1}{14}(x - 2) \qquad [\because \text{ By using } (y - y_1) = m(x - x_1)]$$

$\Rightarrow \quad 14(y - 18) = -x + 2$

$\Rightarrow \quad 14y - 252 = -x + 2$

$\Rightarrow \quad x + 14y - 254 = 0.$

Example 14. *Find the equation of the normal to each of the following at the indicated point :*

(*i*) $y = \cos(3x + 4)$ at $x = \frac{\pi - 8}{6}$

(*ii*) $y = \sin^2 x + \cot^2 x + 3$ at $x = \frac{\pi}{4}$

(*iii*) $y = \frac{\tan x + \sin 2x}{1 + x^2}$ at $x = \pi$.

Solution. (*i*) We have, $y = \cos(3x + 4)$ ...(1)

At $x = \frac{\pi - 8}{6}$, $\quad y = \cos\left[3\left(\frac{\pi - 8}{6}\right) + 4\right] = \cos\left(\frac{\pi - 8}{2} + 4\right)$

$\Rightarrow \quad y = \cos\left(\frac{\pi - 8 + 8}{2}\right) = \cos\frac{\pi}{2} = 0$

$\therefore$ The co-ordinates of point of contact are $\left(\frac{\pi - 8}{6}, 0\right)$.

Differentiating (1) w.r.t. x, we get

$$\frac{dy}{dx} = -\sin(3x + 4) \cdot \frac{d}{dx}(3x + 4) = -\sin(3x + 4).(3 + 0)$$

$$\Rightarrow \quad \frac{dy}{dx} = -3\sin(3x+4) \quad ...(2)$$

$\therefore$ Slope of tangent at $\left(\frac{\pi-8}{6}, 0\right) = m_1 = \left(\frac{dy}{dx}\right)_{\left(\frac{\pi-8}{6}, 0\right)}$

$$= -3\sin\left[3\left(\frac{\pi-8}{6}\right)+4\right] = -3\sin\left[\frac{\pi-8}{2}+4\right]$$

$$= -3\sin\left[\frac{\pi-8+8}{2}\right]$$

$$\Rightarrow \quad m_1 = -3\sin\left(\frac{\pi}{2}\right) = -3(1) = -3.$$

$\therefore$ Slope of normal at $\left(\frac{\pi-8}{6}, 0\right) = m_2 = \frac{-1}{m_1} = \frac{-1}{-3} = \frac{1}{3}$.

$\therefore$ Equation of normal at $\left(\frac{\pi-8}{6}, 0\right)$ is given by

$$(y-0) = \frac{1}{3}\left(x - \frac{\pi-8}{6}\right) \qquad [\because \text{ By using } (y-y_1) = m(x-x_1)]$$

$$\Rightarrow \quad 3y = \left[\frac{6x-(\pi-8)}{6}\right]$$

$$\Rightarrow \quad 18y = 6x - \pi + 8 \quad \Rightarrow \quad 6x - 18y - \pi + 8 = 0$$

(*ii*) We have, $\quad y = \sin^2 x + \cot^2 x + 3 \quad ...(1)$

At $x = \frac{\pi}{4}$, $\quad y = \sin^2\left(\frac{\pi}{4}\right) + \cot^2\left(\frac{\pi}{4}\right) + 3$

$$= \left(\frac{1}{\sqrt{2}}\right)^2 + (1)^2 + 3 = \frac{1}{2} + 1 + 3 = \frac{1}{2} + 4 = \frac{9}{2}.$$

$\therefore$ The co-ordinates of the point are $\left(\frac{\pi}{4}, \frac{9}{2}\right)$.

Differentiating (1) w.r.t. x, we get

$$\frac{dy}{dx} = 2\sin x \cdot \frac{d}{dx}(\sin x) + 2\cot x \cdot \frac{d}{dx}(\cot x) + 0$$

$$\Rightarrow \quad \frac{dy}{dx} = 2\sin x\cos x + 2\cot x(-\operatorname{cosec}^2 x)$$

$$\Rightarrow \quad \frac{dy}{dx} = 2\sin x\cos x - 2\cot x\operatorname{cosec}^2 x \quad ...(2)$$

$\therefore$ Slope of tangent at $\left(\frac{\pi}{4}, \frac{9}{2}\right) = m_1 = \left(\frac{dy}{dx}\right)_{\left(\frac{\pi}{4}, \frac{9}{2}\right)}$

$$= 2 \sin \frac{\pi}{4} \cos \frac{\pi}{4} - 2 \cot \frac{\pi}{4} \operatorname{cosec}^2 \frac{\pi}{4}$$

$$= 2\left(\frac{1}{\sqrt{2}}\right)\left(\frac{1}{\sqrt{2}}\right) - 2(1)\left(\sqrt{2}\right)^2 = 1 - 2(2) = 1 - 4 = -3.$$

$\therefore$ Slope of normal at $\left(\frac{\pi}{4}, \frac{9}{2}\right) = m_2 = \frac{-1}{m_1} = \frac{-1}{-3} = \frac{1}{3}$.

$\therefore$ Equation of normal at $\left(\frac{\pi}{4}, \frac{9}{2}\right)$ is given by

$$\left(y - \frac{9}{2}\right) = \frac{1}{3}\left(x - \frac{\pi}{4}\right) \qquad [\because \text{ By using } (y - y_1) = m(x - x_1)]$$

$\Rightarrow$ $3\left(y - \frac{9}{2}\right) = x - \frac{\pi}{4}$

$\Rightarrow$ $3y - \frac{27}{2} = x - \frac{\pi}{4}$

$\Rightarrow$ $x - 3y + \frac{27}{2} - \frac{\pi}{4} = 0$

$\Rightarrow$ $4x - 12y + 54 - \pi = 0$

(*iii*) We have, $\qquad y = \frac{\tan x + \sin 2x}{1 + x^2}$ ...(1)

At $x = \pi$,

$\therefore$ $\qquad y = \frac{\tan \pi + \sin 2\pi}{1 + \pi^2} = \frac{0 + 0}{1 + \pi^2} = 0.$

$\therefore$ The co-ordinates of the point of contact are $(\pi, 0)$.
Differentiating (1) w.r.t. x, we get

$$\frac{dy}{dx} = \frac{(1 + x^2) \cdot \frac{d}{dx}(\tan x + \sin 2x) - (\tan x + \sin 2x)\frac{d}{dx}(1 + x^2)}{(1 + x^2)^2}$$

$\Rightarrow$ $\qquad \frac{dy}{dx} = \frac{(1 + x^2)(\sec^2 x + 2\cos 2x) - (\tan x + \sin 2x)(0 + 2x)}{(1 + x^2)^2}$ ...(2)

$\therefore$ Slope of tangent at $(\pi, 0) = m_1 = \left(\frac{dy}{dx}\right)_{(\pi, 0)}$

$$= \frac{(1 + \pi^2)[\sec^2 \pi + 2\cos 2\pi] - [\tan \pi + \sin 2\pi](2\pi)}{(1 + \pi^2)^2}$$

$$= \frac{(1+\pi^2)[(-1)^2 + 2(1)] - 0}{(1+\pi^2)^2}$$

$$\Rightarrow \qquad m_1 = \frac{3(1+\pi^2)}{(1+\pi^2)^2} = \frac{3}{(1+\pi^2)}.$$

$\therefore$ Slope of normal at $(\pi, 0) = m_2 = \frac{-1}{m_1} = \frac{-1}{3/(1+\pi^2)} = -\frac{(1+\pi^2)}{3}$.

$\therefore$ Equation of normal at $(\pi, 0)$ is given by

$$(y - 0) = -\frac{(1+\pi^2)}{3}(x - \pi) \qquad [\because \text{ By using } (y - y_1) = m(x - x_1)]$$

$$\Rightarrow \qquad 3y = -(1 + \pi^2)(x - \pi)$$

$$\Rightarrow \qquad 3y = -x(1 + \pi^2) + \pi(1 + \pi^2)$$

$\Rightarrow$ $x(1 + \pi^2) + 3y - \pi(1 + \pi^2) = 0$

Example 15. *(i) Find the equation of the tangent to the curve $y = \sec^4 x - \tan^4 x$ at $x = \frac{\pi}{3}$.*

(ii) Find the equation of the normal to $y = (\sin 2x + \cot x + 2)^2$ at $x = \frac{\pi}{2}$.

Solution. (*i*) We have, $y = \sec^4 x - \tan^4 x$ ...(1)

At $x = \frac{\pi}{3}$, $\qquad y = \sec^4 \frac{\pi}{3} - \tan^4 \frac{\pi}{3} = (2)^4 - \left(\sqrt{3}\right)^4 = 16 - 9 = 7.$

$\therefore$ The co-ordinates of the point of contact are $\left(\frac{\pi}{3}, 7\right)$.

Differentiating (1) w.r.t. x, we get

$$\frac{dy}{dx} = 4\sec^3 x \,.\, \frac{d}{dx}(\sec x) - 4\tan^3 x \,.\, \frac{d}{dx}(\tan x)$$
$$= 4\sec^3 x \,.\, \sec x \tan x - 4\tan^3 x \,.\, \sec^2 x$$
$$= 4\sec^4 x \tan x - 4\sec^2 x \tan^3 x$$
$$= 4\sec^2 x \tan x\,[\sec^2 x - \tan^2 x]$$

$$\Rightarrow \qquad \frac{dy}{dx} = 4\sec^2 x \tan x \qquad ...(2) \quad [\because \sec^2 A - \tan^2 A = 1]$$

$\therefore$ Slope of tangent at $\left(\frac{\pi}{3}, 7\right) = m_1 = \left(\frac{dy}{dx}\right)_{\left(\frac{\pi}{3}, 7\right)}$

$$= 4\sec^2 \frac{\pi}{3} \tan \frac{\pi}{3} = 4(2)^2 - \left(\sqrt{3}\right)$$

$$\therefore \qquad m_1 = 16\sqrt{3}$$

$\therefore$ Equation of tangent at $\left(\frac{\pi}{3}, 7\right)$ is given by

$$(y - 7) = 16\sqrt{3}\left(x - \frac{\pi}{3}\right) \qquad [\because \text{ By using } (y - y_1) = m(x - x_1)]$$

$$\Rightarrow \qquad (y - 7) = 16\sqrt{3}x - \frac{16\pi}{\sqrt{3}}$$

$$\Rightarrow \quad 16\sqrt{3}x - y + 7 - \frac{16\pi}{\sqrt{3}} = 0 \quad \Rightarrow \quad 48x - \sqrt{3}y + 7\sqrt{3} - 16\pi = 0.$$

(*ii*) We have, $\qquad y = (\sin 2x + \cot x + 2)^2 \qquad$ ...(1)

At $x = \dfrac{\pi}{2}$, $\qquad y = \left(\sin \dfrac{2\pi}{2} + \cot \dfrac{\pi}{2} + 2\right)^2 = (0 + 0 + 2)^2 = 4$

$\therefore$ The co-ordinates of point of contact are $\left(\dfrac{\pi}{2}, 4\right)$

Differentiating (1) w.r.t. x, we get

$$\frac{dy}{dx} = 2(\sin 2x + \cot x + 2) \,.\, \frac{d}{dx}(\sin 2x + \cot x + 2)$$

$$\Rightarrow \qquad \frac{dy}{dx} = 2(\sin 2x + \cot x + 2)\,(2\cos 2x - \operatorname{cosec}^2 x) \qquad ...(2)$$

$\therefore$ Slope of tangent at $\left(\dfrac{\pi}{2}, 4\right) = m_1 = \left(\dfrac{dy}{dx}\right)_{\left(\frac{\pi}{2}, 4\right)}$

$$= 2\left(\sin \pi + \cot \frac{\pi}{2} + 2\right)\left(2\cos \pi - \operatorname{cosec}^2 \frac{\pi}{2}\right)$$

$$= 2(0 + 0 + 2)\,(-2 - 1) = 4(-3) = -12.$$

$\therefore$ Slope of normal at $\left(\dfrac{\pi}{2}, 4\right) = m_2 = \dfrac{-1}{m_1} = \dfrac{-1}{-12} = \dfrac{1}{12}$.

$\therefore$ Equation of normal at $\left(\dfrac{\pi}{2}, 4\right)$ is given by

$$(y - 4) = \frac{1}{12}\left(x - \frac{\pi}{2}\right) \qquad [\because \text{ By using } (y - y_1) = m(x - x_1)]$$

$$\Rightarrow \qquad 12(y - 4) = x - \frac{\pi}{2}$$

$$\Rightarrow \qquad 12y - 48 = x - \frac{\pi}{2}$$

$$\Rightarrow \qquad x - 12y + 48 - \frac{\pi}{2} = 0$$

$$\Rightarrow \qquad 2x - 24y + 96 - \pi = 0$$

Example 16. *Find the equations of the tangent and the normal to* $16x^2 + 9y^2 = 144$ *at* (x_1, y_1), *where* $x_1 = 2$ *and* $y_1 > 0$.

Solution. We have, $16x^2 + 9y^2 = 144$ ...(1)

Since the point (x_1, y_1) lies on equation (1) ;

$\therefore \quad 16{x_1}^2 + 9{y_1}^2 = 144$ ...(2)

At $x_1 = 2$, $\therefore \quad 16(2)^2 + 9{y_1}^2 = 144$

$\Rightarrow \quad 64 + 9{y_1}^2 = 144 \quad \Rightarrow \quad 9{y_1}^2 = 144 - 64$

$$\Rightarrow \quad 9{y_1}^2 = 80 \quad \Rightarrow \quad {y_1}^2 = \frac{80}{9} \quad \Rightarrow \quad y_1 = \sqrt{\frac{80}{9}} \qquad [\because \; y > 0]$$

$$\Rightarrow \quad y_1 = \frac{4\sqrt{5}}{3}.$$

$\therefore$ The co-ordinates of the point of contact are $\left(2, \frac{4\sqrt{5}}{3}\right)$.

Differentiating (1) w.r.t. *x*, we get

$$32x + 18y\frac{dy}{dx} = 0 \quad \Rightarrow \quad 18y\frac{dy}{dx} = -32x$$

$$\Rightarrow \quad \frac{dy}{dx} = \frac{-32x}{18y} = \frac{-16x}{9y} \qquad \text{...(3)}$$

$\therefore$ Slope of tangent at $\left(2, \frac{4\sqrt{5}}{3}\right) = m_1 = \left(\frac{dy}{dx}\right)_{\left(2, \frac{4\sqrt{5}}{3}\right)}$

$$m_1 = \frac{-16(2)}{9\left(\frac{4\sqrt{5}}{3}\right)} = \frac{-32}{12\sqrt{5}} = \frac{-8}{3\sqrt{5}}.$$

$\therefore$ The equation of tangent at $\left(2, \frac{4\sqrt{5}}{3}\right)$ is given by

$$\left(y - \frac{4\sqrt{5}}{3}\right) = \frac{-8}{3\sqrt{5}}(x-2) \qquad [\because \text{ By using } (y - y_1) = m(x - x_1)]$$

$$\Rightarrow \quad 3\sqrt{5}\left(y - \frac{4\sqrt{5}}{3}\right) = -8(x-2)$$

$$\Rightarrow \quad 3\sqrt{5}y - 20 = -8x + 16$$

$$\Rightarrow \quad 8x + 3\sqrt{5}y - 36 = 0 \qquad \text{...(4)}$$

Now, the slope of normal at $\left(2, \frac{4\sqrt{5}}{3}\right) = m_2 = \frac{-1}{m_1} = \frac{-1}{\frac{-8}{3\sqrt{5}}} = \frac{3\sqrt{5}}{8}$.

$\therefore$ Equation of normal at $\left(2, \frac{4\sqrt{5}}{3}\right)$ is given by

$$\left(y - \frac{4\sqrt{5}}{3}\right) = \frac{3\sqrt{5}}{8}(x-2) \Rightarrow 8\left(\frac{3y - 4\sqrt{5}}{3}\right) = 3\sqrt{5}\,(x-2)$$

$\Rightarrow$ $$24y - 32\sqrt{5} = 9\sqrt{5}\,(x-2) \Rightarrow 24y - 32\sqrt{5} = 9\sqrt{5}\,x - 18\sqrt{5}$$

$\Rightarrow$ $$9\sqrt{5}\,x - 24y + 14\sqrt{5} = 0.$$

Example 17. (*i*) *Find the equation of the tangent to the curve*

$$x = 1 - \cos\theta \text{ and } y = \theta - \sin\theta \text{ at } \theta = \frac{\pi}{4}.$$

(*ii*) *Find the equation of tangent to the curve*

$$x = a\cos\theta \text{ and } y = b\sin\theta \quad \text{at } \theta = \frac{\pi}{4}.$$

(*iii*) *Find the equation of the tangent and normal to the curve*

$$x = \theta + \sin\theta \text{ and } y = 1 + \cos\theta \quad \text{at } \theta = \frac{\pi}{2}.$$

(*iv*) *Find the equation of tangent and normal to the curve*

$$x = a\sin^3 t \text{ and } y = b\cos^3 t \quad \text{at any point 't'}.$$

Solution. (*i*) We have, $x = 1 - \cos\theta$ and $y = \theta - \sin\theta$

When $\theta = \frac{\pi}{4}$, $\quad x = 1 - \cos\frac{\pi}{4}$ and $y = \frac{\pi}{4} - \sin\frac{\pi}{4}$

$\Rightarrow$ $$x = \left(1 - \frac{1}{\sqrt{2}}\right) \text{ and } y = \left(\frac{\pi}{4} - \frac{1}{\sqrt{2}}\right)$$

$\therefore$ The co-ordinates of point of contact are $\left(1 - \frac{1}{\sqrt{2}}, \frac{\pi}{4} - \frac{1}{\sqrt{2}}\right)$.

Now, $$x = 1 - \cos\theta \text{ and } y = \theta - \sin\theta$$

$\Rightarrow$ $$\frac{dx}{d\theta} = 0 - (-\sin\theta) \text{ and } y = 1 - \cos\theta$$

$\Rightarrow$ $$\frac{dx}{d\theta} = \sin\theta$$

$\therefore$ $$\frac{dy}{dx} = \frac{\frac{dy}{d\theta}}{\frac{dx}{d\theta}} = \frac{1 - \cos\theta}{\sin\theta} \quad ...(1)$$

$\therefore$ Slope of tangent at $\left(\theta = \frac{\pi}{4}\right)$ is $m_1 = \left(\frac{dy}{dx}\right)_{\theta = \frac{\pi}{4}}$

$$m_1 = \frac{1-\cos\frac{\pi}{4}}{\sin\frac{\pi}{4}} = \frac{1-\frac{1}{\sqrt{2}}}{\frac{1}{\sqrt{2}}} = \frac{\sqrt{2}-1}{1}.$$

$\therefore$ The equation of tangent at $\theta = \frac{\pi}{4}$ is given by

$$(y - y_1) = m(x - x_1)$$

$$\Rightarrow \quad \left[y-\left(\frac{\pi}{4}-\frac{1}{\sqrt{2}}\right)\right] = \left(\sqrt{2}-1\right)\left[x-\left(1-\frac{1}{\sqrt{2}}\right)\right]$$

$$\Rightarrow \quad y-\frac{\pi}{4}+\frac{1}{\sqrt{2}} = x\left(\sqrt{2}-1\right)-\left(\sqrt{2}-1\right)\frac{\left(\sqrt{2}-1\right)}{\sqrt{2}}$$

$$\Rightarrow \quad x\left(\sqrt{2}-1\right)-y+\frac{\pi}{4}-\frac{1}{\sqrt{2}}-\frac{\left(\sqrt{2}-1\right)^2}{\sqrt{2}} = 0$$

$$\Rightarrow \quad x\left(\sqrt{2}-1\right)-y+\frac{\pi}{4}-\frac{1}{\sqrt{2}}\left[1-\left(\sqrt{2}-1\right)^2\right] = 0$$

$$\Rightarrow \quad x\left(\sqrt{2}-1\right)-y+\frac{\pi}{4}-\frac{1}{\sqrt{2}}[1-2-1+2\sqrt{2}] = 0$$

$$\Rightarrow \quad x\left(\sqrt{2}-1\right)-y+\frac{\pi}{4}-\frac{2}{\sqrt{2}}\left(\sqrt{2}-1\right) = 0$$

$$\Rightarrow \quad x\left(\sqrt{2}-1\right)-y+\frac{\pi}{4}-\sqrt{2}\left(\sqrt{2}-1\right) = 0.$$

(*ii*) Please try yourself.

[**Hint.** Proceed as in part (*i*) of the same example.] $\left[\textbf{Ans. } \frac{x}{a}+\frac{y}{b}=\sqrt{2}\right]$

(*iii*) We have, $\quad x = \theta + \sin\theta \quad$ and $\quad y = 1 + \cos\theta$

$\therefore$ At $\theta = \frac{\pi}{2}$,

$$\therefore \quad x = \frac{\pi}{2}+\sin\frac{\pi}{2} \quad \text{and} \quad y = 1+\cos\frac{\pi}{2}$$

$$\Rightarrow \quad x = \left(\frac{\pi}{2}+1\right) \quad \text{and} \quad y = 1+0 = 1$$

$\therefore$ The co-ordinates of point of contact are $\left(\frac{\pi}{2}+1, 1\right)$

Now, $\quad x = \theta + \sin\theta \quad$ and $\quad y = 1 + \cos\theta$

$$\Rightarrow \quad \frac{dx}{d\theta} = 1+\cos\theta \quad \text{and} \quad \frac{dy}{d\theta} = -\sin\theta$$

$$\therefore \quad \frac{dy}{dx} = \frac{\frac{dy}{d\theta}}{\frac{dx}{d\theta}} = -\frac{\sin\theta}{1+\cos\theta} \quad ...(1)$$

$\therefore$ Slope of tangent at $\theta = \frac{\pi}{2}$ is $m_1 = \left(\frac{dy}{dx}\right)_{\theta=\frac{\pi}{2}} = -\frac{\sin\frac{\pi}{2}}{1+\cos\frac{\pi}{2}} = \frac{-1}{1+0} = -1$

$\therefore$ The equation of tangent at $\theta = \frac{\pi}{2}$ is given by

$$(y - y_1) = m(x - x_1) \Rightarrow (y-1) = -1\left[x - \left(\frac{\pi}{2}+1\right)\right]$$

$$\Rightarrow \quad y - 1 = -x + \frac{\pi}{2} + 1$$

$$\Rightarrow \quad x + y - \frac{\pi}{2} - 2 = 0$$

$$\Rightarrow \quad 2x + 2y - \pi - 4 = 0 \quad ...(2)$$

Now, the slope of normal at $\theta = \frac{\pi}{2} = m_2 = \frac{-1}{m_1} = \frac{-1}{-1} = 1.$

$\therefore$ The equation of normal at $\theta = \frac{\pi}{2}$ is given by

$$\Rightarrow \quad (y - y_1) = m(x - x_1) \Rightarrow (y-1) = 1\left[x - \left(\frac{\pi}{2}+1\right)\right]$$

$$\Rightarrow \quad (y-1) = x - \frac{\pi}{2} - 1 \Rightarrow x - y - \frac{\pi}{2} = 0$$

$$\Rightarrow \quad 2x - 2y - \pi = 0$$

(*iv*) We have, $\quad x = a\sin^3 t \quad$ and $\quad y = b\cos^3 t$

$$\Rightarrow \quad \frac{dx}{dt} = 3a\sin^2 t\,\frac{d}{dt}(\sin t) \quad \text{and} \quad \frac{dy}{dt} = 3b\cos^2 t \,.\, \frac{d}{dt}(\cos t)$$

$$\Rightarrow \quad \frac{dx}{dt} = 3a\sin^2 t\cos t \quad \text{and} \quad \frac{dy}{dt} = -3b\cos^2 t\sin t$$

$$\therefore \quad \frac{dy}{dx} = \frac{\frac{dy}{dt}}{\frac{dx}{dt}} = \frac{-3b\cos^2 t\sin t}{3a\sin^2 t\cos t} = \frac{-b\cos t}{a\sin t} \quad ...(1)$$

$\therefore$ Slope of tangent at any point t is $= m_1 = \left(\frac{dy}{dx}\right) = \frac{-b\cos t}{a\sin t}$.

$\therefore$ The equation of tangent at any point t is given by

$$(y - y_1) = m(x - x_1)$$

$\Rightarrow$ $(y - b\cos^3 t) = \dfrac{-b\cos t}{a\sin t}(x - a\sin^3 t)$

$\Rightarrow$ $a\sin t(y - b\cos^3 t) = -b\cos t(x - a\sin^3 t)$

$\Rightarrow$ $ay\sin t - ab\cos^3 t\sin t = -bx\cos t + ab\sin^3 t\cos t$

$\Rightarrow$ $bx\cos t + ay\sin t = ab\sin^3 t\cos t + ab\cos^3 t\sin t$

$\Rightarrow$ $bx\cos t + ay\sin t = ab\sin t\cos t(\sin^2 t + \cos^2 t)$

$\Rightarrow$ $bx\cos t + ay\sin t = ab\sin t\cos t$ [$\because$ $\sin^2 A + \cos^2 A = 1$]

Now, the slope of normal at any point t is $= m_2 = \dfrac{-1}{m_1} = \dfrac{-1}{\dfrac{-b\cos t}{a\sin t}} = \dfrac{a\sin t}{b\cos t}$.

$\therefore$ Equation of normal at any point t is given by

$$(y - y_1) = m(x - x_1)$$

$\Rightarrow$ $(y - b\cos^3 t) = \dfrac{a\sin t}{b\cos t}(x - a\sin^3 t)$

$\Rightarrow$ $b\cos t(y - b\cos^3 t) = a\sin t(x - a\sin^3 t)$

$\Rightarrow$ $by\cos t - b^2\cos^4 t = ax\sin t - a^2\sin^4 t$

$\Rightarrow$ $ax\sin t - by\cos t = a^2\sin^4 t - b^2\cos^4 t.$

Example 18. (*i*) *Find the equation of tangent at point (3, 4) to the ellipse* $\dfrac{x^2}{a^2} + \dfrac{y^2}{b^2} = 1$.

(*ii*) *Find the equation of tangent and normal at (– 3, 0) to the curve* $\dfrac{x^2}{9} - \dfrac{y^2}{16} = 1$.

Solution. (*i*) We have, $\dfrac{x^2}{a^2} + \dfrac{y^2}{b^2} = 1$ ...(1)

The co-ordinates of point of contact are (3, 4).
Differentiating (1) w.r.t. x, we get

$$\frac{2x}{a^2} + \frac{2y}{b^2}\cdot\frac{dy}{dx} = 0 \Rightarrow \frac{2y}{b^2}\frac{dy}{dx} = -\frac{2x}{a^2}$$

$\Rightarrow$ $\dfrac{dy}{dx} = -\dfrac{2x}{a^2} \times \dfrac{b^2}{2y} \Rightarrow \dfrac{dy}{dx} = -\dfrac{b^2 x}{a^2 y}$ ...(2)

$\therefore$ Slope of tangent at (3, 4) $= m_1 = \left(\dfrac{dy}{dx}\right)_{(3,4)} = \dfrac{-b^2(3)}{a^2(4)} = \dfrac{-3b^2}{4a^2}$

$\therefore$ Equation of tangent at (3, 4) is given by

$\Rightarrow$ $(y - 4) = \dfrac{-3b^2}{4a^2}(x - 3)$ [$\because$ By using $(y - y_1) = m(x - x_1)$]

$\Rightarrow$ $4a^2(y - 4) = -3b^2(x - 3)$

$\Rightarrow$ $4a^2y - 16a^2 = -3b^2x + 9b^2$

$\Rightarrow$ $3b^2x + 4a^2y = 9b^2 + 16a^2$ [Dividing throughout by a^2b^2]

$$\Rightarrow \qquad \frac{3x}{a^2} + \frac{4y}{b^2} = \frac{9}{a^2} + \frac{16}{b^2} \qquad ...(3)$$

Since the point (3, 4) lies on the ellipse (1)

$$\therefore \qquad \frac{x^2}{a^2} + \frac{y^2}{b^2} = 1$$

$$\Rightarrow \qquad \frac{(3)^2}{a^2} + \frac{(4)^2}{b^2} = 1 \Rightarrow \frac{9}{a^2} + \frac{16}{b^2} = 1 \qquad ...(4)$$

$\therefore$ From equations (3) and (4), we get

$$\frac{3x}{a^2} + \frac{4y}{b^2} = 1.$$

(*ii*) We have, $$\frac{x^2}{9} - \frac{y^2}{16} = 1 \qquad ...(1)$$

The co-ordinates of the point of contact are (– 3, 0).
Differentiating (1) w.r.t. x, we get

$$\frac{2x}{9} - \frac{2y}{16}\frac{dy}{dx} = 0 \Rightarrow -\frac{2y}{16}\frac{dy}{dx} = -\frac{2x}{9}$$

$$\Rightarrow \qquad \frac{dy}{dx} = \frac{16x}{9y} \qquad ...(2)$$

$\therefore$ Slope of tangent at (– 3, 0) is

$$m_1 = \left(\frac{dy}{dx}\right)_{(-3,0)} = \frac{16(-3)}{9(0)} = \infty.$$

$\therefore$ The tangent at (– 3, 0) is parallel to y-axis.
i.e., the equation of tangent is in the form $x = k$.
Since, (– 3, 0) lies on $x = k$.

$$\Rightarrow \qquad -3 = k \qquad ...(3)$$

Now, normal at (– 3, 0) is perpendicular to y-axis.
$\Rightarrow$ Normal at (– 3, 0) is parallel to x-axis.
$\Rightarrow$ The slope of normal at (– 3, 0) is 0.
$\therefore$ The equation of normal at (– 3, 0) is given by

$$(y - 0) = 0\,[x - (-3)]$$

$$\Rightarrow \qquad y = 0.$$

Example 19. (*i*) *Find the equation of normal to the curve $x^2 + 2y^2 - 4x - 6y + 8 = 0$ at the points whose abscissa is 2.*

(*ii*) *Find the equation of the tangent and normal to the curve $y = x^2 + 4x + 1$ at the point whose abscissa is 3.*

Solution. (*i*) We have,

$$x^2 + 2y^2 - 4x - 6y + 8 = 0 \qquad ...(1)$$

Since abscissa is 2 *i.e.*, $x = 2$

$\therefore$ At $x = 2$, $(2)^2 + 2y^2 - 4(2) - 6y + 8 = 0$

$\Rightarrow$ $4 + 2y^2 - 8 - 6y + 8 = 0$

$\Rightarrow$ $2y^2 - 6y + 4 = 0$

$\Rightarrow$ $y^2 - 3y + 2 = 0$

$\Rightarrow$ $(y - 1)(y - 2) = 0$

$\Rightarrow$ $y = 1, 2.$

$\therefore$ The co-ordinates of points of contact are (2, 1) and (2, 2).

Differentiating (1) w.r.t. x, we get

$$2x + 4y\frac{dy}{dx} - 4 - 6\frac{dy}{dx} + 0 = 0$$

$$\Rightarrow \quad (4y - 6)\frac{dy}{dx} = (4 - 2x)$$

$$\Rightarrow \quad \frac{dy}{dx} = \frac{2(2-x)}{2(2y-3)} = \frac{2-x}{2y-3} \qquad ...(2)$$

$\therefore$ Slope of tangent at (2, 1) $= m_1 = \left(\frac{dy}{dx}\right)_{(2,1)} = \frac{2-2}{2-3} = \frac{0}{-1} = 0.$

Also, slope of tangent at (2, 2) $= m_2 = \left(\frac{dy}{dx}\right)_{(2,2)} = \frac{2-2}{4-3} = \frac{0}{1} = 0.$

$\therefore$ The slope of tangent at (2, 1) and (2, 2) is 0.

$\therefore$ Slope of the normal at (2, 1) and (2, 2) is undefined.

$\Rightarrow$ The normal at each point is parallel to y-axis.

$\therefore$ Equation of normal at each point is of the form : $x = k$

It is given that abscissa is 2 *i.e.*, $x = 2 \Rightarrow k = 2$

$\therefore$ Equation of normal at each point is given by

$$x = 2.$$

(*ii*) We have, $y = x^2 + 4x + 1$ ...(1)

Abscissa of point is 3 *i.e.*, $x = 3$.

$\therefore$ $y = (3)^2 + 4(3) + 1 = 9 + 12 + 1 = 22.$

$\therefore$ The co-ordinates of point of contact are (3, 22).

Differentiating (1), w.r.t. x, we get

$$\frac{dy}{dx} = 2x + 4 \qquad ...(2)$$

$\therefore$ Slope of tangent at (3, 22) $= m_1 = \left(\frac{dy}{dx}\right)_{(3,22)} = 2(3) + 4 = 10$

$\therefore$ Equation of tangent at (3, 22) is given by

$(y - 22) = 10(x - 3)$ [$\because$ By using $(y - y_1) = m(x - x_1)$]

$\Rightarrow$ $y - 22 = 10x - 30$

$\Rightarrow$ $10x - y - 8 = 0$ ...(3)

Now, the slope of normal at (3, 22) $= m_2 = \frac{-1}{m_1} = \frac{-1}{10}$

$\therefore$ The equation of normal at (3, 22) is given by

$$(y - 22) = \frac{-1}{10}(x - 3) \Rightarrow 10(y - 22) = -x + 3$$

$$\Rightarrow \quad 10y - 220 = -x + 3 \Rightarrow x + 10y - 223 = 0.$$

Example 20. *Find the equations of tangent and normal to the ellipse* $\frac{x^2}{a^2} + \frac{y^2}{b^2} = 1$ *at* (x_1, y_1).

Solution. We have, $\frac{x^2}{a^2} + \frac{y^2}{b^2} = 1$ ...(1)

The co-ordinates of point of contact are (x_1, y_1).

Differentiating (1) w.r.t. x, we get

$$\frac{2x}{a^2} + \frac{2y}{b^2} \cdot \frac{dy}{dx} = 0 \quad \text{...(2)}$$

$$\Rightarrow \quad \frac{2y}{b^2}\frac{dy}{dx} = -\frac{2x}{a^2} \Rightarrow \frac{dy}{dx} = -\frac{b^2 x}{a^2 y}$$

$\therefore$ Slope of tangent at $(x_1, y_1) = m_1 = \left(\frac{dy}{dx}\right)_{(x_1, y_1)} = \frac{-b^2 x_1}{a^2 y_1}$.

$\therefore$ The equation of tangent at (x_1, y_1) is given by

$$\Rightarrow \quad (y - y_1) = \frac{-b^2 x_1}{a^2 y_1}(x - x_1) \quad [\because \text{ By using } (y - y_1) = m(x - x_1)]$$

$$\Rightarrow \quad a^2y_1(y - y_1) = -b^2x_1(x - x_1)$$

$$\Rightarrow \quad a^2yy_1 - a^2{y_1}^2 = -b^2xx_1 + b^2{x_1}^2$$

$$\Rightarrow \quad b^2xx_1 + a^2yy_1 = a^2{y_1}^2 + b^2{x_1}^2$$

$$\Rightarrow \quad \frac{xx_1}{a^2} + \frac{yy_1}{b^2} = \frac{{x_1}^2}{a^2} + \frac{{y_1}^2}{b^2} \quad \text{...(3) [Dividing throughout by } a^2b^2]$$

Since, the point (x_1, y_1) lies on the curve $\frac{x^2}{a^2} + \frac{y^2}{b^2} = 1$,

$$\therefore \quad \frac{{x_1}^2}{a^2} + \frac{{y_1}^2}{b^2} = 1 \quad \text{...(4)}$$

$\therefore$ From equations (3) and (4), we get

$$\frac{xx_1}{a^2} + \frac{yy_1}{b^2} = 1 \quad \text{...(5)}$$

Now, the slope of normal at $(x_1, y_1) = m_2 = \frac{-1}{m_1} = \frac{-1}{-\frac{b^2 x_1}{a^2 y_1}} = \frac{a^2 y_1}{b^2 x_1}$

$\therefore$ Equation of normal at (x_1, y_1) is given by

$\Rightarrow$ $$(y - y_1) = \frac{a^2 y_1}{b^2 x_1}(x - x_1) \qquad [\because \text{ By using } (y - y_1) = m(x - x_1)]$$

$\Rightarrow$ $$b^2x_1(y - y_1) = a^2y_1(x - x_1)$$

$\Rightarrow$ $$b^2x_1y - b^2x_1y_1 = a^2y_1x - a^2x_1y_1$$

$\Rightarrow$ $$a^2xy_1 - b^2yx_1 = a^2x_1y_1 - b^2x_1y_1$$

$\Rightarrow$ $$\frac{a^2x}{x_1} - \frac{b^2y}{y_1} = a^2 - b^2 \qquad [\text{Dividing throughout by } (x_1y_1)]$$

Example 21. *Find the equation of the normal at (e, 1) to the curve y = log x.*

Solution. We have, $y = \log x$ ...(1)

The co-ordinates of point of contact are $(e, 1)$.

Differentiating (1) w.r.t. x, we get

$$\frac{dy}{dx} = \frac{1}{x} \qquad ...(2)$$

$\therefore$ Slope of tangent at $(e, 1) = m_1 = \left(\frac{dy}{dx}\right)_{(e,1)} = \frac{1}{e}$

$\therefore$ Slope of normal at $(e, 1) = m_2 = \frac{-1}{m_1} = \frac{-1}{1/e} = -e.$

$\therefore$ The equation of the normal at $(e, 1)$ is given by

$\Rightarrow$ $$(y - 1) = -e(x - e) \qquad [\because \text{ By using } (y - y_1) = m(x - x_1)]$$

$\Rightarrow$ $$y - 1 = -ex + e^2$$

$\Rightarrow$ $$ex + y - 1 - e^2 = 0$$

Example 22. *Find the equations of the tangent and the normal to the curve y(x – 2) (x – 3) – x + 7 = 0 at the point, where it cuts x-axis.*

Solution. We have, $y(x - 2)(x - 3) - x + 7 = 0$ ...(1)

Since, the curve cuts the x-axis *i.e.,* at $y = 0$.

$\therefore$ At $y = 0$,

$\Rightarrow$ $0(x - 2)(x - 3) - x + 7 = 0 \Rightarrow x = 7$

$\therefore$ The co-ordinates of point of contact are $(7, 0)$.

Differentiating (1) w.r.t. x, we get

$$y(x-2)\frac{d}{dx}(x-3) + y(x-3)\frac{d}{dx}(x-2) + (x-2)(x-3)\frac{dy}{dx} - 1 + 0 = 0$$

$\Rightarrow$ $$y(x-2)(1-0) + y(x-3)(1-0) + (x-2)(x-3)\frac{dy}{dx} - 1 = 0$$

$\Rightarrow$ $$(x-2)(x-3)\frac{dy}{dx} = 1 - y(x-2) - y(x-3)$$

$\Rightarrow$ $$\frac{dy}{dx} = \frac{1 - y[x - 2 + x - 3]}{(x-2)(x-3)}$$

$$\Rightarrow \quad \frac{dy}{dx} = \frac{1 - y(2x-5)}{(x-2)(x-3)} \quad ...(2)$$

$\therefore$ Slope of tangent at (7, 0) $= m_1 = \left(\frac{dy}{dx}\right)_{(7,0)} = \frac{1 - 0[2(7)-5]}{(7-2)(7-3)} = \frac{1}{(5)(4)} = \frac{1}{20}$

$\therefore$ The equation of tangent at (7, 0) is given by

$$(y - 0) = \frac{1}{20}(x - 7) \quad [\because \text{ By using } (y - y_1) = m(x - x_1)]$$

$$\Rightarrow \quad 20y = x - 7$$

$$\Rightarrow \quad x - 20y - 7 = 0 \quad ...(3)$$

Now, the slope of normal at (7, 0) $= m_2 = \frac{-1}{m_1} = \frac{-1}{1/20} = -20$

$\therefore$ The equation of normal at (7, 0) is given by

$$(y - 0) = -20(x - 7)$$

$$\Rightarrow \quad y = -20x + 140$$

$$\Rightarrow \quad 20x + y - 140 = 0$$

Example 23. *Find the equation of the tangent to the curve $y = (x^3 - 1)(x - 2)$ at the point where the curve cuts the x-axis.*

Solution. Please try yourself.

[**Hint.** Proceed as in example 22.] [**Ans.** $7x - y - 14 = 0$]

Example 24. *Show that the line $\frac{x}{a} + \frac{y}{b} = 1$ touches the curve $y = be^{-x/a}$ at the point where it cuts the y-axis.*

Solution. We have, $y = be^{-x/a}$ ...(1)

Since, it cuts the y-axis *i.e.,* at $x = 0$.

$\therefore$ At $x = 0$ [$\because$ $e^0 = 1$]

$y = be^0 = b$

$\therefore$ The co-ordinates of point of contact are (0, b).

Differentiating (1) w.r.t. x, we get

$$\frac{dy}{dx} = be^{-x/a} \cdot \frac{d}{dx}\left(-\frac{x}{a}\right) = -\frac{b}{a}e^{-x/a} \quad ...(2)$$

$\therefore$ The slope of tangent at (0, b)

$$= m_1 = \left(\frac{dy}{dx}\right)_{(0,b)} = -\frac{b}{a}e^0 = -\frac{b}{a} \quad [\because e^0 = 1]$$

$\therefore$ The equation of the tangent at (0, b) is given by

$$(y - b) = -\frac{b}{a}(x - 0) \quad [\because \text{ By using } (y - y_1) = m(x - x_1)]$$

$$\Rightarrow \quad a(y - b) = -bx$$

$$\Rightarrow \quad ay - ab = -bx$$

$$\Rightarrow \quad bx + ay - ab = 0$$

$\Rightarrow \quad bx + ay = ab$

$\Rightarrow \quad \frac{x}{a} + \frac{y}{b} = 1$ ...(3) [Dividing throughout by ab]

Thus, the line $\frac{x}{a} + \frac{y}{b} = 1$, touches the curve $y = be^{-x/a}$ at the point where it cuts y-axis.

Example 25. *(i) Find the equation of the tangent to the curve $y = x^2$, whose slope is $\frac{1}{2}$.*

(ii) Find the equations of all lines having slope – 1 that are tangents to the curve $y = \frac{1}{x-1}$.

Solution. (*i*) We have, $\quad y = x^2$ ...(1)

Differentiating w.r.t. x, we get

$$\frac{dy}{dx} = 2x$$

$\therefore$ Slope of tangent to the given curve $= \left(\frac{dy}{dx}\right) = 2x$ ...(2)

Also, it is given that, slope of tangent $= \frac{1}{2}$ ...(3)

$\therefore$ From (2) and (3), we get

$$2x = \frac{1}{2} \Rightarrow x = \frac{1}{4}.$$

$\therefore$ At $x = \frac{1}{4}$, $\quad y = \left(\frac{1}{4}\right)^2 = \frac{1}{16}$.

$\therefore$ The co-ordinates of point of contact are $\left(\frac{1}{4}, \frac{1}{16}\right)$.

$\therefore$ The equation of tangent at $\left(\frac{1}{4}, \frac{1}{16}\right)$ is given by

$\left(y - \frac{1}{16}\right) = \frac{1}{2}\left(x - \frac{1}{4}\right)$ [$\because$ By using $(y - y_1) = m(x - x_1)$]

$\Rightarrow \quad \frac{16y - 1}{16} = \frac{1}{2}\left(\frac{4x - 1}{4}\right) \Rightarrow 16y - 1 = 2(4x - 1)$

$\Rightarrow \quad 16y - 1 = 8x - 2 \Rightarrow 8x - 16y - 1 = 0.$

(*ii*) We have, $\quad y = \frac{1}{x-1}$ ...(1)

Differentiating w.r.t. x, we get

$$\frac{dy}{dx} = \frac{-1}{(x-1)^2}$$

$\therefore$ Slope of tangent to the given curve $= \left(\frac{dy}{dx}\right) = \frac{-1}{(x-1)^2}$ ...(2)

Also, it is given that, slope of tangent $= -1$ ...(3)

$\therefore$ From (2) and (3), we have

$$\frac{-1}{(x-1)^2} = -1 \Rightarrow (x-1)^2 = 1$$

$$\Rightarrow \quad x - 1 = \pm 1 \Rightarrow x - 1 = +1, x - 1 = -1$$

$$\Rightarrow \quad x = 0, 2.$$

At $x = 0$, $\quad y = \dfrac{1}{0-1} = -1 \quad \Rightarrow (0, -1)$

At $x = 2$, $\quad y = \dfrac{1}{2-1} = \dfrac{1}{1} = 1 \Rightarrow (2, 1)$

$\therefore$ The co-ordinates of the points of contact are $(0, -1)$ and $(2, 1)$.

$\therefore$ The equation of tangent at $(0, -1)$ is given by

$$[y - (-1)] = -1(x - 0) \qquad [\because \text{ By using } (y - y_1) = m(x - x_1)]$$

$$\Rightarrow \quad y + 1 = -x$$

$$\Rightarrow \quad x + y + 1 = 0 \qquad ...(4)$$

The equation of tangent at $(2, 1)$ is given by

$$(y - 1) = -1(x - 2)$$

$$\Rightarrow \quad y - 1 = -x + 2$$

$$\Rightarrow \quad x + y - 3 = 0$$

Example 26. (*i*) *Find the equation of tangent to the curve* $y = 2x^2 + 7$ *which is parallel to the line* $4x - y + 3 = 0$.

(*ii*) *Find the equation of the tangent to the curve* $x^2 + 3y = 3$ *which is parallel to the line* $y - 4x + 5 = 0$.

(*iii*) *Find the equation of the tangent to the curve* $y = \sqrt{5x-3} - 2$ *which is parallel to the line* $4x - 2y + 3 = 0$.

(*iv*) *Find the equation of the tangent to the curve* $y = \sqrt{3x-2}$ *which is parallel to the line* $4x - 2y + 5 = 0$.

Solution. (*i*) The given equation of the curve is

$$y = 2x^2 + 7 \qquad ...(1)$$

Differentiating w.r.t. x, we get

$$\frac{dy}{dx} = 4x$$

$\therefore$ Slope of tangent at $(x, y) = m_1 = \left(\dfrac{dy}{dx}\right) = 4x$ ...(2)

Now, the given equation of line is

$$4x - y + 3 = 0 \qquad ...(3)$$

$\therefore$ Slope of line $= m_2 = \dfrac{-4}{-1} = 4$ ...(4) $\left[\begin{array}{l}\because \text{ For the line } ax + by + c = 0 \\ \Rightarrow \quad \text{slope} = -\dfrac{\text{co-eff. of } x}{\text{co-eff. of } y}\end{array}\right]$

Since, the tangent to the given curve is parallel to the given line.

$\therefore$ Their slopes must be equal *i.e.*, $m_1 = m_2$

$$\Rightarrow \quad 4x = 4$$

$$\Rightarrow \quad x = 1$$

$\therefore$ At $x = 1$,

From (1),

$$y = 2(1)^2 + 7 = 9$$

$\therefore$ The co-ordinates of point of contact are (1, 9).

$\therefore$ The equation of tangent at (1, 9) is given by

$$(y - 9) = 4(x - 1) \qquad [\because \text{ By using } (y - y_1) = m(x - x_1)]$$

$$\Rightarrow \quad y - 9 = 4x - 4$$

$$\Rightarrow \quad 4x - y + 5 = 0.$$

(*ii*) Please try yourself.

[**Hint.** Proceed as in part (*i*)] [**Ans.** $4x - y + 13 = 0$]

(*iii*) The given equation of the curve is

$$y = \sqrt{5x - 3} - 2 \qquad ...(1)$$

Differentiating w.r.t. x, we get

$$\frac{dy}{dx} = \frac{1}{2\sqrt{5x-3}} \cdot \frac{d}{dx}(5x - 3) - 0 = \frac{1}{2\sqrt{5x-3}}(5 - 0)$$

$$\Rightarrow \quad \frac{dy}{dx} = \frac{5}{2\sqrt{5x-3}}$$

$$\therefore \text{ Slope of tangent at } (x, y) = m_1 = \left(\frac{dy}{dx}\right) = \frac{5}{2\sqrt{5x-3}} \qquad ...(2)$$

Now, the given equation of line is

$$4x - 2y + 3 = 0 \qquad ...(3)$$

$$\therefore \text{ Slope of line} \quad = m_2 = \frac{-4}{-2} = 2 \qquad \left[\begin{array}{l}\because \text{ For the line } ax + by + c = 0 \\ \Rightarrow \quad \text{slope} = -\dfrac{\text{co-eff. of } x}{\text{co-eff. of } y}\end{array}\right]$$

Since, the tangent to the given curve is parallel to the given line.

$\therefore$ Their slopes must be equal *i.e.*, $m_1 = m_2$

$$\Rightarrow \quad \frac{5}{2\sqrt{5x-3}} = 2 \quad \Rightarrow \quad 5 = 4\sqrt{5x-3} \qquad \text{[On squaring]}$$

$$\Rightarrow \quad 25 = 16(5x - 3) \quad \Rightarrow \quad 25 = 80x - 48$$

$$\Rightarrow \quad 80x = 73 \quad \Rightarrow \quad x = \frac{73}{80}$$

$\therefore$ At $x = \dfrac{73}{80}$,

From (1),

$$y = \sqrt{5\left(\frac{73}{80}\right) - 3} - 2 = \sqrt{\frac{73}{16} - 3} - 2$$

$$= \sqrt{\frac{73-48}{16}} - 2 = \sqrt{\frac{25}{16}} - 2 = \frac{5}{4} - 2 = \frac{5-8}{4} = -\frac{3}{4}$$

$\therefore$ The co-ordinates of point of contact are $\left(\frac{73}{80}, \frac{-3}{4}\right)$.

$\therefore$ The equation of the tangent at $\left(\frac{73}{80}, \frac{-3}{4}\right)$ is given by

$$\left[y - \left(-\frac{3}{4}\right)\right] = 2\left(x - \frac{73}{80}\right) \qquad [\because \text{ By using } (y - y_1) = m(x - x_1)]$$

$\Rightarrow$ $$\left(y + \frac{3}{4}\right) = 2\left(\frac{80x - 73}{80}\right)$$

$\Rightarrow$ $$\frac{4y+3}{4} = \frac{80x-73}{40}$$

$\Rightarrow$ $$4y + 3 = \frac{80x - 73}{10}$$

$\Rightarrow$ $$10(4y + 3) = 80x - 73$$

$\Rightarrow$ $$40y + 30 = 80x - 73$$

$\Rightarrow$ $$80x - 40y - 103 = 0.$$

(*iv*) Please try yourself.

[**Hint.** Proceed as in part (*iii*)] [**Ans.** $48x - 24y - 23 = 0$]

Example 27. *Find the equation of the tangent to the curve $y = x^2 + 4x - 16$ which is parallel to the line $3x - y + 1 = 0$.*

Solution. The given equation of the curve is

$$y = x^2 + 4x - 16 \qquad ...(1)$$

Differentiating w.r.t. x, we get

$$\frac{dy}{dx} = 2x + 4$$

$\therefore$ Slope of tangent at $(x, y) = m_1 = \left(\frac{dy}{dx}\right) = 2x + 4$ ...(2)

Now, the equation of the given line is

$$3x - y + 1 = 0 \qquad ...(3)$$

$\therefore$ Slope of line $= m_2 = -\left(\frac{3}{-1}\right) = 3$ $\left[\begin{array}{l}\because \text{ For the line } ax + by + c = 0 \\ \Rightarrow \quad \text{slope} = -\frac{\text{co-eff. of } x}{\text{co-eff. of } y}\end{array}\right]$

Since, the tangent to the given curve is parallel to the given line.

∴ Their slopes must be equal *i.e.,*

$$m_1 = m_2 \quad \Rightarrow \quad 2x + 4 = 3$$

$$\Rightarrow \qquad 2x = -1 \quad \Rightarrow \quad x = \frac{-1}{2}.$$

∴ At $x = \frac{-1}{2}$,

From (1),

$$y = \left(\frac{-1}{2}\right)^2 + 4\left(\frac{-1}{2}\right) - 16$$

$$= \frac{1}{4} - 2 - 16 = \frac{1}{4} - 18 = \frac{1-72}{4} = \frac{-71}{4}$$

∴ The co-ordinates of point of contact are $\left(-\frac{1}{2}, \frac{-71}{4}\right)$.

∴ The equation of tangent at $\left(-\frac{1}{2}, \frac{-71}{4}\right)$ is given by

$$\Rightarrow \qquad \left[y - \left(\frac{-71}{4}\right)\right] = 3\left[x - \left(\frac{-1}{2}\right)\right] \qquad [\because \text{ By using } (y - y_1) = m(x - x_1)]$$

$$\Rightarrow \qquad \left(y + \frac{71}{4}\right) = 3\left(x + \frac{1}{2}\right)$$

$$\Rightarrow \qquad \left(\frac{4y + 71}{4}\right) = 3\left(\frac{2x + 1}{2}\right) \quad \Rightarrow \quad 4y + 71 = 6(2x + 1)$$

$$\Rightarrow \qquad 4y + 71 = 12x + 6 \quad \Rightarrow \quad 12x - 4y - 65 = 0.$$

Example 28. (*i*) *Determine the equation of the normal to the curve* $y = 4x^3 - 3x + 5$ *which are parallel to the line* $9y + x + 3 = 0$.

(*ii*) *Find the equation of the normal to the curve* $3x^2 - y^2 = 8$ *which is parallel to the line* $x + 3y = 4$.

(*iii*) *Find the equation of the normal to the curve* $2x^2 - y^2 = 14$ *which is parallel to the line* $x + 3y = 4$.

Solution. (*i*) The given equation of the curve is

$$y = 4x^3 - 3x + 5 \qquad ...(1)$$

Differentiating w.r.t. x, we get

$$\frac{dy}{dx} = 12x^2 - 3$$

∴ Slope of tangent at $(x, y) = m_1 = \left(\frac{dy}{dx}\right) = 12x^2 - 3$

∴ Slope of normal at $(x, y) = m_2 = \frac{-1}{m_1} = \frac{-1}{12x^2 - 3}$ ...(2)

Now, the given equation of line is

$$9y + x + 3 = 0 \qquad ...(3)$$

$\therefore$ Slope of line $= m_3 = \dfrac{-1}{9}$ $\left[\begin{array}{l}\because \text{ For the line } ax + by + c = 0 \\ \Rightarrow \quad \text{slope} = -\dfrac{\text{co-eff. of } x}{\text{co-eff. of } y}\end{array}\right]$

Since, the normal to the given curve is parallel to the given line

$\therefore$ Their slopes must be equal *i.e.*, $m_2 = m_3$

$$\Rightarrow \quad \frac{-1}{12x^2 - 3} = \frac{-1}{9} \quad \Rightarrow \quad 12x^2 - 3 = 9$$

$$\Rightarrow \quad 12x^2 = 12 \quad \Rightarrow \quad x^2 = 1 \quad \Rightarrow \quad x = \pm 1$$

$\therefore$ At $x = 1$,

From (1), $y = 4(1)^3 - 3(1) + 5 = 6$

At $x = -1$,

From (1), $y = 4(-1)^3 - 3(-1) + 5 = -4 + 3 + 5 = 4$

$\therefore$ We have two normal, one at (1, 6) and the other at (– 1, 4) which are parallel to the given line.

$\therefore$ The equation of normal at (1, 6) is given by

$$(y - 6) = -\frac{1}{9}(x - 1) \qquad [\because \text{ By using } (y - y_1) = m(x - x_1)]$$

$$\Rightarrow \quad 9(y - 6) = -x + 1 \quad \Rightarrow \quad 9y - 54 = -x + 1$$

$$\Rightarrow \quad x + 9y - 55 = 0 \qquad ...(4)$$

$\therefore$ The equation of normal at (– 1, 4) is given by

$$\Rightarrow \quad (y - 4) = -\frac{1}{9}[x - (-1)] \quad \Rightarrow \quad 9(y - 4) = -1(x + 1)$$

$$\Rightarrow \quad 9y - 36 = -x - 1 \quad \Rightarrow \quad x + 9y - 35 = 0$$

(*ii*) The given equation of the curve is

$$3x^2 - y^2 = 8 \qquad ...(1)$$

Differentiating w.r.t. x, we get

$$6x - 2y\frac{dy}{dx} = 0$$

$$\Rightarrow \quad -2y\frac{dy}{dx} = -6x \quad \Rightarrow \quad \frac{dy}{dx} = \frac{3x}{y}$$

$\therefore$ The slope of tangent at $(x, y) = m_1 = \left(\dfrac{dy}{dx}\right) = \dfrac{3x}{y}$

$\therefore$ The slope of normal at $(x, y) = m_2 = \dfrac{-1}{m_1} = \dfrac{-1}{3x/y} = \dfrac{-y}{3x}$ ...(2)

Now, the given equation of the line is

$$x + 3y - 4 = 0 \qquad ...(3)$$

$\therefore$ Slope of the line $= m_3 = -\dfrac{1}{3}$ $\left[\begin{array}{l}\because \text{ For the line } ax+by+c=0 \\ \Rightarrow \quad \text{slope} = -\dfrac{\text{co-eff. of } x}{\text{co-eff. of } y}\end{array}\right]$

Since, the normal to the given curve is parallel to the given line.

$\therefore$ Their slopes must be equal *i.e.*, $m_2 = m_3$

$$\Rightarrow \quad \frac{-y}{3x} = \frac{-1}{3} \Rightarrow x = y \quad ...(4)$$

$\therefore$ At $x = y$,

From (1),

$$\Rightarrow \quad 3x^2 - x^2 = 8$$

$$\Rightarrow \quad 2x^2 = 8 \Rightarrow x^2 = 4 \Rightarrow x = \pm 2.$$

$\therefore$ We have, two normals, one at (2, 2) and the other at (– 2, – 2) which are parallel to the given line.

$\therefore$ Equation of the normal at (2, 2) is given by

$$(y-2) = \frac{-1}{3}(x-2) \quad [\because \text{ By using } (y - y_1) = m(x - x_1)]$$

$$\Rightarrow \quad 3(y-2) = -1(x-2)$$

$$\Rightarrow \quad 3y - 6 = -x + 2$$

$$\Rightarrow \quad x + 3y - 8 = 0 \quad ...(5)$$

Equation of the normal at (– 2, – 2) is given by

$$[y-(-2)] = \frac{-1}{3}[x-(-2)]$$

$$\Rightarrow \quad 3(y+2) = -1(x+2)$$

$$\Rightarrow \quad 3y + 6 = -x - 2$$

$$\Rightarrow \quad x + 3y + 8 = 0$$

(*iii*) Please try yourself.

[Hint. Proceed as in part (*ii*)] **[Ans.** $x + 3y - 9 = 0$, $x + 3y + 9 = 0$**]**

Example 29. *Find the equation of the tangent line to the curve* $y = x^2 - 2x + 7$, *which is*

(*i*) *Parallel to the line* $2x - y + 9 = 0$

(*ii*) *Perpendicular to the line* $5y - 15x = 13$.

Solution. (*i*) The given equation of the curve is

$$y = x^2 - 2x + 7 \quad ...(1)$$

Differentiating w.r.t. x, we get

$$\frac{dy}{dx} = 2x - 2$$

$$\therefore \quad \text{Slope of tangent at } (x, y) = m_1 = \left(\frac{dy}{dx}\right) = 2x - 2 \quad ...(2)$$

Now, the equation of the given line is

$$2x - y + 9 = 0 \quad ...(3)$$

$\therefore$ Slope of line $= m_2 = -\left(\dfrac{2}{-1}\right) = 2$ $\left[\begin{array}{l}\because \text{ For the line } ax+by+c=0 \\ \Rightarrow \quad \text{slope} = -\dfrac{\text{co-eff. of } x}{\text{co-eff. of } y}\end{array}\right]$

Since, the tangent to the given curve is parallel to the given line.

$\therefore$ Their slopes must be equal. *i.e.,* $m_1 = m_2$.

$\Rightarrow \quad 2x - 2 = 2 \Rightarrow 2x = 4 \Rightarrow x = 2$

$\therefore$ At $x = 2$,

From (1),

$$y = (2)^2 - 2(2) + 7 = 4 - 4 + 7 = 7.$$

$\therefore$ The co-ordinates of point of contact are (2, 7).

$\therefore$ The equation of tangent at (2, 7) is given by

$\Rightarrow \quad (y - 7) = 2(x - 2)$ $\quad [\because$ By using $(y - y_1) = m(x - x_1)]$

$\Rightarrow \quad y - 7 = 2x - 4$

$\Rightarrow \quad 2x - y + 3 = 0$ ...(4)

(*ii*) The given equation of line is

$5y - 15x - 13 = 0$ ...(5)

$\therefore$ Slope of the line $= (m_3) = -\left(\frac{-15}{5}\right) = 3$

$\therefore$ Slope of the tangent line which is perpendicular to given line $= \frac{-1}{3}$

$\left[\because \text{ For the line } ax + by + c = 0 \Rightarrow \text{slope} = -\frac{\text{co-eff. of } x}{\text{co-eff. of } y}\right]$

Since, the tangent to the given curve is perpendicular to the given line.

$\therefore$ Their product must be equal to -1. *i.e.,* $m_1 m_3 = -1$

$\Rightarrow \quad (2x - 2)(3) = -1 \Rightarrow 6x - 6 = -1$

$\Rightarrow \quad 6x = -1 + 6 \Rightarrow 6x = 5 \Rightarrow x = \frac{5}{6}$

$\therefore$ At $x = \frac{5}{6}$,

From (1), $y = \left(\frac{5}{6}\right)^2 - 2\left(\frac{5}{6}\right) + 7 = \frac{25}{36} - \frac{5}{3} + 7 = \frac{25 - 60 + 252}{36} = \frac{217}{36}$

$\therefore$ The co-ordinates of the point of contact are $\left(\frac{5}{6}, \frac{217}{36}\right)$

$\therefore$ The equation of tangent at $\left(\frac{5}{6}, \frac{217}{36}\right)$ is given by

$\left(y - \frac{217}{36}\right) = \frac{-1}{3}\left(x - \frac{5}{6}\right)$ $\quad [\because$ By using $(y - y_1) = m(x - x_1)]$

$\Rightarrow \quad 3\left(\frac{36y - 217}{36}\right) = -1\left(\frac{6x - 5}{6}\right)$

$\Rightarrow \quad \frac{36y - 217}{12} = \frac{-6x + 5}{6}$

$\Rightarrow \quad 36y - 217 = 2(-6x + 5)$

$\Rightarrow \quad 36y - 217 = -12x + 10$

$\Rightarrow \quad 12x + 36y - 217 - 10 = 0$

$\Rightarrow \quad 12x + 36y - 227 = 0.$

Example 30. *Find the equation of normal at any point 't' to the curve x = a (cos t + t sin t), y = a(sin t – t cos t). Also, show that the normal to the curve is at a constant distance from the origin.*

Solution. We have, $x = a(\cos t + t \sin t)$ and $y = a(\sin t - t \cos t)$

$$\therefore \quad \frac{dx}{dt} = a\left[-\sin t + t \cdot \frac{d}{dt}(\sin t) + \sin t \cdot \frac{d}{dt}(t)\right]$$

$$= a[-\sin t + t\cos t + \sin t \cdot 1]$$

$$\Rightarrow \quad \frac{dx}{dt} = at\cos t \qquad \text{...(1)}$$

Also,
$$\frac{dy}{dt} = a\left[\cos t - t \cdot \frac{d}{dt}(\cos t) - \cos t \cdot \frac{d}{dt}(t)\right]$$

$$= a[\cos t - t(-\sin t) - \cos t \cdot 1] = a[t \sin t]$$

$$\Rightarrow \quad \frac{dy}{dt} = at\sin t \qquad \text{...(2)}$$

$$\Rightarrow \quad \frac{dy}{dx} = \frac{dy/dt}{dx/dt} = \frac{at\sin t}{at\cos t} = \tan t \qquad \text{...(3)}$$

$\therefore$ Slope of the tangent at any point $t = m_1 = \left(\dfrac{dy}{dx}\right) = \tan t.$

$\therefore$ Slope of the normal at any point $= m_2 = \dfrac{-1}{m_1} = \dfrac{-1}{\tan t} = -\cot t.$

$\therefore$ The equation of normal at any point t is given by

$$[y - a(\sin t - t\cos t)] = -\cot t\,[x - a(\cos t + t\sin t)]$$

[$\because$ By using $(y - y_1) = m(x - x_1)$]

$$\Rightarrow \quad [y - a\sin t + at\cos t] = -\frac{\cos t}{\sin t}[x - a\cos t - at\sin t]$$

$$\Rightarrow \quad \sin t\,[y - a\sin t + at\cos t] = -\cos t\,[x - a\cos t - at\sin t]$$

$$\Rightarrow \quad y\sin t - a\sin^2 t + at\sin t\cos t = -x\cos t + a\cos^2 t + at\sin t\cos t$$

$$\Rightarrow \quad x\cos t + y\sin t = a(\sin^2 t + \cos^2 t) \qquad [\because \sin^2 A + \cos^2 A = 1]$$

$$\Rightarrow \quad x\cos t + y\sin t = a \qquad \text{...(4)}$$

Now, the length of the perpendicular from the origin (0, 0) to the line $x \cos t + y \sin t = a$ is given by

$$= \left|\frac{0 \cdot \cos t + 0 \cdot \sin t - a}{\sqrt{\cos^2 t + \sin^2 t}}\right| \qquad \left[\because \text{Perpendicular distance} = \frac{Ax + By - C}{\sqrt{A^2 + B^2}}\right]$$

$= |-a| = a$, which is a constant. [$\because \sin^2 A + \cos^2 A = 1$]

$\therefore$ The normal at any point t to the given curve is at a constant distance from the origin.

Example 31. *Prove that the curve* $\left(\frac{x}{a}\right)^n + \left(\frac{y}{b}\right)^n = 2$ *touches the straight line* $\frac{x}{a} + \frac{y}{b} = 2$ *at the point (a, b), whatever be the value of n.*

Solution. The given equation of the curve is

$$\Rightarrow \quad \left(\frac{x}{a}\right)^n + \left(\frac{y}{b}\right)^n = 2 \qquad ...(1)$$

Differentiating w.r.t. x, we get

$$n\left(\frac{x}{a}\right)^{n-1} \cdot \frac{1}{a} + n\left(\frac{y}{b}\right)^{n-1} \cdot \frac{1}{b}\frac{dy}{dx} = 0$$

$$\Rightarrow \quad \frac{nx^{n-1}}{a^{n-1} \cdot a} + \frac{ny^{n-1}}{b^{n-1} \cdot b}\frac{dy}{dx} = 0$$

$$\Rightarrow \quad \frac{ny^{n-1}}{b^n}\frac{dy}{dx} = -\frac{nx^{n-1}}{a^n}$$

$$\Rightarrow \quad \frac{dy}{dx} = -\frac{nx^{n-1}}{a^n} \times \frac{b^n}{ny^{n-1}}$$

$$\Rightarrow \quad \frac{dy}{dx} = -\frac{b^n x^{n-1}}{a^n y^{n-1}} \qquad ...(2)$$

$\therefore$ The slope of the tangent at $(a, b) = m_1 = \left(\frac{dy}{dx}\right)_{(a,b)} = -\frac{b^n \cdot a^{n-1}}{a^n \cdot b^{n-1}} = -\frac{b}{a}$

Now, the equation of the tangent at (a, b) is given by

$$(y - b) = -\frac{b}{a}(x - a) \qquad [\because \text{ By using } (y - y_1) = m(x - x_1)]$$

$$\Rightarrow \quad a(y - b) = -b(x - a)$$

$$\Rightarrow \quad ay - ab = -bx + ab$$

$$\Rightarrow \quad bx + ay = 2ab \qquad [\text{Dividing throughout by } ab]$$

$$\Rightarrow \quad \frac{x}{a} + \frac{y}{b} = 2$$

$\therefore$ The line $\frac{x}{a} + \frac{y}{b} = 2$, touches the curve $\left(\frac{x}{a}\right)^n + \left(\frac{y}{b}\right)^n = 2$, whatever be the value of n.

Example 32. *Tangents are drawn from the origin to the curve, $y = \sin x$. Prove that their points of contact lie on the curve $x^2y^2 = (x^2 - y^2)$.*

Solution. The given equation of the curve is

$$y = \sin x \qquad ...(1)$$

Differentiating w.r.t. x, we get

$$\frac{dy}{dx} = \cos x \qquad ...(2)$$

Let the co-ordinates of point of contact are (x_1, y_1).

$\therefore$ Slope of tangent at $(x_1, y_1) = m_1 = \left(\frac{dy}{dx}\right)_{(x_1, y_1)} = \cos x_1$

$\therefore$ The equation of tangent at (x_1, y_1) is given by

$\Rightarrow$ $(y - y_1) = \cos x_1 (x - x_1)$ ...(3) [$\because$ By using $(y - y_1) = m(x - x_1)$]

$\Rightarrow$ Since, this tangent passes through the origin. *i.e.*, (0, 0).

$\therefore$ $0 - y_1 = \cos x_1(0 - x_1)$

$\Rightarrow$ $-y_1 = -x_1 \cos x_1 \Rightarrow \frac{y_1}{x_1} = \cos x_1$ ...(4)

Also, the point (x_1, y_1) lies on the given curve.

$\therefore$ $y_1 = \sin x_1$ ...(5)

Squaring (4) and (5) and adding, we get

$$\frac{y_1^2}{x_1^2} + y_1^2 = (\cos x_1)^2 + (\sin x_1)^2$$

$\Rightarrow$ $\frac{y_1^2 + x_1^2 y_1^2}{x_1^2} = \cos^2 x_1 + \sin^2 x_1$

$\Rightarrow$ $y_1^2 + x_1^2 y_1^2 = x_1^2$ [$\because$ $\cos^2 A + \sin^2 A = 1$]

$\Rightarrow$ $x_1^2 y_1^2 = x_1^2 - y_1^2$

$\therefore$ The point of contact (x_1, y_1) lies on the curve $x^2y^2 = (x^2 - y^2)$.

Example 33. *If $x \cos \alpha + y \sin \alpha = p$ touches the curve $\frac{x^2}{a^2} + \frac{y^2}{b^2} = 1$, then show that*

$$a^2 \cos^2 \alpha + b^2 \sin^2 \alpha = p^2.$$

Solution. Let the line $x \cos \alpha + y \sin \alpha = p$ ...(1)

touches the given curve $\frac{x^2}{a^2} + \frac{y^2}{b^2} = 1$ ...(2)

at any point (x_1, y_1).

Then, the equation of tangent to the given curve at (x_1, y_1) is given by

$$\frac{xx_1}{a^2} + \frac{yy_1}{b^2} = 1 \qquad ...(3) \quad \text{[For Solution See Ex. 20]}$$

Since, the equations (1) and (3) identical.

$\therefore$ Comparing the co-efficients, we get

$$\frac{x_1/a^2}{\cos \alpha} = \frac{y_1/b^2}{\sin \alpha} = \frac{1}{p}$$

$\Rightarrow$ $\frac{x_1/a}{a \cos \alpha} = \frac{y_1/b}{b \sin \alpha} = \frac{1}{p}$

$$\Rightarrow \quad \frac{x_1}{a} = \frac{a \cos \alpha}{p} \quad \text{...(4)}$$

and
$$\frac{y_1}{b} = \frac{b \sin \alpha}{p} \quad \text{...(5)}$$

Squaring (4) and (5) and adding, we get

$$\frac{x_1^2}{a^2} + \frac{y_1^2}{b^2} = \frac{a^2 \cos^2 \alpha}{p^2} + \frac{b^2 \sin^2 \alpha}{p^2}$$

$$\Rightarrow \quad \frac{x_1^2}{a^2} + \frac{y_1^2}{b^2} = \frac{a^2 \cos^2 \alpha + b^2 \sin^2 \alpha}{p^2}$$

$$\Rightarrow \quad 1 = \frac{a^2 \cos^2 \alpha + b^2 \sin^2 \alpha}{p^2} \quad [\because \text{ By using (2)}]$$

$\Rightarrow \quad a^2 \cos^2 \alpha + b^2 \sin^2 \alpha = p^2$.

Example 34. *If $x \cos \alpha + y \sin \alpha = p$ touches the curve, $x^m y^n = a^{m+n}$. Prove that :*

$$p^{m+n} \cdot m^m \cdot n^n = (m+n)^{m+n} \cdot \cos^m \alpha \sin^n \alpha.$$

Solution. Let the equation of the given curve is

$$x^m y^n = a^{m+n} \quad \text{....(1)}$$

Taking logarithms on both sides, we get

$$\log (x^m y^n) = \log (a^{m+n})$$

$$\Rightarrow \quad \log x^m + \log y^n = \log (a^{m+n}) \quad [\because \log (m.n) = \log m + \log n]$$

$$\Rightarrow \quad m \log x + n \log y = (m+n) \log a \quad [\because \log m^n = n \log m]$$

Differentiating w.r.t. x, we get

$$\frac{m}{x} + \frac{n}{y} \cdot \frac{dy}{dx} = 0 \Rightarrow \frac{n}{y}\frac{dy}{dx} = -\frac{m}{x}$$

$$\Rightarrow \quad \frac{dy}{dx} = -\frac{my}{nx} \quad \text{...(2)}$$

Let the co-ordinates of point of contact are (x_1, y_1).

$\therefore$ Slope of tangent at $(x_1, y_1) = m_1 = \left(\frac{dy}{dx}\right)_{(x_1, y_1)} = -\frac{my_1}{nx_1}$.

$\therefore$ Equation of the tangent to the given curve at (x_1, y_1) is given by

$$(y - y_1) = -\frac{my_1}{nx_1}(x - x_1) \quad [\because \text{ By using } (y - y_1) = m(x - x_1)]$$

$$\Rightarrow \quad nx_1(y - y_1) = -my_1(x - x_1)$$

$$\Rightarrow \quad nx_1 y - nx_1 y_1 = -my_1 x + my_1 x_1$$

$$\Rightarrow \quad my_1 x + nx_1 y = my_1 x_1 + nx_1 y_1$$

$$\Rightarrow \quad (my_1)x + (nx_1)y = (m+n)x_1 y_1 \quad \text{...(3)}$$

Since, the given equation of the line is

$$x \cos \alpha + y \sin \alpha = p \quad \text{...(4)}$$

As the equations (3) and (4) are identical.

$\therefore$ Comparing the co-efficients, we get

$$\frac{my_1}{\cos\alpha} = \frac{nx_1}{\sin\alpha} = \frac{(m+n)x_1y_1}{p}$$

$$\therefore \quad \frac{my_1}{\cos\alpha} = \frac{(m+n)x_1y_1}{p} \quad \text{and} \quad \frac{nx_1}{\sin\alpha} = \frac{(m+n)x_1y_1}{p}$$

$$\Rightarrow \quad \frac{mp}{(m+n)\cos\alpha} = x_1 \quad \text{and} \quad \frac{np}{(m+n)\sin\alpha} = y_1$$

$$\Rightarrow \quad x_1 = \frac{mp}{(m+n)\cos\alpha} \qquad ...(5)$$

and $$y_1 = \frac{np}{(m+n)\sin\alpha} \qquad ...(6)$$

Also, the (x_1, y_1) lies on the given curve, therefore, we have

$$x_1{}^m y_1{}^n = a^{m+n} \qquad ...(7)$$

Now, putting the values of x_1, y_1 from equations (5) and (6) in equation (7), we get

$$\left[\frac{mp}{(m+n)\cos\alpha}\right]^m \cdot \left[\frac{np}{(m+n)\sin\alpha}\right]^n = a^{m+n}$$

$$\Rightarrow \quad \frac{m^m p^m}{(m+n)^m \cos^m\alpha} \times \frac{n^n p^n}{(m+n)^n \sin^n\alpha} = a^{m+n}$$

$$\Rightarrow \quad \frac{m^m n^n p^{(m+n)}}{(m+n)^{m+n} \cdot \cos^m\alpha \sin^n\alpha} = a^{m+n}$$

$$\Rightarrow \quad p^{m+n} \cdot m^m \cdot n^n = a^{m+n} \cdot (m+n)^{m+n} \cdot \cos^m\alpha \cdot \sin^n\alpha.$$

Example 35. *Find the angle of intersection between the curves*

(*i*) $y^2 = 2x$ *and* $x^2 + y^2 = 8$ (*ii*) $y^2 = ax$ *and* $x^2 + y^2 = 2a^2$

(*iii*) $y^2 = 4x$ *and* $x^2 = 4y$ (*iv*) $xy = 6$ *and* $x^2y = 12$.

Solution. (*i*) The equations of two curves are

$$y^2 = 2x \qquad ...(1)$$

and $$x^2 + y^2 = 8 \qquad ...(2)$$

Now, to find the point of intersection of the two curves, we have to solve equations (1) and (2).

$$\Rightarrow \quad x^2 + 2x = 8 \qquad [\because \text{ By using (1)}]$$

$$\Rightarrow \quad x^2 + 2x - 8 = 0 \quad \Rightarrow \quad (x+4)(x-2) = 0$$

$$\Rightarrow \quad x = -4, 2$$

$\therefore$ When $x = -4$,

From (1), we have

$$y^2 = 2(-4) = -8$$

which gives imaginary values of y and is rejected.

When $x = 2$,

From (1), we have

$$y^2 = 2(2) = 4 \quad \Rightarrow \quad y = \pm 2.$$

$\therefore$ The points of intersection are (2, 2) and (2, – 2).

Now, Differentiating (1) w.r.t. x, we get

$$2y\frac{dy}{dx} = 2 \quad \Rightarrow \quad \frac{dy}{dx} = \frac{1}{y} \qquad ...(3)$$

Differentiating (2) w.r.t. x, we get

$$2x + 2y\frac{dy}{dx} = 0 \quad \Rightarrow \quad 2y\frac{dy}{dx} = -2x$$

$$\Rightarrow \quad \frac{dy}{dx} = -\frac{x}{y} \qquad ...(4)$$

Point of intersection (2, 2) :

$\therefore$ Slope of tangent to the curve $y^2 = 2x$ at (2, 2) $= m_1 = \left(\frac{dy}{dx}\right)_{(2,2)} = \frac{1}{2}$

Slope of tangent to the curve $x^2 + y^2 = 8$ at (2, 2) $= m_2 = \left(\frac{dy}{dx}\right)_{(2,2)} = -\frac{2}{2} = -1.$

If θ is the acute angle between the curves at (2, 2), then

$$\tan\theta = \left|\frac{m_1 - m_2}{1 + m_1 m_2}\right| = \left|\frac{\frac{1}{2} - (-1)}{1 + \left(\frac{1}{2}\right)(-1)}\right| = \left|\frac{\frac{1}{2} + 1}{1 - \frac{1}{2}}\right| = \frac{\frac{3}{2}}{\frac{1}{2}} = 3$$

$$\Rightarrow \quad \theta = \tan^{-1} 3.$$

Point of intersection (2, – 2) :

$\therefore$ Slope of tangent to the curve $y^2 = 2x$ at (2, – 2) $= m_1 = \left(\frac{dy}{dx}\right)_{(2,-2)} = -\frac{1}{2}$

Slope of tangent to the curve $x^2 + y^2 = 8$ at (2, – 2) $= m_2 = \left(\frac{dy}{dx}\right)_{(2,-2)} = -\frac{2}{-2} = 1$

If ϕ is the acute angle between the curves at (2, – 2), then

$$\tan\phi = \left|\frac{m_1 - m_2}{1 + m_1 m_2}\right| = \left|\frac{-\frac{1}{2} - 1}{1 + \left(-\frac{1}{2}\right)(1)}\right| = \left|\frac{-\frac{3}{2}}{\frac{1}{2}}\right| = |-3| = 3.$$

$$\Rightarrow \quad \phi = \tan^{-1} 3.$$

(*ii*) The equations of the two curves are

$$y^2 = ax \qquad ...(1)$$

and

$$x^2 + y^2 = 2a^2 \qquad ...(2)$$

Now, to find the point of intersection of two curves, we have to solve equations (1) and (2).

$$\Rightarrow \quad x^2 + ax = 2a^2 \qquad [\because \text{ By using (1)}]$$

$\Rightarrow \qquad x^2 + ax - 2a^2 = 0$

$\Rightarrow \qquad (x + 2a)(x - a) = 0$

$\Rightarrow \qquad x = -2a, a$

When $x = -2a$,

From (1), we have

$$y^2 = a(-2a) = -2a^2$$

which gives imaginary values of y and is rejected.

When $x = a$,

From (1), we have

$$y^2 = a(a) = a^2 \Rightarrow y = \pm a$$

$\therefore$ The points of intersections are (a, a) and $(a, -a)$.

Now, Differentiating (1) w.r.t. x, we get

$$2y\frac{dy}{dx} = a \qquad \Rightarrow \qquad \frac{dy}{dx} = \frac{a}{2y} \qquad ...(3)$$

Differentiating (2) w.r.t. x, we get

$$2x + 2y\frac{dy}{dx} = 0$$

$$\Rightarrow \qquad 2y\frac{dy}{dx} = -2x \Rightarrow \frac{dy}{dx} = -\frac{x}{y} \qquad ...(4)$$

Point of intersection (a, a) :

$\therefore$ Slope of tangent to the curve $y^2 = ax$ at $(a, a) = m_1 = \left(\frac{dy}{dx}\right)_{(a,a)} = \frac{a}{2a} = \frac{1}{2}$

Slope of tangent to the curve $x^2 + y^2 = 2a^2$ at $(a, a) = m_2 = \left(\frac{dy}{dx}\right)_{(a,a)} = -\frac{a}{a} = -1.$

If θ is the acute angle between the curves at (a, a), then

$$\tan\theta = \left|\frac{m_1 - m_2}{1 + m_1 m_2}\right| = \left|\frac{\frac{1}{2} - (-1)}{1 + \frac{1}{2}(-1)}\right| = \left|\frac{\frac{1}{2} + 1}{1 - \frac{1}{2}}\right| = \frac{\frac{3}{2}}{\frac{1}{2}} = 3.$$

$\Rightarrow \qquad \theta = \tan^{-1} 3.$

Point of intersection (a, – a) :

$\therefore$ Slope of tangent to the curve $y^2 = ax$ at $(a, -a) = m_1 = \left(\frac{dy}{dx}\right)_{(a,-a)} = \frac{a}{-2a} = -\frac{1}{2}$

Slope of tangent to the curve $x^2 + y^2 = 2a^2$ at $(a, -a) = m_2 = \left(\frac{dy}{dx}\right)_{(a,-a)} = -\left(\frac{a}{-a}\right) = 1.$

If ϕ is the acute angle between the curves at $(a, -a)$, then

$$\tan \phi = \left| \frac{m_1 - m_2}{1 + m_1 m_2} \right| = \left| \frac{-\frac{1}{2} - 1}{1 + \left(-\frac{1}{2}\right)(1)} \right| = \left| \frac{-\frac{3}{2}}{\frac{1}{2}} \right| = |-3| = 3.$$

$\Rightarrow \qquad \phi = \tan^{-1} 3.$

(*iii*) The equations of two curves are

$$y^2 = 4x \qquad ...(1)$$

and

$$x^2 = 4y \qquad ...(2)$$

Now, to find the point of intersection of the two curves, we have to solve equations (1) and (2).

$$\Rightarrow \quad \left(\frac{y^2}{4}\right)^2 = 4y \qquad \left[\because \text{ From (1), } y^2 = 4x \Rightarrow x = \frac{y^2}{4}\right]$$

$$\Rightarrow \quad \frac{y^4}{16} = 4y$$

$$\Rightarrow \quad y^4 = 64y \Rightarrow y^4 - 64y = 0$$

$$\Rightarrow \quad y(y^3 - 64) = 0$$

$$\Rightarrow \quad y = 0, \quad y^3 - 64 = 0 \Rightarrow y = 4$$

$\therefore$ When $y = 0$, from (1)

$$0 = 4x \Rightarrow x = 0$$

When $y = 4$, from (1)

$$(4)^2 = 4x \Rightarrow 16 = 4x \Rightarrow x = 4.$$

$\therefore$ The points of intersection are (0, 0) are (4, 4).

Now, Differentiating (1) w.r.t. x, we get

$$\Rightarrow \quad 2y \cdot \frac{dy}{dx} = 4 \Rightarrow \frac{dy}{dx} = \frac{4}{2y} = \frac{2}{y} \qquad ...(3)$$

Differentiating (2) w.r.t. x, we get

$$2x = 4\frac{dy}{dx} \Rightarrow \frac{dy}{dx} = \frac{2x}{4} = \frac{x}{2} \qquad ...(4)$$

Point of intersection (0, 0) :

$\therefore$ Slope of tangent to the curve $y^2 = 4x$ at $(0, 0) = m_1 = \left(\frac{dy}{dx}\right)_{(0,0)} = \frac{2}{0} = \infty.$

$\therefore$ Tangent to the curve $y^2 = 4x$ at (0, 0) is parallel to y-axis.

Slope of tangent to the curve $x^2 = 4y$ at $(0, 0) = m_2 = \left(\frac{dy}{dx}\right)_{(0,0)} = \frac{0}{2} = 0$

$\therefore$ Tangent to the curve $x^2 = 4y$ at (0, 0) is parallel to x-axis.

Thus, the angle between the tangents to the given curves at (0, 0) is a right angle.

Hence, the two curves intersect at right angle at (0, 0).

Point of intersection (4, 4) :

$\therefore$ Slope of tangent to the curve $y^2 = 4x$ at (4, 4) $= m_1 = \left(\frac{dy}{dx}\right)_{(4,4)} = \frac{2}{4} = \frac{1}{2}$

Slope of tangent to the curve $x^2 = 4y$ at (4, 4) $= m_2 = \left(\frac{dy}{dx}\right)_{(4,4)} = \frac{4}{2} = 2$

If θ is the acute angle between the curves at (4, 4), then

$$\tan\theta = \left|\frac{m_1 - m_2}{1 + m_1 m_2}\right| = \left|\frac{2 - \left(\frac{1}{2}\right)}{1 + 2\left(\frac{1}{2}\right)}\right| = \left|\frac{\frac{3}{2}}{2}\right| = \frac{3}{4}$$

$$\Rightarrow \qquad \theta = \tan^{-1}\left(\frac{3}{4}\right).$$

(*iv*) Please try yourself. $\left[\textbf{Ans. } \tan^{-1}\left(\frac{3}{11}\right)\right]$

Example 36. *Show that the curves $xy = a^2$ and $x^2 + y^2 = 2a^2$ touch each other.*

Solution. The equations of two curves are

$$x^2 + y^2 = 2a^2 \qquad ...(1)$$

and

$$xy = a^2 \qquad ...(2)$$

Now, to find the point of intersection of the two curves, we have to solve equations (1) and (2).

$$\Rightarrow \qquad x^2 + \left(\frac{a^2}{x}\right)^2 = 2a^2 \qquad \left[\because \text{ From (2), } xy = a^2 \Rightarrow y = \frac{a^2}{x}\right]$$

$$\Rightarrow \qquad x^2 + \frac{a^4}{x^2} = 2a^2 \qquad \Rightarrow \frac{x^4 + a^4}{x^2} = 2a^2$$

$$\Rightarrow \qquad x^4 + a^4 = 2a^2x^2 \qquad \Rightarrow x^4 - 2a^2x^2 + a^4 = 0$$

$$\Rightarrow \qquad (x^2 - a^2)^2 = 0 \qquad \Rightarrow x^2 - a^2 = 0$$

$$\Rightarrow \qquad x^2 = a^2 \qquad \Rightarrow x = \pm a$$

$\therefore$ When $x = a$,

From (2),

$$ay = a^2 \qquad \Rightarrow y = a$$

When $x = -a$,

From (2),

$$(-a)y = a^2 \qquad \Rightarrow y = -a.$$

$\therefore$ The points of intersection are (a, a) and $(-a, -a)$.

Now, Differentiating (1) w.r.t. x, we get

$$2x + 2y\frac{dy}{dx} = 0$$

$\Rightarrow \quad 2y \dfrac{dy}{dx} = -2x \quad \Rightarrow \quad \dfrac{dy}{dx} = -\dfrac{x}{y}$ ...(3)

Differentiating (2) w.r.t. x, we get

$$x \cdot \frac{dy}{dx} + y \cdot 1 = 0 \quad \Rightarrow \quad x \frac{dy}{dx} = -y$$

$\Rightarrow \quad \dfrac{dy}{dx} = -\dfrac{y}{x}$ ...(4)

Point of intersection (a, a) :

$\therefore$ Slope of tangent to the curve $x^2 + y^2 = 2a^2$ at $(a, a) = m_1 = \left(\dfrac{dy}{dx}\right)_{(a,a)} = -\dfrac{a}{a} = -1$

Slope of tangent to the curve $xy = a^2$ at $(a, a) = m_2 = \left(\dfrac{dy}{dx}\right)_{(a,a)} = -\dfrac{a}{a} = -1$

Since, the slopes of tangents at (a, a) are equal, therefore, the two curves touches each other at (a, a).

Point of intersection (– a, – a) :

$\therefore$ Slope of tangent to the curve $x^2 + y^2 = 2a^2$ at $(-a, -a)$

$$= m_1 = \left(\frac{dy}{dx}\right)_{(-a,-a)} = -\frac{(-a)}{(-a)} = -1.$$

Slope of tangent to the curve $xy = a^2$ at $(-a, -a)$

$$= m_2 = \left(\frac{dy}{dx}\right)_{(-a,-a)} = -\frac{(-a)}{(-a)} = -1$$

Since, the slopes of tangents at $(-a, -a)$ are equal, therefore, the two curves touches each other at $(-a, -a)$.

Example 37. *Show that the condition that the curves $ax^2 + by^2 = 1$ and $a'x^2 + b'y^2 = 1$ should intersect orthogonally is that* $\dfrac{1}{a} - \dfrac{1}{b} = \dfrac{1}{a'} - \dfrac{1}{b'}$.

Solution. The equations of two curves are

$ax^2 + by^2 = 1$ ...(1)

and $a'x^2 + b'y^2 = 1$ ...(2)

Let (x_1, y_1) be the point of intersection of the two curves.

Therefore, (x_1, y_1) lies on the given curves

$\Rightarrow \quad ax_1^2 + by_1^2 = 1$ ...(3)

and $a'x_1^2 + b'y_1^2 = 1$ ...(4)

Now, Differentiating (1) w.r.t. x, we get

$$2ax + 2by \frac{dy}{dx} = 0 \quad \Rightarrow \quad 2by \frac{dy}{dx} = -2ax$$

$\Rightarrow \quad \dfrac{dy}{dx} = -\dfrac{ax}{by}$ ...(5)

Differentiating (2) w.r.t. x, we get

$$2a'x + 2b'y\frac{dy}{dx} = 0 \quad \Rightarrow \quad 2b'y\frac{dy}{dx} = -2a'x$$

$$\Rightarrow \quad \frac{dy}{dx} = -\frac{a'x}{b'y} \quad \text{...(6)}$$

$\therefore$ Slope of tangent to the curve (1) at $(x_1, y_1) = m_1 = \left(\frac{dy}{dx}\right)_{(x_1, y_1)} = -\frac{ax_1}{by_1}$ ...(7)

Slope of tangent to the curve (2) at (x_1, y_1) $= m_2 = \left(\frac{dy}{dx}\right)_{(x_1, y_1)} = -\frac{a'x_1}{b'y_1}$ ...(8)

Now, the two curves intersects orthogonally if, $m_1m_2 = -1$

[$\because$ By using (7) and (8)]

$$\Rightarrow \quad \left(-\frac{ax_1}{by_1}\right)\left(-\frac{a'x_1}{b'y_1}\right) = -1$$

$$\Rightarrow \quad \frac{aa'x_1^2}{bb'y_1^2} = -1$$

$$\Rightarrow \quad aa'x_1^2 = -bb'y_1^2 \quad \text{...(9)}$$

Subtracting (4) from (3), we get

$$(a - a')x_1^2 + (b - b')y_1^2 = 0$$

$$\Rightarrow \quad (a - a')x_1^2 = -(b - b')y_1^2 \quad \text{...(10)}$$

Dividing equation (10) by (9), we get

$$\frac{(a - a')x_1^2}{aa'x_1^2} = -\frac{(b - b')y_1^2}{-bb'y_1^2}$$

$$\Rightarrow \quad \frac{a - a'}{aa'} = \frac{b - b'}{bb'}$$

$$\Rightarrow \quad \frac{1}{a'} - \frac{1}{a} = \frac{1}{b'} - \frac{1}{b}$$

$$\Rightarrow \quad \frac{1}{a} - \frac{1}{b} = \frac{1}{a'} - \frac{1}{b'}.$$

Example 38. *Prove that the curves* $x = y^2$ *and* $xy = k$ *cut at right angle if* $8k^2 = 1$.

Solution. The equations of two curves are

$$x = y^2 \quad \text{...(1)}$$

$$xy = k \quad \text{...(2)}$$

Now, to find the point of intersection of the two curves, we have to solve equations (1) and (2).

$\Rightarrow \quad (y^2)y = k \Rightarrow y^3 = k \Rightarrow y = k^{1/3}$ [$\because$ From (1), $y^2 = x$]

$\therefore$ When $y = k^{1/3}$, From (1),

$$x = (k^{1/3})^2 = k^{2/3}$$

$\therefore$ The point of intersection is $(k^{2/3}, k^{1/3})$.

Differentiating (1) w.r.t. x, we get

$$1 = 2y\frac{dy}{dx} \Rightarrow \frac{dy}{dx} = \frac{1}{2y} \quad ...(3)$$

Differentiating (2) w.r.t. x, we get

$$x \cdot \frac{dy}{dx} + y.1 = 0$$

$$\Rightarrow \quad x\frac{dy}{dx} = -y \Rightarrow \frac{dy}{dx} = \frac{-y}{x} \quad ...(4)$$

Point of intersection ($\mathbf{k^{2/3}, k^{1/3}}$) :

Slope of tangent to the curve $x = y^2$ at $(k^{2/3}, k^{1/3})$

$$= m_1 = \left(\frac{dy}{dx}\right)_{(k^{2/3},\, k^{1/3})} = \frac{1}{2k^{1/3}} \quad ...(5)$$

Slope of tangent to the curve $xy = k$ at $(k^{2/3}, k^{1/3})$

$$= m_2 = \left(\frac{dy}{dx}\right)_{(k^{2/3},\, k^{1/3})} = -\frac{k^{1/3}}{k^{2/3}} = -\frac{1}{k^{1/3}} \quad ...(6)$$

Now, the two curves intersects at right angle if $m_1m_2 = -1$

$$\Rightarrow \quad \left(\frac{1}{2k^{1/3}}\right)\left(\frac{-1}{k^{1/3}}\right) = -1$$

$$\Rightarrow \quad \frac{1}{2k^{2/3}} = 1 \Rightarrow 2k^{2/3} = 1 \quad \text{[Cubing both sides]}$$

$$\Rightarrow \quad 8k^2 = 1$$

$\therefore$ The curves cut at right angle if $8k^2 = 1$.

Example 39. *Find the angle of intersection of the curves $x^3 - 3xy^2 = -2$ and $3x^2y - y^3 = 2$.*

Solution. The equations of two curves are

$$x^3 - 3xy^2 = -2 \quad ...(1)$$

and

$$3x^2y - y^3 = 2 \quad ...(2)$$

Now, to find the point of intersection of the two curves, we have to solve equations (1) and (2).

Adding equations (1) and (2), we have

$$x^3 - 3xy^2 + 3x^2y - y^3 = 0$$

$$\Rightarrow \quad x^3 - y^3 - 3xy(y - x) = 0$$

$$\Rightarrow \quad (x - y)(x^2 + xy + y^2) + 3xy(x - y) = 0 \quad [\because (a^3 - b^3) = (a - b)(a^2 + ab + b^2)]$$

$$\Rightarrow \quad (x - y)[x^2 + xy + y^2 + 3xy] = 0$$

$$\Rightarrow \quad (x - y)(x^2 + y^2 + 4xy) = 0$$

$$\Rightarrow \quad (x - y) = 0 \quad [\because (x^2 + y^2 + 4xy) \neq 0]$$

$$\Rightarrow \quad x = y$$

$\therefore$ When $x = y$, From (1),

$$\Rightarrow \quad x^3 - 3x(x^2) = -2 \Rightarrow x^3 - 3x^3 = -2$$

$$\Rightarrow \quad -2x^3 = -2 \Rightarrow x^3 = 1 \Rightarrow x = 1$$

$\therefore$ When $x = 1, y = 1$

$\therefore$ The point of intersection is (1, 1).

Differentiating (1) w.r.t. x, we get

$$3x^2 - 3\left[x \cdot \frac{d}{dx}(y^2) + y^2 \cdot \frac{d}{dx}(x)\right] = 0$$

$$\Rightarrow \quad 3x^2 - 3\left[x\left(2y\frac{dy}{dx}\right) + y^2 \cdot 1\right] = 0$$

$$\Rightarrow \quad 3x^2 - 6xy\frac{dy}{dx} - 3y^2 = 0$$

$$\Rightarrow \quad 2xy\frac{dy}{dx} = x^2 - y^2$$

$$\Rightarrow \quad \frac{dy}{dx} = \frac{x^2 - y^2}{2xy} \qquad ...(3)$$

Differentiating (2) w.r.t. x, we get

$$\Rightarrow \quad 3\left(x^2 \cdot \frac{dy}{dx} + 2xy\right) - 3y^2 \cdot \frac{dy}{dx} = 0$$

$$\Rightarrow \quad 3x^2\frac{dy}{dx} + 6xy - 3y^2\frac{dy}{dx} = 0$$

$$\Rightarrow \quad (x^2 - y^2)\frac{dy}{dx} + 2xy = 0$$

$$\Rightarrow \quad (x^2 - y^2)\frac{dy}{dx} = -2xy$$

$$\Rightarrow \quad \frac{dy}{dx} = \frac{2xy}{y^2 - x^2} \qquad ...(4)$$

Point of intersection (1, 1) :

$\therefore$ Slope of tangent to the curve (1) at (1, 1) $= m_1 = \left(\frac{dy}{dx}\right)_{(1,1)} = \frac{1-1}{2(1)(1)} = 0$

Slope of tangent to the curve (2) at (1, 1) $= m_2 = \left(\frac{dy}{dx}\right)_{(1,1)} = \frac{2(1)(1)}{1-1} = \infty$

$\therefore$ Tangent to the curve (1) at (1, 1) is parallel to x-axis and tangent to the curve (2) at (1, 1) is parallel to y-axis.

$\therefore$ Angle between the two tangents is 90°.

$\therefore$ The two curves intersects each other at right angles.

EXERCISE FOR PRACTICE

1. Show that the tangents to the curve $y = 2x^3 - 4$ at the points where $x = 2$ and $x = -2$ are parallel.
2. At what point on the circle $x^2 + y^2 - 2x - 4y + 1 = 0$, the tangent is parallel to x-axis.
3. At what point on the curve $x^2 + y^2 - 2x - 4y + 1 = 0$, the tangent is parallel to y-axis.
4. Find the equation of the normal to the curve $y = 2x^3 - x^2 + 3$ at (1, 4).
5. Find the equations of tangent and normal to the curve $y = x^4 - bx^3 + 13x^2 - 10x + 5$ at (0, 5).
6. Find the equation of the tangent to the curve $y = x - \sin x \cos x$ at $x = \dfrac{\pi}{2}$.
7. Find the equation of the normal to the curve $y = 2x^2 + 3 \sin x$ at $x = 0$.
8. Find the points on the curve $y = x^3 - 2x^2 - x$ at which the tangent lines are parallel to the line $y = 3x - 2$.
9. Find the co-ordinates of the points on the curve $y = x^2 + 3x + 4$, the tangent at which passes through the origin.
10. Show that the curves $2x^2 + y^2 = 20$ and $4y^2 - x^2 = 8$ are orthogonal.

Answers

2. (1, 0), (1, 4) **3.** (– 1, 2), (3, 2) **4.** $x + 4y - 17 = 0$

5. $y + 10x - 5 = 0$; $x - 10y + 50 = 0$ **6.** $4x - 2y - \pi = 0$

7. $x + 3y = 0$ **8.** $(2, -2)$; $\left(-\dfrac{2}{3}, -\dfrac{14}{27}\right)$ **9.** (– 2, 2) ; (2, 14).

3

Increasing and Decreasing Functions

3.1 INTRODUCTION

In this chapter, we shall study the use of derivative in examining the behaviour of functions *i.e.*, the monotonicity of functions. A function $f(x)$ is said to be a monotonic function, if it is either increasing or decreasing. And a function $f(x)$ is said to be strictly monotonic function if it is either strictly increasing or strictly decreasing.

3.2 INCREASING FUNCTION

A function $f(x)$ is said to be increasing on an open interval (a, b), if $f(x)$ increases as x increases in the open interval (a, b).

i.e., $\quad x_1 < x_2 \quad \Rightarrow \quad f(x_1) \le f(x_2) \quad$ for all $x_1, x_2 \in (a, b)$.

Also, $f(x)$ is said to be strictly increasing on an open interval (a, b), if

$$x_1 < x_2 \quad \Rightarrow \quad f(x_1) < f(x_2) \quad \text{for all } x_1, x_2 \in (a, b).$$

Since, $\quad f(x_1) < f(x_2) \quad \Rightarrow \quad f(x_1) \le f(x_2)$.

***i.e.*, A strictly increasing function on an open interval is always increasing on that interval, but the converse is not true. *i.e.*, An increasing function on an open interval may not be strictly increasing on that interval.**

In the figure :

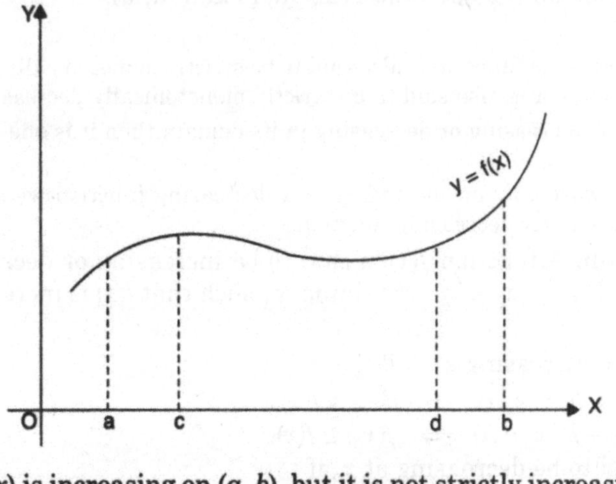

The function $f(x)$ is increasing on (a, b), but it is not strictly increasing on (a, b), whereas $f(x)$ is strictly increasing on the open intervals (a, c) and (d, b).

3.3 DECREASING FUNCTION

A function $f(x)$ is said to be decreasing on an open interval (a, b) if $f(x)$ decreases as x increases in the open interval (a, b).

i.e., $$x_1 < x_2 \Rightarrow f(x_1) \geq f(x_2) \quad \text{for all } x_1, x_2 \in (a, b).$$

Also, $f(x)$ is said to be strictly decreasing on an open interval (a, b) if,

$$x_1 < x_2 \Rightarrow f(x_1) > f(x_2) \quad \text{for all } x_1, x_2 \in (a, b).$$

Since, $f(x_1) > f(x_2)$ implies $f(x_1) \geq f(x_2)$, a strictly decreasing function on an open interval is always decreasing on that interval but the converse is not true. *i.e.,* A decreasing function on an open interval may not be strictly decreasing on that interval.

In the figure :

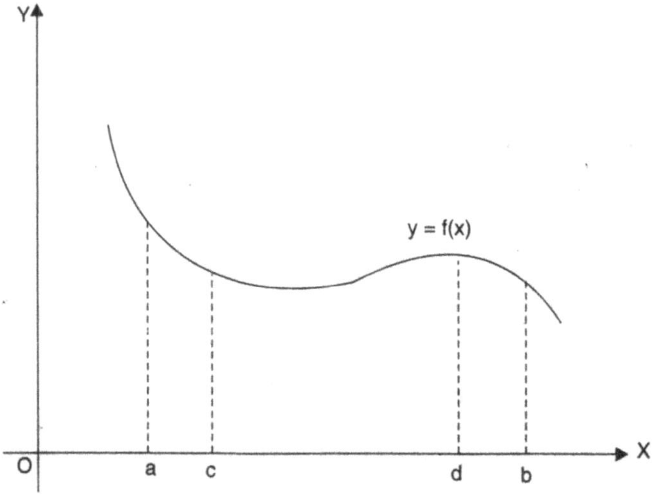

The function $f(x)$ is decreasing on (a, b), but it is not strictly decreasing on (a, b), whereas $f(x)$ is strictly decreasing on the open intervals (a, c) and (d, b).

Remark :

(*i*) A strictly increasing function is also said to be strictly monotonically increasing and a strictly decreasing function is also said to be strictly monotonically decreasing.

(*ii*) If a function is increasing or decreasing in its domain then it is one-one and onto and hence, invertible.

(*iii*) From now onwards, by an increasing or a decreasing function we shall mean a strictly increasing or a strictly decreasing function.

3.3.1 Definition. A function $f(x)$ is said to be increasing or decreasing at a point x_0 if there is an interval $(x_0 - h, x_0 + h)$ containing x_0 such that $f(x)$ is increasing or decreasing on $(x_0 - h, x_0 + h)$. *i.e.,*

$f(x)$ is said to be increasing at x_0 if :

(*i*) $x_0 < x$ in $(x_0 - h, x_0 + h) \Rightarrow f(x_0) \leq f(x)$

(*ii*) $x_0 > x$ in $(x_0 - h, x_0 + h) \Rightarrow f(x_0) \geq f(x)$.

And, $f(x)$ is said to be decreasing at x_0 if :

(*i*) $x_0 < x$ in $(x_0 - h, x_0 + h) \Rightarrow f(x_0) \geq f(x)$

(*ii*) $x_0 > x$ in $(x_0 - h, x_0 + h) \Rightarrow f(x_0) \leq f(x)$.

3.3.2 Definition. A function $f(x)$ is said to be increasing on $[a, b]$ if it is increasing or decreasing on (a, b) and it is also increasing or decreasing at $x = a$ and $x = b$.

3.4 THEOREM

Let $f(x)$ be continuous on $[a, b]$ and differentiable on the open interval (a, b).

Then :

(i) $f(x)$ is increasing on $[a, b]$ if $f'(x) > 0$ for each $x \in (a, b)$.

(ii) $f(x)$ is decreasing on $[a, b]$ if $f'(x) < 0$ for each $x \in (a, b)$.

Proof. *(i)* It is given that : $f'(x) > 0$ for each $x \in (a, b)$.

Let $x_1, x_2 \in [a, b]$ such that $x_1 < x_2$.

Then, By Lagrange's mean value theorem, there exists *a* point *c* between x_1 and x_2 such that :

$$f'(c) = \frac{f(x_2) - f(x_1)}{x_2 - x_1}$$

$$\Rightarrow \quad f(x_2) - f(x_1) = f'(c)\,(x_2 - x_1) \qquad \left[\begin{array}{l} \because\ f'(c) > 0 \text{ (given)} \\ \text{and } x_1 < x_2 \ \Rightarrow\ x_2 - x_1 > 0 \end{array}\right]$$

$$\Rightarrow \quad f(x_2) - f(x_1) > 0$$

$$\Rightarrow \quad f(x_2) > f(x_1)$$

Thus, we have, $x_2 > x_1 \Rightarrow f(x_2) > f(x_1)$ for all x_1, x_2 in $[a, b]$

Hence, $f(x)$ is an increasing function on $[a, b]$.

(ii) It is given that : $f'(x) < 0$ for each $x \in (a, b)$.

Let $x_1, x_2 \in [a, b]$ such that $x_1 < x_2$.

Then, By Lagrange's mean value theorem, there exists a point *c* between x_1 and x_2 such that

$$f'(c) = \frac{f(x_2) - f(x_1)}{x_2 - x_1}$$

$$\Rightarrow \quad f(x_2) - f(x_1) = f'(c)\,(x_2 - x_1)$$

$$\Rightarrow \quad f(x_2) - f(x_1) < 0 \qquad \left[\begin{array}{l} \because\ f'(c) < 0 \text{ (given)} \\ \text{and } x_1 < x_2 \ \Rightarrow\ x_2 - x_1 > 0 \end{array}\right]$$

$$\Rightarrow \quad f(x_2) < f(x_1)$$

Thus, we have $x_2 > x_1 \Rightarrow f(x_2) < f(x_1)$ for all x_1, x_2 in $[a, b]$

Hence, $f(x)$ is a decreasing function on $[a, b]$.

3.5 CRITICAL VALUE

A value $x = x_0$ is called a critical value for the function $f(x)$ if either $f'(x) = 0$ or $f'(x_0)$ does not exist. The point $(x_0, f(x_0))$ is called a critical point on the curve of $y = f(x)$.

If $f'(x_0) = 0$, then $f(x)$ is said to be stationary at $x = x_0$.

Note. *(i)* The critical values of a function generally separates the increasing and decreasing portions of the given function.

(ii) If the critical values for a function are more than one, then all critical values must be arranged in the ascending order of magnitude.

Some Useful Results :

The following results are very helpful in solving the inequations involved in the problems of increasing or decreasing functions :

(*i*) $ab > 0 \Rightarrow$ either $a > 0$ and $b > 0$ or $a < 0$ and $b < 0$

(*ii*) $ab < 0 \Rightarrow$ either $a > 0$ and $b < 0$ or $a < 0$ and $b > 0$

(*iii*) $ab > 0$ and $a < 0 \Rightarrow b < 0$

(*iv*) $ab > 0$ and $a > 0 \Rightarrow b > 0$

(*v*) Let $\alpha < \beta$ then, $(x - \alpha)(x - \beta) < 0 \Rightarrow \alpha < x < \beta$.

(*vi*) Let $\alpha < \beta$ then, $(x - \alpha)(x - \beta) > 0 \Rightarrow x < \alpha$ or $x > \beta$.

i.e., x does not lie between α and β.

i.e., the solution set is $(-\infty, \alpha) \cup (\beta, \infty)$.

(*vii*) Let $a > 0$ and $x^2 < a^2 \Rightarrow -a < x < a$.

$$[\because x^2 < a^2 \Rightarrow x^2 - a^2 < 0 \Rightarrow (x + a)(x - a) < 0 \Rightarrow -a < x < a]$$

(*viii*) Let $a > 0$ and $x^2 > a^2 \Rightarrow x < -a$ or $x > a$

(*ix*) $f'(x) > 0 \ \forall x \in (a, b) \Leftrightarrow f$ is increasing on (a, b)

(*x*) $f'(x) < 0 \ \forall x \in (a, b) \Leftrightarrow f$ is decreasing on (a, b).

(*xi*) $|x| < \alpha \Rightarrow -\alpha < x < \alpha$

(*xii*) $|x| > \alpha \Rightarrow x > \alpha$ or $x < -\alpha$.

SOME SOLVED EXAMPLES

Example 1. *Without using derivatives, show that :*

(*i*) *$f(x) = 4x + 7$ is strictly increasing on R.*

(*ii*) *$f(x) = 2 - 3x$ is strictly decreasing on R.*

(*iii*) *$f(x) = 1 - 6x$ is strictly decreasing on R.*

(*iv*) *$f(x) = 2x + 3$ is strictly increasing on R.*

(*v*) *$f(x) = a^x$; $a > 1$ is strictly increasing on R.*

(*vi*) *$f(x) = a^x$; $0 < a < 1$ is strictly decreasing on R.*

Solution. (*i*) We have $f(x) = 4x + 7$ ...(1)

Let x_1, x_2 be real numbers such that

$$x_1 < x_2$$

$$\therefore \quad 4x_1 < 4x_2$$

$$\Rightarrow \quad 4x_1 + 7 < 4x_2 + 7$$

$$\Rightarrow \quad f(x_1) < f(x_2)$$

$$\therefore \quad x_1 < x_2 \Rightarrow f(x_1) < f(x_2) \quad \forall x_1, x_2 \in \text{R}.$$

Hence, $f(x)$ is strictly increasing on R.

(*ii*) We have $f(x) = 2 - 3x$ ...(1)

Let x_1, x_2 be real numbers such that

$$x_1 < x_2$$

$$\Rightarrow \quad 3x_1 < 3x_2$$

$$\Rightarrow \quad -3x_1 > -3x_2$$

$$\Rightarrow \quad 2 - 3x_1 > 2 - 3x_2$$

$$\Rightarrow \quad f(x_1) > f(x_2)$$

$\therefore \quad x_1 < x_2 \Rightarrow f(x_1) > f(x_2) \;\; \forall\, x_1, x_2 \in R.$

Hence, $f(x)$ is strictly decreasing on R.

(*iii*) We have $\quad f(x) = 1 - 6x \quad$ ...(1)

Let x_1, x_2 be real numbers such that

$$x_1 < x_2$$
$$\Rightarrow \quad 6x_1 < 6x_2$$
$$\Rightarrow \quad -6x_1 > -6x_2$$
$$\Rightarrow \quad 1 - 6x_1 > 1 - 6x_2$$
$$\Rightarrow \quad f(x_1) > f(x_2)$$
$$\therefore \quad x_1 < x_2 \Rightarrow f(x_1) > f(x_2) \quad \text{for all } x_1, x_2 \in R$$

Hence, $f(x)$ is strictly decreasing on R.

(*iv*) We have $\quad f(x) = 2x + 3 \quad$ ...(1)

Let x_1, x_2 be real numbers such that

$$x_1 < x_2$$
$$\Rightarrow \quad 2x_1 < 2x_2$$
$$\Rightarrow \quad 2x_1 + 3 < 2x_2 + 3$$
$$\Rightarrow \quad f(x_1) < f(x_2)$$
$$\therefore \quad x_1 < x_2 \Rightarrow f(x_1) < f(x_2) \quad \text{for all } x_1, x_2 \in R.$$

Hence, $f(x)$ is strictly increasing on R.

(*v*) We have $\quad f(x) = a^x \,;\, a > 1 \quad$ ...(1)

Let x_1, x_2 be real numbers such that

$$x_1 < x_2$$
$$\Rightarrow \quad a^{x_1} < a^{x_2} \quad \left[\begin{array}{l} \because \; a > 1 \text{ (given)} \\ \therefore \; x_1 < x_2 \Rightarrow a^{x_1} < a^{x_2} \end{array}\right]$$
$$\Rightarrow \quad f(x_1) < f(x_2)$$
$$\therefore \quad x_1 < x_2 \Rightarrow f(x_1) < f(x_2) \quad \text{for all } x_1, x_2 \in R.$$

Hence, $f(x)$ is strictly increasing on R.

(*vi*) We have $\quad f(x) = a^x \,;\, 0 < a < 1 \quad$ ...(1)

Let x_1, x_2 be real numbers such that

$$x_1 < x_2$$
$$\Rightarrow \quad a^{x_1} > a^{x_2} \quad \left[\begin{array}{l} \because \; 0 < a < 1 \text{ (given)} \\ \therefore \; x_1 < x_2 \Rightarrow a^{x_1} > a^{x_2} \end{array}\right]$$
$$\Rightarrow \quad f(x_1) > f(x_2)$$
$$\therefore \quad x_1 < x_2 \Rightarrow f(x_1) > f(x_2) \quad \text{for all } x_1, x_2 \in R.$$

Hence $f(x)$ is strictly decreasing on R.

Example 2. (*i*) *Show that the function $f(x) = x^2$ is a strictly increasing function on $[0, \infty)$.*

(*ii*) *Show that the function $f(x) = x^2$ is a strictly decreasing function on $(-\infty, 0]$.*

Solution. (*i*) We have

$$f(x) = x^2 \quad \text{...(1)}$$

Let x_1, x_2 be positive real numbers such that

$$x_1 < x_2$$
$$\Rightarrow \quad x_1^2 < x_1 x_2 \quad \text{...(2) [Multiplying both sides by } x_1 > 0]$$

Also, $x_1 < x_2$

$$\Rightarrow \quad x_1 x_2 < x_2^2 \quad \text{...(3) [Multiplying both sides by } x_2 > 0]$$

$\therefore$ From (2) and (3), we have

$$x_1^2 < x_2^2 \Rightarrow f(x_1) < f(x_2)$$

$\therefore$ $x_1 < x_2 \Rightarrow f(x_1) < f(x_2)$ for all $x_1, x_2 \in [0, \infty)$.

Hence, $f(x)$ is strictly increasing on $[0, \infty)$.

(*ii*) We have $f(x) = x^2$ ...(1)

Let x_1, x_2 be negative real numbers such that

$$x_1 < x_2$$

$\Rightarrow$ $x_1^2 > x_1 x_2$ ...(2) [Multiplying both sides by $x_1 < 0$]

Also, $x_1 < x_2$

$\Rightarrow$ $x_1 x_2 > x_2^2$ ...(3) [Multiplying both sides by $x_2 < 0$]

$\therefore$ From (2) and (3), we have

$$x_1^2 > x_2^2$$

$\Rightarrow$ $f(x_1) > f(x_2)$

$\therefore$ $x_1 < x_2 \Rightarrow f(x_1) > f(x_2)$ for all $x_1, x_2 \in (-\infty, 0]$

Hence, $f(x)$ is strictly decreasing on $(-\infty, 0]$.

Example 3. (*i*) *Prove that the function $x^2 - x + 1$ is neither strictly increasing nor strictly decreasing on (0, 1).*

(*ii*) *Prove that the function $x^2 - x + 1$ is neither strictly increasing nor strictly decreasing on (– 1, 1).*

Solution. (*i*) Let $f(x) = x^2 - x + 1$ ...(1)

$\Rightarrow$ $f'(x) = 2x - 1$

Now, $f(x)$ is strictly increasing if $f'(x) > 0$

$\therefore$ $f'(x) > 0 \Rightarrow 2x - 1 > 0 \Rightarrow 2x > 1 \Rightarrow x > \frac{1}{2}$

$\Rightarrow$ $x \in \left(\frac{1}{2}, 1\right)$

$\therefore$ $f(x)$ is strictly increasing on $\left(\frac{1}{2}, 1\right)$

Also, $f(x)$ is strictly decreasing if $f'(x) < 0$

$\therefore$ $f'(x) < 0 \Rightarrow 2x - 1 < 0 \Rightarrow 2x < 1 \Rightarrow x < \frac{1}{2}$

$\Rightarrow$ $x \in \left(0, \frac{1}{2}\right)$

$\therefore$ $f(x)$ is strictly decreasing on $\left(0, \frac{1}{2}\right)$.

Hence, the given function is neither strictly increasing nor strictly decreasing on the entire interval (0, 1).

(*ii*) Let $f(x) = x^2 - x + 1$ ...(1)

$\therefore$ $f'(x) = 2x - 1$

Now, $f(x)$ is strictly increasing if $f'(x) > 0$

$$\therefore \quad f'(x) > 0 \Rightarrow 2x - 1 > 0 \Rightarrow 2x > 1 \Rightarrow x > \frac{1}{2}$$

$$\Rightarrow \quad x \in \left(\frac{1}{2}, 1\right)$$

$\therefore$ $f(x)$ is strictly increasing on $\left(\frac{1}{2}, 0\right)$.

Also, $f(x)$ is strictly decreasing if $f'(x) < 0$.

$$\therefore \quad f'(x) < 0 \Rightarrow 2x - 1 < 0 \Rightarrow 2x < 1 \Rightarrow x < \frac{1}{2}$$

$$\Rightarrow \quad x \in \left(-1, \frac{1}{2}\right)$$

$\therefore$ $f(x)$ is strictly decreasing on $\left(-1, \frac{1}{2}\right)$.

Hence, the given function is neither strictly increasing nor strictly decreasing on the entire interval (–1, 1).

Example 4. *Show that :*

(*i*) $f(x) = x^2 - 8x$ *is strictly decreasing on* $(-\infty, 4)$.

(*ii*) $f(x) = x^3 - 6x^2 + 12x - 18$ *is strictly increasing on R.*

(*iii*) $f(x) = x^3 - 3x^2 + 3x - 100$ *is increasing on R.*

(*iv*) $f(x) = \left(\frac{1}{2}\right)^x$ *is strictly decreasing on R.*

Solution. (*i*) We have

$$f(x) = x^2 - 8x \quad ...(1)$$

$$\Rightarrow \quad f'(x) = 2x - 8 = 2(x - 4)$$

Since, $x \in (-\infty, 4) \Rightarrow (x - 4) < 0$

$$\Rightarrow \quad 2(x - 4) < 0$$

$$\Rightarrow \quad f'(x) < 0$$

$\therefore$ $f(x)$ is strictly decreasing on $(-\infty, 4)$.

(*ii*) We have $f(x) = x^3 - 6x^2 + 12x - 18$ ...(1)

$$\Rightarrow \quad f'(x) = 3x^2 - 12x + 12$$

$$= 3(x^2 - 4x + 4) = 3(x - 2)^2$$

Since, $x \in R \Rightarrow (x - 2)^2 \geq 0$

$$\Rightarrow \quad 3(x - 2)^2 > 0$$

$$\Rightarrow \quad f'(x) > 0$$

$\therefore$ $f(x)$ is strictly increasing on R.

(*iii*) We have $f(x) = x^3 - 3x^2 + 3x - 100$ ...(1)

$$\Rightarrow \quad f'(x) = 3x^2 - 6x + 3$$

$$= 3(x^2 - 2x + 1) = 3(x - 1)^2$$

Since, $x \in R \Rightarrow (x - 1)^2 \geq 0$

$\Rightarrow \quad 3(x-1)^2 \geq 0 \Rightarrow f'(x) \geq 0$

$\therefore$ $f(x)$ is increasing on R.

(iv) We have $\quad f(x) = \left(\frac{1}{2}\right)^x$ ...(1)

$\Rightarrow \quad f'(x) = \left(\frac{1}{2}\right)^x \log \frac{1}{2}$ $\quad \left[\because \frac{d}{dx}(a^x) = a^x \cdot \log a\right]$

$= \frac{1}{2^x} \log 2^{-1}$ $\quad [\because \log n^m = m \log n]$

$= \frac{1}{2^x} (-1) \log 2 = -\frac{\log 2}{2^x}$

Since, $\quad x \in R \Rightarrow 2^x > 0$

$\Rightarrow \quad \frac{\log 2}{2^x} > 0$

$\Rightarrow \quad -\frac{\log 2}{2^x} < 0 \Rightarrow f'(x) < 0$

$\therefore$ $f(x)$ is strictly decreasing on R.

Example 5. *(i) Show that the function $f(x) = \frac{3}{x} + 7$ is decreasing for $x \in R$ $(x \neq 0)$.*

(ii) Show that the function $x + \frac{1}{x}$; $x \geq 1$ is increasing.

Solution. *(i)* Let $\quad f(x) = \frac{3}{x} + 7$ ...(1)

$\Rightarrow \quad f'(x) = -\frac{3}{x^2} + 0 = -\frac{3}{x^2}$

Since, $\quad x \in R, x \neq 0 \Rightarrow \frac{1}{x^2} > 0$

$\Rightarrow \quad -\frac{1}{x^2} < 0$

$\Rightarrow \quad -\frac{3}{x^2} < 0$

$\Rightarrow \quad f'(x) < 0$

$\therefore$ $f(x)$ is decreasing for $x \in R$, $x \neq 0$.

(ii) Let $\quad f(x) = x + \frac{1}{x}$ ... (1)

$\Rightarrow \quad f'(x) = 1 - \frac{1}{x^2} = \frac{x^2 - 1}{x^2}$

Since, $\quad x \geq 1$

$\Rightarrow \quad x^2 \geq 1$

$\Rightarrow \quad x^2 - 1 \geq 0 \Rightarrow \dfrac{x^2 - 1}{x^2} \geq 0$

$\Rightarrow \quad f'(x) \geq 0$

$\therefore$ $f(x)$ is increasing for $x \geq 1$.

Example 6. (*i*) *Find the value of 'a' for which the function $f(x) = x^2 - 2ax + 6$ is strictly increasing when $x > 0$.*

(*ii*) *Find the least value of a such that the function $x^2 + ax + 1$ is increasing on $[1, 2]$.*

Solution. (*i*) We have

$$f(x) = x^2 - 2ax + 6 \quad ...(1)$$

$\Rightarrow \quad f'(x) = 2x - 2a = 2(x - a)$

Since, $f(x)$ is strictly increasing when $x > 0$. *i.e.*, $f'(x) > 0$

$\Rightarrow \quad 2(x - a) > 0$

$\Rightarrow \quad x - a > 0 \Rightarrow x > a$

$\therefore \quad a \leq 0$, because $x > 0$.

(*ii*) Let $\quad f(x) = x^2 + ax + 1 \quad ...(1)$

$\Rightarrow \quad f'(x) = 2x + a$

Since, $f(x)$ is increasing on $[1, 2]$

$\therefore \quad f'(x) \geq 0$ on $[1, 2]$

Now, we have $\quad 1 \leq x \leq 2$

$\Rightarrow \quad 2 \leq 2x \leq 4$ [Multiplying throughout by 2]

$\Rightarrow \quad 2 + a \leq 2x + a \leq 4 + a$ [Adding 'a' throughout]

$\Rightarrow \quad 2 + a \leq f'(x) \leq 4 + a.$

$\Rightarrow \quad 2 + a \geq 0$ [$\because \; f'(x) \geq 0$]

$\Rightarrow \quad a \geq -2$

$\therefore$ The least value of a is -2.

Example 7. *Find the intervals on which the following functions are :*

(*a*) *strictly increasing* (*b*) *strictly decreasing*

(*i*) $6x^2 - 72x + 1$ (*ii*) $-x^2 - 2x + 15$

(*iii*) $2x^3 - 24x + 107$ (*iv*) $2x^3 - 9x^2 + 12x + 15$

(*v*) $2x^3 + x^2 - 20x$ (*vi*) $-2x^3 + 3x^2 + 12x + 6$

(*vii*) $x^3 - 12x^2 + 36x + 17$.

Solution. (*i*) Let $\quad f(x) = 6x^2 - 72x + 1 \quad ...(1)$

$\Rightarrow \quad f'(x) = 12x - 72 = 12(x - 6).$

Now, $\quad f'(x) = 0 \Rightarrow 12(x - 6) = 0$

$\Rightarrow \quad x - 6 = 0 \Rightarrow x = 6.$

$\therefore$ The critical value for $f(x)$ is $x = 6$.

When $x < 6$

$\Rightarrow \quad f'(x) = 12(-\text{ve}) = -\text{ve}$

$\Rightarrow$ $f(x)$ is strictly decreasing for $x < 6$

$\therefore$ $f(x)$ is strictly decreasing on $(-\infty, 6)$

When $x > 6$

$\Rightarrow \quad f'(x) = 12(+ve) = +ve$

$\Rightarrow$ $f(x)$ is strictly increasing for $x > 6$

$\therefore$ $f(x)$ is strictly increasing on $(6, \infty)$.

(*ii*) Let $\quad f(x) = -x^2 - 2x + 15 \qquad$ ...(1)

$\Rightarrow \quad f'(x) = -2x - 2 = -2(x+1)$

Now, $\quad f'(x) = 0 \Rightarrow -2(x+1) = 0$

$\Rightarrow \quad x + 1 = 0 \Rightarrow x = -1$

$\therefore$ The critical value for $f(x)$ is $x = -1$

When $x < -1$

$\Rightarrow \quad f'(x) = -2(-ve) = +ve$

$\Rightarrow$ $f(x)$ is strictly increasing for $x < -1$

$\therefore$ $f(x)$ is strictly increasing on $(-\infty, -1)$

When $x > -1$

$\Rightarrow \quad f'(x) = -2(+ve) = -ve$

$\Rightarrow$ $f(x)$ is strictly decreasing for $x > -1$

$\therefore$ $f(x)$ is strictly decreasing on $(-1, \infty)$.

(*iii*) Let $\quad f(x) = 2x^3 - 24x + 107 \qquad$ ...(1)

$\Rightarrow \quad f'(x) = 6x^2 - 24 = 6(x^2 - 4) = 6(x-2)(x+2)$

Now, $\quad f'(x) = 0 \Rightarrow 6(x-2)(x+2) = 0$

$\Rightarrow \quad (x-2)(x+2) = 0$

$\Rightarrow \quad x = -2, 2.$

$\therefore$ The critical values for $f(x)$ are $x = -2, 2$.

When $x < -2$

$\Rightarrow \quad f'(x) = 6(-ve)(-ve) = +ve$

$\Rightarrow$ $f(x)$ is strictly increasing for $x < -2$

$\therefore$ $f(x)$ is strictly increasing on $(-\infty, -2)$

When $-2 < x < 2$

$\Rightarrow \quad f'(x) = 6(-ve)(+ve) = -ve$

$\Rightarrow$ $f(x)$ is strictly decreasing for $-2 < x < 2$

$\therefore$ $f(x)$ is strictly increasing on $(-2, 2)$.

When $x > 2$

$\Rightarrow$ $f'(x) = 6(+ve)(+ve) = +ve$

$\Rightarrow$ $f(x)$ is strictly increasing for $x > 2$

$\therefore$ $f(x)$ is strictly increasing on $(2, \infty)$

(*iv*) Let $\quad f(x) = 2x^3 - 9x^2 + 12x + 15 \qquad$ ...(1)

$\Rightarrow \quad f'(x) = 6x^2 - 18x + 12 = 6(x^2 - 3x + 2)$

$\qquad = 6(x-1)(x-2)$

Now, $\quad f'(x) = 0 \Rightarrow 6(x-1)(x-2) = 0$

$\Rightarrow \quad (x-1)(x-2) = 0$

$\Rightarrow \quad x = 1, 2.$

$\therefore$ The critical values for $f(x)$ are $x = 1, 2$

When $x < 1$,

$\Rightarrow$ $$f'(x) = 6(-\text{ve})(-\text{ve}) = +\text{ve}$$

$\Rightarrow$ $f(x)$ is strictly increasing for $x < 1$

$\therefore$ $f(x)$ is strictly increasing on $(-\infty, 1)$

When $1 < x < 2$,

$\Rightarrow$ $$f'(x) = 6(+\text{ve})(-\text{ve}) = -\text{ve}$$

$\Rightarrow$ $f(x)$ is strictly decreasing for $1 < x < 2$

$\therefore$ $f(x)$ is strictly decreasing on $(1, 2)$.

When $x > 2$,

$\Rightarrow$ $$f'(x) = 6(+\text{ve})(+\text{ve}) = +\text{ve}$$

$\Rightarrow$ $f(x)$ is strictly increasing for $x > 2$

$\therefore$ $f(x)$ is strictly increasing on $(2, \infty)$.

(v) Let $$f(x) = 2x^3 + x^2 - 20x \qquad \text{...(1)}$$

$\Rightarrow$ $$f'(x) = 6x^2 + 2x - 20 = 2(3x^2 + x - 10)$$
$$= 2(x + 2)(3x - 5)$$

Now, $$f'(x) = 0 \Rightarrow 2(x + 2)(3x - 5) = 0$$
$$\Rightarrow (x + 2)(3x - 5) = 0$$
$$\Rightarrow x = -2, \frac{5}{3}$$

$\therefore$ The critical values for $f(x)$ are $x = -2, \frac{5}{3}$

When $x < -2$,

$\Rightarrow$ $$f'(x) = 2(-\text{ve})(-\text{ve}) = +\text{ve}$$

$\Rightarrow$ $f(x)$ is strictly increasing for $x < -2$

$\therefore$ $f(x)$ is strictly increasing on $(-\infty, -2)$

When $-2 < x < \frac{5}{3}$,

$\Rightarrow$ $$f'(x) = 2(+\text{ve})(-\text{ve}) = -\text{ve}$$

$\Rightarrow$ $f(x)$ is strictly decreasing for $-2 < x < \frac{5}{3}$,

$\therefore$ $f(x)$ is strictly decreasing on $\left(-2, \frac{5}{3}\right)$.

When $x > \frac{5}{3}$,

$\Rightarrow$ $$f'(x) = 2(+\text{ve})(+\text{ve}) = +\text{ve}$$

$\Rightarrow$ $f(x)$ is strictly increasing for $x > \frac{5}{3}$.

$\therefore$ $f(x)$ is strictly increasing on $\left(\frac{5}{3}, \infty\right)$.

(*vi*) Let $f(x) = -2x^3 + 3x^2 + 12x + 6$ ...(1)

$\Rightarrow$ $f'(x) = -6x^2 + 6x + 12$

$= -6(x^2 - x - 2) = -6(x-2)(x+1)$

Now, $f'(x) = 0 \Rightarrow -6(x-2)(x+1) = 0$

$\Rightarrow$ $(x-2)(x+1) = 0 \Rightarrow x = 2, -1$

$\therefore$ The critical values for $f(x)$ are $x = -1, 2$.

When $x < -1$,

$\Rightarrow$ $f'(x) = -6(-\text{ve})(-\text{ve}) = -\text{ve}$

$\Rightarrow$ $f(x)$ is strictly decreasing for $x < -1$

$\therefore$ $f(x)$ is strictly decreasing on $(-\infty, -1)$

When $-1 < x < 2$,

$\Rightarrow$ $f'(x) = -6(-\text{ve})(+\text{ve}) = +\text{ve}$

$\Rightarrow$ $f(x)$ is strictly increasing for $-1 < x < 2$

$\therefore$ $f(x)$ is strictly increasing on $(-1, 2)$.

When $x > 2$,

$\Rightarrow$ $f'(x) = -6(+\text{ve})(+\text{ve}) = -\text{ve}$

$\Rightarrow$ $f(x)$ is strictly decreasing for $x > 2$

$\therefore$ $f(x)$ is strictly decreasing on $(2, \infty)$.

(*vii*) Let $f(x) = x^3 - 12x^2 + 36x + 17$

Please try yourself.

[**Hint.** Critical values are $x = 2, 6$]

[**Ans.** $f(x)$ is strictly increasing on $(-\infty, 2)$, $(6, \infty)$ and strictly decreasing on $(2, 6)$.]

Example 8. *Find the intervals on which the following functions are (a) strictly increasing (b) strictly decreasing.*

(*i*) $-2x^3 + 3x^2 + 36x + 8$ (*ii*) $(x+1)^3 (x-3)^3$

(*iii*) $-x^2 - 9x + 6$ (*iv*) $-2x^2 - 6x + 10$

(*v*) $x^3 - 8$.

Solution. (*i*) Let $f(x) = -2x^3 + 3x^2 + 36x + 8$ ...(1)

$\Rightarrow$ $f'(x) = -6x^2 + 6x + 36$

$= -6(x^2 - x - 6) = -6(x-3)(x+2)$

Now, $f'(x) = 0 \Rightarrow -6(x-3)(x+2) = 0$

$\Rightarrow$ $(x-3)(x+2) = 0$

$\Rightarrow$ $x = 3, -2$

$\therefore$ The critical values of $f(x)$ are $x = -2, 3$.

When $x < -2$,

$\Rightarrow$ $f'(x) = -6(-\text{ve})(-\text{ve}) = -\text{ve}$

$\Rightarrow$ $f(x)$ is strictly decreasing for $x < -2$

$\therefore$ $f(x)$ is strictly decreasing on $(-\infty, -2)$

When $-2 < x < 3$,

$\Rightarrow$ $f'(x) = -6(-\text{ve})(+\text{ve}) = +\text{ve}$

$\Rightarrow$ $f(x)$ is strictly increasing for $-2 < x < 3$

$\therefore$ $f(x)$ is strictly increasing on $(-2, 3)$.

When $x > 3$,

$\Rightarrow \quad f'(x) = -6(+ve)(+ve) = -ve$

$\Rightarrow$ $f(x)$ is strictly decreasing for $x > 3$

$\therefore$ $f(x)$ is strictly decreasing on $(3, \infty)$.

(*ii*) Let $\quad f(x) = (x+1)^3 (x-3)^3 \quad$ ...(1)

$$\Rightarrow \quad f'(x) = (x+1)^3 \cdot \frac{d}{dx}(x-3)^3 + (x-3)^3 \cdot \frac{d}{dx}(x+1)^3$$

$$= (x+1)^3 \cdot 3(x-3)^2 \frac{d}{dx}(x-3) + (x-3)^3 \cdot 3(x+1)^2 \frac{d}{dx}(x+1)$$

$$= 3(x+1)^3 (x-3)^2 \cdot 1 + 3(x-3)^3 (x+1)^2 \cdot 1$$

$$= 3(x+1)^2 (x-3)^2 [x+1+x-3]$$

$$= 3(x+1)^2 (x-3)^2 (2x-2)$$

$$= 6(x+1)^2 (x-3)^2 (x-1)$$

Now, $\quad f'(x) = 0 \Rightarrow 6(x+1)^2 (x-3)^2 (x-1) = 0$

$\Rightarrow \quad (x+1)^2 (x-3)^2 (x-1) = 0$

$\Rightarrow \quad x = -1, 3, 1$

$\therefore$ The critical values of $f(x)$ are $x = -1, 1, 3$.

When $x < -1$,

$\Rightarrow \quad f'(x) = 6(+ve)(+ve)(-ve) = -ve$

$\Rightarrow$ $f(x)$ is strictly decreasing for $x < -1$

$\therefore$ $f(x)$ is strictly decreasing on $(-\infty, -1)$

When $-1 < x < 1$,

$\Rightarrow \quad f'(x) = 6(+ve)(+ve)(-ve) = -ve$

$\Rightarrow$ $f(x)$ is strictly decreasing for $-1 < x < 1$

$\Rightarrow$ $f(x)$ is strictly decreasing on $(-1, 1)$.

When $1 < x < 3$,

$\Rightarrow \quad f'(x) = 6(+ve)(+ve)(+ve) = +ve$

$\Rightarrow$ $f(x)$ is strictly increasing for $1 < x < 3$

$\therefore$ $f(x)$ is strictly increasing on $(1, 3)$.

When $x > 3$,

$\Rightarrow \quad f'(x) = 6(+ve)(+ve)(+ve) = +ve$

$\Rightarrow$ $f(x)$ is strictly increasing for $x > 3$.

$\therefore$ $f(x)$ is strictly increasing on $(3, \infty)$.

(*iii*) Let $\quad f(x) = -x^2 - 9x + 6 \quad$ ...(1)

$\Rightarrow \quad f'(x) = -2x - 9 = -1(2x+9)$

Now, $\quad f'(x) = 0 \Rightarrow -2x - 9 = 0$

$$\Rightarrow \quad -2x = 9 \Rightarrow x = -\frac{9}{2}$$

$\therefore$ The critical value of $f(x)$ is $x = -\dfrac{9}{2}$.

When $x < -\frac{9}{2}$,

$\Rightarrow$ $f'(x) = -1\ (-\text{ve}) = +\text{ve}$

$\Rightarrow$ $f(x)$ is strictly increasing for $x < -\frac{9}{2}$

$\therefore$ $f(x)$ is strictly increasing on $\left(-\infty, -\frac{9}{2}\right)$.

When $x > -\frac{9}{2}$,

$\Rightarrow$ $f'(x) = -1\ (+\text{ve}) = -\text{ve}$

$\Rightarrow$ $f(x)$ is strictly decreasing for $x > -\frac{9}{2}$

$\therefore$ $f(x)$ is strictly decreasing on $\left(-\frac{9}{2}, \infty\right)$.

(*iv*) Let $f(x) = -2x^2 - 6x + 10$

Please try yourself.

$\left[\textbf{Hint.} \text{ The critical value of } f(x) \text{ is } x = -\frac{3}{2}\right]$

$\left[\textbf{Ans.}\ f(x) \text{ is strictly increasing on } \left(-\infty, -\frac{3}{2}\right) \text{ and strictly decreasing on } \left(-\frac{3}{2}, \infty\right)\right]$

(*v*) Let $f(x) = x^3 - 8$ ...(1)

$\Rightarrow$ $f'(x) = 3x^2$

Now, $f'(x) = 0 \Rightarrow 3x^2 = 0$

$\Rightarrow$ $x^2 = 0 \Rightarrow x = 0$

$\therefore$ The critical value of $f(x)$ is $x = 0$.

When $x < 0$,

$\Rightarrow$ $f'(x) = 3\ (+\text{ve}) = +\text{ve}$

$\Rightarrow$ $f(x)$ is strictly increasing for $x < 0$

$\therefore$ $f(x)$ is strictly increasing on $(-\infty, 0)$

When $x > 0$,

$\Rightarrow$ $f'(x) = 3\ (+\text{ve}) = +\text{ve}$

$\Rightarrow$ $f(x)$ is strictly increasing for $x > 0$.

$\Rightarrow$ $f(x)$ is strictly increasing on $(0, \infty)$

Now, on combining the two results :

$f(x)$ is increasing on $R - \{0\}$.

Example 9. *Determine the intervals in which the following functions are strictly increasing or strictly decreasing.*

(*i*) $\frac{x}{2} + \frac{2}{x}, x \neq 0$ (*ii*) $\frac{x-2}{x+1}, x \neq -1$

(*iii*) $\frac{x}{x^2+1}$ (*iv*) $5x^{3/2} - 3x^{5/2}, x > 0.$

Solution. (*i*) Let $f(x) = \dfrac{x}{2} + \dfrac{2}{x}$, $x \neq 0$

$$\Rightarrow \quad f'(x) = \frac{1}{2} - \frac{2}{x^2} = \frac{x^2 - 4}{2x^2} = \frac{(x+2)(x-2)}{2x^2}$$

Now, $f'(x) = 0 \Rightarrow \dfrac{(x+2)(x-2)}{2x^2} = 0$

$\Rightarrow \quad (x + 2)(x - 2) = 0$

$\Rightarrow \quad x = -2, 2$

$\therefore$ The critical values of $f(x)$ are $x = -2, 2$.

When $x < -2$,

$$\Rightarrow \quad f'(x) = \frac{(-ve)(-ve)}{2(+ve)} = +ve$$

$\Rightarrow$ $f(x)$ is strictly increasing for $x < -2$

$\therefore$ $f(x)$ is strictly increasing on $(-\infty, -2)$.

When $-2 < x < 2$,

$$\Rightarrow \quad f'(x) = \frac{(-ve)(+ve)}{2(+ve)} = -ve$$

$\Rightarrow$ $f(x)$ is strictly decreasing for $-2 < x < 2$

$\therefore$ $f(x)$ is strictly decreasing on $(-2, 2)$.

When $x > 2$,

$$\Rightarrow \quad f'(x) = \frac{(+ve)(+ve)}{2(+ve)} = +ve$$

$\Rightarrow$ $f(x)$ is strictly increasing for $x > 2$.

$\therefore$ $f(x)$ is strictly increasing on $(2, \infty)$.

(*ii*) Let $f(x) = \dfrac{x-2}{x+1}$, $x \neq -1$

$$\Rightarrow \quad f'(x) = \frac{(x+1)\frac{d}{dx}(x-2) - (x-2)\frac{d}{dx}(x+1)}{(x+1)^2}$$

$$= \frac{(x+1).1 - (x-2).1}{(x+1)^2} = \frac{x+1-x+2}{(x+1)^2}$$

$$= \frac{3}{(x+1)^2}\ ;\ x \neq -1$$

$\Rightarrow$ $f'(x) > 0$ for all real values of x except $x = -1$.

$\therefore$ $f(x)$ is increasing on, $R - \{-1\}$.

(*iii*) Let $f(x) = \dfrac{x}{1+x^2}$ ...(1)

$$\Rightarrow \qquad f'(x) = \frac{(1+x^2)\,.\,\frac{d}{dx}(x) - x\,.\,\frac{d}{dx}(1+x^2)}{(1+x^2)^2}$$

$$= \frac{(1+x^2)\,.\,1 - x\,(2x)}{(1+x^2)^2} = \frac{1+x^2-2x^2}{(1+x^2)^2}$$

$$= \frac{1-x^2}{(1+x^2)^2} = \frac{(1-x)\,(1+x)}{(1+x^2)^2}$$

Now, $\qquad f'(x) = 0 \;\Rightarrow\; \dfrac{(1-x)\,(1+x)}{(1+x^2)^2} = 0$

$\Rightarrow \qquad (1-x)\,(1+x) = 0$

$\Rightarrow \qquad x = 1, -1$

$\therefore$ The critical values of $f(x)$ are $x = -1, 1$.

When $x < -1$

$$\Rightarrow \qquad f'(x) = \frac{(+\text{ve})\,(-\text{ve})}{(+\text{ve})} = -\text{ve}$$

$\Rightarrow$ $f(x)$ is strictly decreasing for $x < -1$

$\therefore$ $f(x)$ is strictly decreasing on $(-\infty, -1)$.

When $-1 < x < 1$

$$\Rightarrow \qquad f'(x) = \frac{(+\text{ve})\,(+\text{ve})}{(+\text{ve})} = +\text{ve}$$

$\Rightarrow$ $f(x)$ is strictly increasing for $-1 < x < 1$

$\therefore$ $f(x)$ is strictly increasing on $(-1, 1)$.

When $x > 1$

$$\Rightarrow \qquad f'(x) = \frac{(-\text{ve})\,(+\text{ve})}{(+\text{ve})} = -\text{ve}$$

$\Rightarrow$ $f(x)$ is strictly decreasing for $x > 1$

$\therefore$ $f(x)$ is strictly decreasing on $(1, \infty)$.

(*iv*) Let $\qquad f(x) = 5x^{3/2} - 3x^{5/2},\ x > 0 \qquad$ ...(1)

$$\Rightarrow \qquad f'(x) = 5\,.\,\frac{3}{2}x^{1/2} - 3\,.\,\frac{5}{2}x^{3/2}$$

$$= \frac{15}{2}\sqrt{x} - \frac{15}{2}x\sqrt{x} = \frac{15}{2}\sqrt{x}\,(1-x)$$

Now, $\qquad f'(x) = 0 \;\Rightarrow\; \dfrac{15}{2}\sqrt{x}\,(1-x) = 0$

$\Rightarrow \qquad \sqrt{x}\,(1-x) = 0$

$\Rightarrow \qquad x = 1 \qquad [\because\ x > 0]$

$\therefore$ The critical value of $f(x)$ is $x = 1$.

When $0 < x < 1$,

$$\Rightarrow \qquad f'(x) = \frac{15}{2}\,(+\text{ve})\,(+\text{ve}) = +\text{ve}$$

$\Rightarrow$ $f(x)$ is strictly increasing for $0 < x < 1$

$\therefore$ $f(x)$ is strictly increasing on $(0, 1)$

When $x > 1$

$$\Rightarrow \qquad f'(x) = \frac{15}{2} (+\text{ve})\,(-\text{ve}) = -\text{ve}$$

$\Rightarrow$ $f(x)$ is strictly decreasing for $x > 1$

$\therefore$ $f(x)$ is strictly decreasing on $(1, \infty)$.

Example 10. *Find the intervals on which the following functions are strictly increasing or strictly decreasing.*

(i) $\dfrac{x}{\log x}$ (ii) $\log x,\ x > 0$

(iii) $\log(1 + x) - \dfrac{2x}{2 + x}$ (iv) $2\log(x - 2) - x^2 + 4x + 1.$

Solution. (i) Let $f(x) = \dfrac{x}{\log x}$ ...(1)

$$\Rightarrow \qquad f'(x) = \frac{\log x \cdot \frac{d}{dx}(x) - x \cdot \frac{d}{dx}(\log x)}{(\log x)^2} = \frac{\log x \cdot 1 - x \cdot \frac{1}{x}}{(\log x)^2} = \frac{\log x - 1}{(\log x)^2}$$

Now, $f'(x) = 0 \Rightarrow \dfrac{\log x - 1}{(\log x)^2} = 0$

$\Rightarrow$ $\log x - 1 = 0$

$\Rightarrow$ $\log x = 1$

$\Rightarrow$ $x = e^1 = e$

Also $f'(x)$ is not defined at $x = 1$.

$\therefore$ The critical values for $f(x)$ are $x = 1, e$.

When $0 < x < 1$,

$$\Rightarrow \qquad f'(x) = \frac{(-\text{ve})}{(+\text{ve})} = -\text{ve}$$

$\Rightarrow$ $f(x)$ is strictly decreasing for $0 < x < 1$

$\therefore$ $f(x)$ is strictly decreasing on $(0, 1)$.

When $1 < x < e$,

$$\Rightarrow \qquad f'(x) = \frac{(-\text{ve})}{(+\text{ve})} = -\text{ve}$$

$\Rightarrow$ $f(x)$ is strictly decreasing for $1 < x < e$

$\therefore$ $f(x)$ is strictly decreasing on $(1, e)$.

When $x > e$,

$$\Rightarrow \qquad f'(x) = \frac{(+\text{ve})}{(+\text{ve})} = +\text{ve}$$

$\Rightarrow$ $f(x)$ is strictly increasing for $x > e$

$\therefore$ $f(x)$ is strictly increasing on $(1, e)$.

(*ii*) Let $f(x) = \log x\ ,\ (x > 0)$ ...(1)

$$\Rightarrow \quad f'(x) = \frac{1}{x}$$

Since, $x > 0 \Rightarrow f'(x) > 0$ *i.e.,* +ve

$\therefore$ $f(x)$ is strictly increasing for all (+ve) real values of x.

(*iii*) Let $f(x) = \log(1 + x) - \dfrac{2x}{2 + x}$ ...(1)

$$\therefore \quad f'(x) = \frac{1}{(1+x)} \cdot \frac{d}{dx}(1 + x) - \frac{(2+x)\,.\,\frac{d}{dx}(2x) - 2x\,.\,\frac{d}{dx}(2+x)}{(2+x)^2}$$

$$= \frac{1}{1+x} - \frac{(2+x)\,.\,2 - 2x(1)}{(2+x)^2} = \frac{1}{1+x} - \frac{4 + 2x - 2x}{(2+x)^2}$$

$$= \frac{1}{1+x} - \frac{4}{(2+x)^2} = \frac{(2+x)^2 - 4(1+x)}{(1+x)\,(2+x)^2}$$

$$= \frac{4 + x^2 + 4x - 4 - 4x}{(1+x)\,(2+x)^2} = \frac{x^2}{(1+x)\,(2+x)^2}$$

The function $f(x)$ is strictly increasing only if $f'(x) > 0$ *i.e.,* $\dfrac{x^2}{(1+x)\,(2+x)^2} > 0$

$$\Rightarrow \quad \left(\frac{x}{2+x}\right)^2 \left(\frac{1}{1+x}\right) > 0$$

$$\Rightarrow \quad \frac{1}{1+x} > 0 \qquad \left[\because \left(\frac{x}{2+x}\right)^2 > 0\right]$$

$$\Rightarrow \quad 1 + x > 0 \Rightarrow x > -1$$

$$\Rightarrow \quad x \in (-1, \infty).$$

$\therefore$ $f(x)$ is strictly increasing on $(-1, \infty)$.

(*iv*) Let $f(x) = 2\log(x - 2) - x^2 + 4x + 1$ ...(1)

$$\Rightarrow \quad f'(x) = \frac{2}{x-2} - 2x + 4 \qquad [\because \text{ Clearly, } f(x) \text{ is defined only for all } x > 2]$$

$$= \frac{2 - 2x(x-2) + 4(x-2)}{x-2} = \frac{2 - 2x^2 + 4x + 4x - 8}{x-2}$$

$$= \frac{-2x^2 + 8x - 6}{x-2} = \frac{-2(x^2 - 4x + 3)}{(x-2)} = \frac{-2\,(x-1)\,(x-3)}{(x-2)}$$

Now, $f'(x) = 0 \Rightarrow \dfrac{-2(x-1)(x-3)}{(x-2)} = 0$

$\Rightarrow -2(x-1)(x-3) = 0 \Rightarrow (x-1)(x-3) = 0$

$\Rightarrow x = 1, 3$

$\therefore$ The critical values of $f(x)$ is 3. [$\because$ $f(x)$ is defined only for all $x > 2$]

When $2 < x < 3$,

$$\Rightarrow f'(x) = \frac{-2\,(+\text{ve})\,(-\text{ve})}{(+\text{ve})} = +\text{ve}$$

$\Rightarrow$ $f(x)$ is strictly increasing for $2 < x < 3$

$\therefore$ $f(x)$ is strictly increasing on (2, 3)

When $x > 3$,

$$\Rightarrow f'(x) = \frac{-2\,(+\text{ve})\,(+\text{ve})}{(+\text{ve})} = -\text{ve}$$

$\Rightarrow$ $f(x)$ is strictly decreasing for $x > 3$

$\therefore$ $f(x)$ is strictly decreasing on $(3, \infty)$.

Example 11. *Find the intervals on which the following functions are strictly increasing or strictly decreasing.*

(i) x^x ; $x > 0$ *(ii)* $(x-1)\,e^x + 1$; $x > 0$.

Solution. *(i)* Let $f(x) = x^x$; $x > 0$ ...(1)

Taking logarithms on both sides, we get

$\Rightarrow \log f(x) = \log x^x$

$\Rightarrow \log f(x) = x \log x$ [$\because$ $\log n^m = m \log n$]

Differentiating w.r.t. x, we get

$$\frac{1}{f(x)} \cdot \frac{d}{dx}(f(x)) = x \cdot \frac{1}{x} + \log x \,.1$$

$$\Rightarrow \frac{f'(x)}{f(x)} = 1 + \log x$$

$\Rightarrow f'(x) = f(x)\,[1 + \log x]$

$\Rightarrow f'(x) = x^x\,(1 + \log x)$ [$\because$ $f(x) = x^x$]

Now, $f'(x) = 0 \Rightarrow x^x(1 + \log x) = 0$

$\Rightarrow 1 + \log x = 0 \Rightarrow \log x = -1$

$\Rightarrow x = e^{-1} \Rightarrow x = \dfrac{1}{e}$

$\therefore$ The critical value for $f(x)$ is $x = \dfrac{1}{e}$

When $0 < x < \dfrac{1}{e}$,

$\Rightarrow f'(x) = (+\text{ve})\,(-\text{ve}) = -\text{ve}$

$\Rightarrow$ $f(x)$ is strictly decreasing for $0 < x < \dfrac{1}{e}$

$\therefore$ $f(x)$ is strictly decreasing on $\left(0, \frac{1}{e}\right)$.

$\because \quad 0 < x < \frac{1}{e}$

$\Rightarrow \quad \log x < \log \frac{1}{e}$

$\Rightarrow \quad \log x < \log e^{-1}$

$\Rightarrow \quad \log x < (-1) \log e \quad [\because \log e = 1]$

$\Rightarrow \quad \log x < -1$

$\Rightarrow \quad 1 + \log x < 0 \quad$ *i.e.,* (–ve)

When $x > \frac{1}{e}$,

$\Rightarrow \quad f'(x) = (+\text{ve})\ (+\text{ve}) = +\text{ve}$

$\Rightarrow$ $f(x)$ is strictly increasing for $x > \frac{1}{e}$

$\therefore$ $f(x)$ is strictly increasing on $\left(\frac{1}{e}, \infty\right)$.

$\because \quad \frac{1}{e} < x < \infty$

$\Rightarrow \quad \log \frac{1}{e} < \log x$

$\Rightarrow \quad \log e^{-1} < \log x$

$\Rightarrow \quad (-1) \log e < \log x$

$\Rightarrow \quad -1 < \log x$

$\Rightarrow \quad 0 < 1 + \log x$

$\Rightarrow \quad 1 + \log x > 0$

i.e., (+ve)

(*ii*) Let $\quad f(x) = (x-1)\, e^x + 1 \; ; x > 0 \quad \ldots(1)$

$$\Rightarrow \quad f'(x) = (x-1) \cdot \frac{d}{dx}(e^x) + e^x \cdot \frac{d}{dx}(x-1)$$

$$= (x-1)\, e^x + e^x \cdot 1 = e^x \cdot x - e^x + e^x = x\, e^x$$

Now, for all $x > 0$, $x\, e^x > 0$.

$\therefore \quad f'(x) > 0$ for all $x > 0$

$\therefore$ $f(x)$ is strictly increasing for all $x > 0$.

Example 12. (*i*) *Show that the function $f(x) = e^x$ is strictly increasing on R*

(*ii*) *Show that the function $f(x) = e^{1/x}$ is strictly decreasing for all $x > 0$.*

(*iii*) *Show that the function $(x + 2)\, e^{-x}$ is strictly increasing on $(-\infty, -1)$.*

Solution. (*i*) We have

$$f(x) = e^x \quad \ldots(1)$$

$$\Rightarrow \quad f'(x) = e^x$$

Since, $e^x > 0$ for all real values of x.

When $x > 0$,

$$\Rightarrow \quad f'(x) = e^x = \left(1 + x + \frac{x^2}{2!} + \frac{x^3}{3!} + \ldots\ldots\right) > 0$$

$\Rightarrow$ $f'(x) > 0$ for all $x > 0$ $\left[\because\ e^x = 1 + x + \dfrac{x^2}{2!} + \dfrac{x^3}{3!} + \ldots\ldots\right]$

When $x = 0$,

$\Rightarrow$ $f'(x) = e^x = e^0 = 1 > 0$

$\Rightarrow$ $f'(x) > 0$ for all $x = 0$.

When $x < 0$,

Let $x = -y$ [where y is positive]

$\Rightarrow$ $f'(x) = e^x = e^{-y}$

$= \dfrac{1}{e^y} > 0$ [$\because$ y is positive]

$\Rightarrow$ $f'(x) > 0$ for all $x < 0$

$\Rightarrow$ $f'(x) > 0$ for all real values of x.

$\therefore$ $f(x)$ is strictly increasing on $x \in R$.

(*ii*) We have $f(x) = e^{1/x}$ ...(1)

$\Rightarrow$ $f'(x) = e^{1/x} \cdot \dfrac{d}{dx}\left(\dfrac{1}{x}\right) = -\dfrac{1}{x^2} \cdot e^{1/x} < 0$

[$\because$ $x^2 > 0$, and $e^{1/x} > 0$ for all $x > 0$]

$\Rightarrow$ $f'(x) < 0$ for all $x > 0$

$\therefore$ $f(x)$ is strictly decreasing for all $x > 0$.

(*iii*) Let $f(x) = (x + 2)\,e^{-x}$ ...(1)

$$f'(x) = (x + 2)\,.\,\frac{d}{dx}(e^{-x}) + e^{-x}\,.\,\frac{d}{dx}(x + 2)$$

$$= (x + 2)\,e^{-x}(-1) + e^{-x}\,.\,1 = e^{-x}(-x - 2 + 1)$$

$$= e^{-x}(-x - 1) = -e^{-x}(x + 1)$$

$$= -\frac{(x + 1)}{e^x}$$

Since, $x \in (-\infty, -1)$ *i.e.*, $x < -1$

$\Rightarrow$ $f'(x) > 0$ for all $x < -1$ [$\because$ $(x + 1) < 0$, $e^x > 0$]

$\therefore$ $f(x)$ is strictly decreasing on $(-\infty, -1)$.

Example 13. *Show that :*

(*i*) *sin x is strictly increasing on* $\left(-\dfrac{\pi}{2}, \dfrac{\pi}{2}\right)$.

(*ii*) *cos* $\left(2x + \dfrac{\pi}{4}\right)$ *is strictly increasing on* $\left(\dfrac{3\pi}{8}, \dfrac{7\pi}{8}\right)$.

(*iii*) *(x − sin x) is strictly increasing for all x ∈ R.*

(*iv*) $\cos^2 x$ *is strictly decreasing on* $\left(0, \dfrac{\pi}{2}\right)$.

(*v*) *tan x is strictly increasing on* $\left(-\dfrac{\pi}{2}, \dfrac{\pi}{2}\right)$

(*vi*) *cos x is strictly decreasing on (0, π).*

Solution. (*i*) Let $f(x) = \sin x$ ...(1)

$\Rightarrow$ $f'(x) = \cos x$

Since, $x \in \left(-\frac{\pi}{2}, \frac{\pi}{2}\right) \Rightarrow f'(x) = \cos x > 0$ [$\because$ cosine function is positive in Ist and IVth quadrants]

$\Rightarrow$ $f'(x) > 0$ for $x \in \left(-\frac{\pi}{2}, \frac{\pi}{2}\right)$

$\therefore$ $f(x)$ is strictly increasing on $\left(-\frac{\pi}{2}, \frac{\pi}{2}\right)$.

(*ii*) Let $f(x) = \cos\left(2x + \frac{\pi}{4}\right)$ ...(1)

$$\Rightarrow \quad f'(x) = -\sin\left(2x + \frac{\pi}{4}\right) \cdot \frac{d}{dx}\left(2x + \frac{\pi}{4}\right)$$

$$= -2\sin\left(2x + \frac{\pi}{4}\right).$$

Since, $x \in \left(\frac{3\pi}{8}, \frac{7\pi}{8}\right)$

$$\Rightarrow \quad \frac{3\pi}{8} < x < \frac{7\pi}{8}$$

$$\Rightarrow \quad \frac{3\pi}{4} < 2x < \frac{7\pi}{4}$$ [Multiplying throughout by 2]

$$\Rightarrow \quad \left(\frac{3\pi}{4} + \frac{\pi}{4}\right) < \left(2x + \frac{\pi}{4}\right) < \left(\frac{7\pi}{4} + \frac{\pi}{4}\right)$$ $\left[\text{Adding } \frac{\pi}{4} \text{ throughout}\right]$

$$\Rightarrow \quad \pi < \left(2x + \frac{\pi}{4}\right) < 2\pi$$

[$\because$ sine function is negative in IIIrd and IVth quadrants]

$$\Rightarrow \quad \sin\left(2x + \frac{\pi}{4}\right) < 0 \quad \Rightarrow \quad -2\sin\left(2x + \frac{\pi}{4}\right) > 0$$

$$\Rightarrow \quad f'(x) > 0$$

$\Rightarrow$ $f(x)$ is strictly increasing on $\left(\frac{3\pi}{8}, \frac{7\pi}{8}\right)$.

(*iii*) Let $f(x) = x - \sin x$ ...(1)

$\Rightarrow$ $f'(x) = 1 - \cos x$

Since, the value of cosine function always lies between – 1 and + 1.

$\therefore$ $f'(x) \geq 0$ for all $x \in$ R

$\therefore$ $f(x)$ is strictly increasing for all $x \in$ R.

(*iv*) Let $f(x) = \cos^2 x$ ...(1)

$\Rightarrow$ $f'(x) = 2 \cos x \cdot \dfrac{d}{dx}(\cos x)$

$= - 2 \cos x \sin x$ [$\because$ $\sin 2A = 2 \sin A \cos A$]

$= - \sin 2x$

Since, $x \in \left(0, \dfrac{\pi}{2}\right) \Rightarrow 0 < x < \dfrac{\pi}{2}$ [Multiplying throughout by 2]

$\Rightarrow$ $0 < 2x < \pi$

[$\because$ sine function is positive in Ist and IInd quadrants]

$\Rightarrow$ $\sin 2x > 0 \Rightarrow - \sin 2x < 0$

$\Rightarrow$ $f'(x) < 0$

$\Rightarrow$ $f(x)$ is strictly decreasing on $\left(0, \dfrac{\pi}{2}\right)$.

(*v*) Let $f(x) = \tan x$ ...(1)

$\Rightarrow$ $f'(x) = \sec^2 x$

Since, $x \in \left(-\dfrac{\pi}{2}, \dfrac{\pi}{2}\right)$

$\Rightarrow$ $\sec x > 0$ [$\because$ secant function is positive in Ist and IVth quadrants]

$\Rightarrow$ $\sec^2 x > 0$

$\Rightarrow$ $f'(x) > 0$

$\Rightarrow$ $f(x)$ is strictly increasing on $\left(-\dfrac{\pi}{2}, \dfrac{\pi}{2}\right)$.

(*vi*) Let $f(x) = \cos x$ ...(1)

$\Rightarrow$ $f'(x) = - \sin x$

Since, $x \in (0, \pi)$

$\Rightarrow$ $0 < x < \pi$

[$\because$ sine function is positive in Ist and IInd quadrants]

$\Rightarrow$ $\sin x > 0$

$\Rightarrow$ $- \sin x < 0$

$\Rightarrow$ $f'(x) < 0$

$\Rightarrow$ $f(x)$ is strictly decreasing on $(0, \pi)$.

Example 14. *Show that :*

(*i*) *(tan x – x) is strictly increasing on* $\left(0, \dfrac{\pi}{2}\right)$.

(*ii*) *tan*$^{-1}$ *(sin x + cos x) is strictly increasing on* $\left(0, \dfrac{\pi}{4}\right)$.

(*iii*) $\dfrac{x}{\sin x}$ *is strictly increasing on* $\left(0, \dfrac{\pi}{2}\right)$.

(*iv*) *(tan x – 4x) is strictly decreasing on* $\left(-\dfrac{\pi}{3}, \dfrac{\pi}{3}\right)$.

(v) $\left(-\frac{x}{2}+\sin x\right)$ *is strictly increasing on* $\left(-\frac{\pi}{3}, \frac{\pi}{3}\right)$.

(vi) $\cot^{-1}(\sin x+\cos x)$ *is strictly decreasing on* $\left(0, \frac{\pi}{4}\right)$.

Solution. (i) Let $f(x)=\tan x-x$ ...(1)

$\Rightarrow \quad f'(x)=\sec^2 x-1$

Since, $x \in \left(0, \frac{\pi}{2}\right) \Rightarrow \sec x>0$

$\Rightarrow \quad \sec^2 x>0 \Rightarrow \sec^2 x-1>0$

[$\because$ sec x is positive in Ist quadrant]

$\Rightarrow \quad f'(x)>0$

$\therefore$ $f(x)$ is strictly increasing on $\left(0, \frac{\pi}{2}\right)$.

(ii) Let $f(x)=\tan^{-1}(\sin x+\cos x)$ ...(1)

$$\Rightarrow \quad f'(x)=\frac{1}{1+(\sin x+\cos x)^2} \cdot \frac{d}{dx}(\sin x+\cos x)$$

$$\left[\because \frac{d}{dx}(\tan^{-1} x)=\frac{1}{1+x^2}\right]$$

$$=\frac{1}{1+(\sin x+\cos x)^2} \cdot (\cos x-\sin x)$$

$$=\frac{(\cos x-\sin x)}{1+(\sin x+\cos x)^2}=\frac{\cos x(1-\tan x)}{1+(\sin x+\cos x)^2}$$

Since, $x \in \left(0, \frac{\pi}{4}\right)$.

$$\cos x>0, \frac{1}{1+(\sin x+\cos x)^2}>0$$

[$\because$ cos x is positive in Ist quadrant]

Also, $\tan x<\tan \frac{\pi}{4}$ [$\because$ $1+(\sin x+\cos x)^2$ being the sum of two positive numbers is also positive]

$\tan x<1$

$\Rightarrow \quad (1-\tan x)>0$

$$\therefore \quad f'(x)=\frac{\cos x(1-\tan x)}{1+(\sin x+\cos x)^2}>0$$

$\Rightarrow \quad f'(x)>0$ for all $x \in \left(0, \frac{\pi}{4}\right)$.

$\therefore$ $f(x)$ is strictly increasing on $\left(0, \frac{\pi}{4}\right)$.

(*iii*) Let $$f(x) = \frac{x}{\sin x} \quad ...(1)$$

$$\Rightarrow \quad f'(x) = \frac{\sin x . \frac{d}{dx}(x) - x . \frac{d}{dx}(\sin x)}{(\sin x)^2}$$

$$= \frac{\sin x . 1 - x \cos x}{\sin^2 x} = \frac{\cos x (\tan x - x)}{\sin^2 x}$$

Since, $x \in \left(0, \frac{\pi}{2}\right)$

$\Rightarrow \quad \cos x > 0$ [$\because$ sin x and cos x are positive in Ist quadrant]

$\Rightarrow \quad \sin^2 x > 0$

$\Rightarrow \quad (\tan x - x) > 0$ [Proved in part (*i*)]

$$\Rightarrow \quad f'(x) = \frac{\cos x (\tan x - x)}{\sin^2 x} > 0$$

$$\Rightarrow \quad f'(x) > 0 \text{ for all } x \in \left(0, \frac{\pi}{2}\right)$$

$\therefore$ $f(x)$ is strictly increasing on $\left(0, \frac{\pi}{2}\right)$.

(*iv*) Let $$f(x) = \tan x - 4x \quad ...(1)$$

$$\Rightarrow \quad f'(x) = \sec^2 x - 4$$

$$= \frac{1}{\cos^2 x} - 4 = \frac{1 - 4\cos^2 x}{\cos^2 x}$$

$$= \frac{(1 - 2\cos x)(1 + 2\cos x)}{\cos^2 x}$$

$$= \sec^2 x (1 - 2\cos x)(1 + 2\cos x)$$

Since, $x \in \left(-\frac{\pi}{3}, \frac{\pi}{3}\right)$

$$= \sec^2 x > 0, \cos x > \frac{1}{2}$$

$\therefore \quad f'(x) = (+\text{ve})(-\text{ve})(+\text{ve}) = -\text{ve}$

$\therefore$ $f(x)$ is strictly decreasing on $\left(-\frac{\pi}{3}, \frac{\pi}{3}\right)$.

(*v*) Let $$f(x) = -\frac{x}{2} + \sin x \quad ...(1)$$

$\Rightarrow \quad f'(x) = -\frac{1}{2} + \cos x$

Since, $\quad x \in \left(-\frac{\pi}{3}, \frac{\pi}{3}\right)$

$\Rightarrow \quad \cos x > \frac{1}{2}$

$\Rightarrow \quad -\frac{1}{2} + \cos x > -\frac{1}{2} + \frac{1}{2}$

$\Rightarrow \quad -\frac{1}{2} + \cos x > 0$

$\Rightarrow \quad f'(x) > 0$ for all $x \in \left(-\frac{\pi}{3}, \frac{\pi}{3}\right)$

$\therefore$ $f(x)$ is strictly increasing on $\left(-\frac{\pi}{3}, \frac{\pi}{3}\right)$.

(*vi*) Let $\quad f(x) = \cot^{-1} (\sin x + \cos x)$ ...(1)

$\Rightarrow \quad f'(x) = \frac{-1}{1 + (\sin x + \cos x)^2} \cdot \frac{d}{dx} (\sin x + \cos x)$

$\left[\because \frac{d}{dx}(\cot^{-1} x) = \frac{-1}{1+x^2}\right]$

$$= \frac{-1}{1 + (\sin x + \cos x)^2} \cdot (\cos x - \sin x)$$

$$= \frac{-1(\cos x - \sin x)}{1 + (\sin x + \cos x)^2} = \frac{-\cos x (1 - \tan x)}{1 + (\sin x + \cos x)^2}$$

Since $\quad x \in \left(0, \frac{\pi}{4}\right)$

$\Rightarrow \quad \cos x > 0$ $\quad$ [$\because$ cos x in positive in Ist quadrant]

$\Rightarrow \quad \tan x < \tan \frac{\pi}{4}$

$\Rightarrow \quad \tan x < 1$ $\quad$ [$\because$ $1 + (\sin x + \cos x)^2 > 0$ being the sum of two positive numbers is also positive]

$\Rightarrow \quad 1 - \tan x > 0$

Also, $1 + (\sin x + \cos x)^2 > 0$

$\therefore \quad f'(x) = \frac{(-\text{ve})\,(+\text{ve})\,(+\text{ve})}{(+\text{ve})} = -\text{ve}$

$\Rightarrow$ $f(x)$ is strictly decreasing on $\left(0, \frac{\pi}{4}\right)$.

Example 15. *Separate the interval* $\left[0, \frac{\pi}{2}\right]$ *into sub-intervals in which* $f(x) = \sin^4 x + \cos^4 x$ *is increasing or decreasing.*

Or

Find the interval on which the function $f(x) = \sin^4 x + \cos^4 x$; $0 \le x \le \frac{\pi}{2}$ *is increasing or decreasing.*

Solution. We have

$$f(x) = \sin^4 x + \cos^4 x \quad ...(1)$$

$$\Rightarrow \quad f'(x) = 4 \sin^3 x . \frac{d}{dx}(\sin x) + 4\cos^3 x . \frac{d}{dx}(\cos x)$$

$$= 4\sin^3 x(\cos x) + 4\cos^3 x(-\sin x)$$

$$= 4\sin^3 x \cos x - 4\cos^3 x \sin x$$

$$= 4\sin x \cos x(\sin^2 x - \cos^2 x)$$

$$= -2(2\sin x \cos x)(\cos^2 x - \sin^2 x)$$

$$= -2\sin 2x \cos 2x \quad \left[\because \begin{array}{l}\sin 2A = 2\sin A\cos A \\ \cos 2A = \cos^2 A - \sin^2 A\end{array}\right]$$

$$= -\sin 4x$$

Now, $f'(x) = 0$

$\Rightarrow$ $-\sin 4x = 0 \Rightarrow \sin 4x = 0$

$\Rightarrow$ $4x = n\pi$; $n \in z$

$\Rightarrow$ $x = \frac{n\pi}{4}$; $n \in z$

Put $n = 1$, $\left[\because \text{ only } n = 1, \text{ can give the values of } x \text{ lying between 0 and } \frac{\pi}{2}.\right]$

$\Rightarrow$ $x = \frac{\pi}{4}$

$\therefore$ The critical value of $f(x)$ is $x = \frac{\pi}{4}$.

When $0 < x < \frac{\pi}{4}$ $\Rightarrow$ $0 < 4x < \pi$

$\Rightarrow$ $\sin 4x > 0$ [$\because$ $\sin x$ is positive in Ist and IInd quadrants]

$\Rightarrow$ $-\sin 4x < 0$

$\Rightarrow$ $f'(x) < 0$ for all $x \in \left(0, \frac{\pi}{4}\right)$.

$\therefore$ $f(x)$ is strictly decreasing on $\left(0, \frac{\pi}{4}\right)$.

When $\frac{\pi}{4} < x < \frac{\pi}{2} \Rightarrow \pi < 4x < 2\pi$

$\Rightarrow \quad \sin 4x < 0$ [$\because$ sin x is negative in IIIrd and IVth quadrants]

$\Rightarrow \quad -\sin 4x > 0$

$\Rightarrow \quad f'(x) > 0$ for all $x \in \left(\frac{\pi}{4}, \frac{\pi}{2}\right)$.

$\therefore$ $f(x)$ is strictly increasing on $\left(\frac{\pi}{4}, \frac{\pi}{2}\right)$.

Example 16. *Separate* $\left[0, \frac{\pi}{2}\right]$ *into sub-intervals in which* $f(x) = \sin 3x$ *is strictly increasing or strictly decreasing.*

Solution. We have

$$f(x) = \sin 3x \quad \text{...(1)}$$

$$\Rightarrow \quad f'(x) = 3 \cos 3x$$

Since, $x \in \left(0, \frac{\pi}{2}\right) \Rightarrow 0 < x < \frac{\pi}{2}$

$$\Rightarrow \quad 0 < 3x < \frac{3\pi}{2}$$

Since, cosine function is positive in Ist quadrant and negative in the IInd and IIIrd quadrants.

$\therefore$ We have two cases.

Case I. When $0 < 3x < \frac{\pi}{2}$

$\Rightarrow \quad \cos 3x > 0$

$\Rightarrow \quad 3 \cos 3x > 0$

$\Rightarrow \quad f'(x) > 0$ for $0 < 3x < \frac{\pi}{2}$ *i.e.,* $0 < x < \frac{\pi}{6}$

$\therefore$ $f(x)$ is strictly increasing on $\left(0, \frac{\pi}{6}\right)$.

Case II. When $\frac{\pi}{2} < 3x < \frac{3\pi}{2}$

$\Rightarrow \quad \cos 3x < 0$

$\Rightarrow \quad 3 \cos 3x < 0$

$\Rightarrow \quad f'(x) < 0$ for $\frac{\pi}{2} < 3x < \frac{3\pi}{2}$ *i.e.,* $\frac{\pi}{6} < x < \frac{\pi}{2}$

$\therefore$ $f(x)$ is strictly decreasing on $\left(\frac{\pi}{6}, \frac{\pi}{2}\right)$.

Example 17. (*i*) *Determine the intervals in which $f(x) = \sin x - \cos x$; $0 \le x \le 2\pi$ is strictly increasing or strictly decreasing.*

(*ii*) *Determine the intervals in which the function $f(x) = \sin x + \sqrt{3} \cos x$; $0 \le x \le 2\pi$ is increasing or decreasing.*

Solution. We have

(*i*) $$f(x) = \sin x - \cos x \quad ...(1)$$

$\Rightarrow$ $$f'(x) = \cos x + \sin x$$

$$= \sqrt{2}\left(\frac{1}{\sqrt{2}}\cos x + \frac{1}{\sqrt{2}}\sin x\right) \quad \text{[Multiply and divide by } \sqrt{2}]$$

$$= \sqrt{2}\left(\cos x . \sin\frac{\pi}{4} + \sin x \cos\frac{\pi}{4}\right)$$

$$= \sqrt{2}\sin\left(\frac{\pi}{4} + x\right) \quad [\because \sin(A + B) = \sin A \cos B + \cos A \sin B]$$

Since, $$0 \le x \le 2\pi \Rightarrow \frac{\pi}{4} \le \frac{\pi}{4} + x \le \frac{\pi}{4} + 2\pi \quad \left[\text{Adding } \frac{\pi}{4} \text{ throughout}\right]$$

$\Rightarrow$ $$\frac{\pi}{4} \le \left(\frac{\pi}{4} + x\right) \le \frac{9\pi}{4}$$

Now, the function $f(x)$ is strictly increasing, when $f'(x) > 0$

$\Rightarrow$ $$f'(x) > 0$$

$\Rightarrow$ $$\sqrt{2}\sin\left(\frac{\pi}{4} + x\right) > 0$$

$\Rightarrow$ $$\sin\left(\frac{\pi}{4} + x\right) > 0 \quad [\because \text{ sine function is positive in Ist and IInd quadrants}]$$

$\Rightarrow$ $$\frac{\pi}{4} < \frac{\pi}{4} + x < \pi \,;\, 2\pi < \frac{\pi}{4} + x < \frac{9\pi}{4}$$

$\Rightarrow$ $$\frac{\pi}{4} - \frac{\pi}{4} < x < \pi - \frac{\pi}{4} \,;\, 2\pi - \frac{\pi}{4} < x < \frac{9\pi}{4} - \frac{\pi}{4}$$

$\Rightarrow$ $$0 < x < \frac{3\pi}{4} \,;\, \frac{7\pi}{4} < x < 2\pi$$

$\therefore$ $f(x)$ is strictly increasing on $\left(0, \frac{3\pi}{4}\right)$ and $\left(\frac{7\pi}{4}, 2\pi\right)$.

Now, the function $f(x)$ is strictly decreasing

When $$f'(x) < 0$$

$\Rightarrow$ $$f'(x) < 0$$

$\Rightarrow$ $$\sqrt{2}\sin\left(\frac{\pi}{4} + x\right) < 0$$

$\Rightarrow$ $$\sin\left(\frac{\pi}{4} + x\right) < 0 \quad [\because \text{ sine function is negative in IIIrd and IVth quadrant}]$$

$\Rightarrow \quad \pi < \frac{\pi}{4} + x < 2\pi$

$\Rightarrow \quad \pi - \frac{\pi}{4} < x < 2\pi - \frac{\pi}{4}$

$\Rightarrow \quad \frac{3\pi}{4} < x < \frac{7\pi}{4}$

$\therefore$ $f(x)$ is strictly decreasing on $\left(\frac{3\pi}{4}, \frac{7\pi}{4}\right)$.

(*ii*) We have $f(x) = \sin x + \sqrt{3} \cos x$ ...(1)

$\Rightarrow \quad f'(x) = \cos x + \sqrt{3} \sin x$

$= 2\left(\frac{1}{2}\cos x + \frac{\sqrt{3}}{2}\sin x\right)$ [Multiply and divide by 2]

$= 2\left(\frac{\sqrt{3}}{2}\sin x + \frac{1}{2}\cos x\right)$

$= 2\left(\cos\frac{\pi}{6}\sin x + \sin\frac{\pi}{6}\cos x\right)$

$= 2\sin\left(x + \frac{\pi}{6}\right)$ [$\because$ $\sin(A + B) = \sin A \cos B + \cos A \sin B$]

Since, $0 \le x \le 2\pi \quad \Rightarrow \quad \frac{\pi}{6} \le x + \frac{\pi}{6} \le 2\pi + \frac{\pi}{6}$

$\Rightarrow \quad \frac{\pi}{6} \le x + \frac{\pi}{6} \le \frac{13\pi}{6}$

Now, the function $f(x)$ is strictly increasing,
When $f'(x) > 0$

$\Rightarrow \quad f'(x) > 0$

$\Rightarrow \quad 2\sin\left(x + \frac{\pi}{6}\right) > 0$

$\Rightarrow \quad \sin\left(x + \frac{\pi}{6}\right) > 0$ [$\because$ sine function is positive in Ist and IInd quadrants]

$\Rightarrow \quad \frac{\pi}{6} < x + \frac{\pi}{6} < \pi \; ; \; 2\pi < x + \frac{\pi}{6} < \frac{13\pi}{6}$

$\Rightarrow \quad \frac{\pi}{6} - \frac{\pi}{6} < x < \pi - \frac{\pi}{6} \; ; \; 2\pi - \frac{\pi}{6} < x < \frac{13\pi}{6} - \frac{\pi}{6}$

$\Rightarrow \quad 0 < x < \frac{5\pi}{6} \; ; \; \frac{11\pi}{6} < x < 2\pi$

$\therefore$ $f(x)$ is strictly increasing on $\left(0, \frac{5\pi}{6}\right)$ and $\left(\frac{11\pi}{6}, 2\pi\right)$.

Now, the function $f(x)$ is strictly decreasing.

When $\quad f'(x) < 0$

$\Rightarrow \quad f'(x) < 0$

$\Rightarrow \quad 2 \sin\left(x + \frac{\pi}{6}\right) < 0$

$\Rightarrow \quad \sin\left(x + \frac{\pi}{6}\right) < 0$ [$\because$ sine function is negative in IIIrd and IVth quadrants]

$\Rightarrow \quad \pi < x + \frac{\pi}{6} < 2\pi$

$\Rightarrow \quad \pi - \frac{\pi}{6} < x < 2\pi - \frac{\pi}{6}$

$\Rightarrow \quad \frac{5\pi}{6} < x < \frac{11\pi}{6}$

$\Rightarrow$ $f(x)$ is strictly decreasing on $\left(\frac{5\pi}{6}, \frac{11\pi}{6}\right)$.

Example 18. *Show that* $\log(x + 1) \geq \frac{x}{x+1}$ *as* x *increases from 0 to* ∞.

Solution. Let $\quad f(x) = \log(x + 1) - \frac{x}{x+1}$ ...(1)

$$\Rightarrow \quad f'(x) = \frac{1}{x+1} - \frac{(x+1).1 - x(1+0)}{(x+1)^2}$$

$$= \frac{1}{x+1} - \frac{x+1-x}{(x+1)^2} = \frac{1}{x+1} - \frac{1}{(x+1)^2}$$

$$= \frac{x+1-1}{(x+1)^2} = \frac{x}{(x+1)^2} \geq 0 \quad \text{as } x \geq 0$$

$\Rightarrow \quad f'(x) \geq 0$ as $x \geq 0$.

$\Rightarrow$ $f(x)$ is an increasing function for $x \geq 0$

$\Rightarrow \quad f(x) \geq f(0)$

$\Rightarrow \quad \log(x + 1) - \frac{x}{x+1} \geq \log 1 - \frac{0}{0+1}$

$\Rightarrow \quad \log(x + 1) - \frac{x}{x+1} \geq 0$ [$\because$ $\log 1 = 0$]

$\Rightarrow \quad \log(x + 1) \geq \frac{x}{(x+1)}$.

Example 19. *Find the intervals in which the function*

$$f(x) = \frac{4 \sin x - 2x - x \cos x}{2 + \cos x} \text{ ; } 0 \le x \le 2\pi \text{ is (a) increasing (b) decreasing.}$$

Solution. We have

$$f(x) = \frac{4 \sin x - 2x - x \cos x}{2 + \cos x} \quad ...(1)$$

$$\Rightarrow \quad f'(x) = \frac{(2 + \cos x)\dfrac{d}{dx}(4 \sin x - 2x - x \cos x) - (4 \sin x - 2x - x \cos x)\dfrac{d}{dx}(2 + \cos x)}{(2 + \cos x)^2}$$

$$= \frac{(2 + \cos x)(4 \cos x - 2 - \cos x + x \sin x) - (4 \sin x - 2x - x \cos x)(-\sin x)}{(2 + \cos x)^2}$$

$$= \frac{\begin{array}{c} 8 \cos x + 4 \cos^2 x - 4 - 2 \cos x - 2 \cos x - \cos^2 x + 2x \sin x \\ + x \sin x \cos x + 4 \sin^2 x - 2x \sin x - x \sin x \cos x \end{array}}{(2 + \cos x)^2}$$

$$= \frac{4 \cos x + 4 \cos^2 x + 4 \sin^2 x - \cos^2 x - 4}{(2 + \cos x)^2}$$

$$= \frac{4 \cos x + 4 (\cos^2 x + \sin^2 x) - \cos^2 x - 4}{(2 + \cos x)^2} \quad [\because \sin^2 A + \cos^2 A = 1]$$

$$= \frac{4 \cos x + 4 - \cos^2 x - 4}{(2 + \cos x)^2} = \frac{\cos x (4 - \cos x)}{(2 + \cos x)^2}$$

Now, $f(x)$ is an increasing function, when $f'(x) > 0$

$$\Rightarrow \quad f'(x) > 0$$

$$\Rightarrow \quad \frac{\cos x(4 - \cos x)}{(2 + \cos x)^2} > 0 \quad \left[\because \frac{4 - \cos x}{(2 + \cos x)^2} > 0\right]$$

$\Rightarrow \quad \cos x > 0$ $\quad$ [$\because$ cosine function is positive in Ist and IVth quadrants]

$$\Rightarrow \quad 0 < x < \frac{\pi}{2} \text{ ; } \frac{3\pi}{2} < x < 2\pi$$

$\Rightarrow$ $f(x)$ is increasing on $\left(0, \frac{\pi}{2}\right)$ and $\left(\frac{3\pi}{2}, 2\pi\right)$.

Now, $f(x)$ is a decreasing function, when $f'(x) < 0$

$$\Rightarrow \quad f'(x) < 0$$

$$\Rightarrow \quad \frac{\cos x(4 - \cos x)}{(2 + \cos x)^2} < 0 \quad \left[\because \frac{4 - \cos x}{(2 + \cos x)^2} > 0\right]$$

$$\Rightarrow \quad \cos x < 0$$

$$\Rightarrow \quad \frac{\pi}{2} < x < \frac{3\pi}{2}$$

[$\because$ cosine function is negative in IInd and IIIrd quadrants]

$\therefore$ $f(x)$ is decreasing on $\left(\frac{\pi}{2}, \frac{3\pi}{2}\right)$.

Example 20. *Which of the following functions are strictly decreasing on* $\left(0, \frac{\pi}{2}\right)$.

(i) cos x (ii) cos $2x$

(iii) cos $3x$ (iv) tan x.

Solution. (i) Let $f(x) = \cos x$

$\Rightarrow$ $f'(x) = -\sin x$

Since, $x \in \left(0, \frac{\pi}{2}\right) \Rightarrow \sin x > 0$ [$\because$ sin x is positive in Ist quadrant]

$\Rightarrow -\sin x < 0$

$\Rightarrow f'(x) < 0$ for all $x \in \left(0, \frac{\pi}{2}\right)$.

$\therefore$ $f(x)$ is strictly decreasing on $\left(0, \frac{\pi}{2}\right)$.

(ii) Let $f(x) = \cos 2x$

$\Rightarrow$ $f'(x) = -2 \sin 2x$

Since, $x \in \left(0, \frac{\pi}{2}\right) \Rightarrow 0 < x < \frac{\pi}{2}$

$\Rightarrow$ $0 < 2x < \pi$ [$\because$ sin x is positive in Ist and IInd quadrants]

$\Rightarrow$ $\sin 2x > 0$

$\Rightarrow$ $-2 \sin 2x < 0$

$\Rightarrow$ $f'(x) < 0$ for all $x \in \left(0, \frac{\pi}{2}\right)$

$\therefore$ $f(x)$ is strictly decreasing on $\left(0, \frac{\pi}{2}\right)$.

(iii) Let $f(x) = \cos 3x$

$\Rightarrow$ $f'(x) = -3 \sin 3x$

Since, $x \in \left(0, \frac{\pi}{2}\right) \Rightarrow 0 < x < \frac{\pi}{2}$

$\Rightarrow$ $0 < 3x < \frac{3\pi}{2}$

$\Rightarrow$ sin $3x$ can be positive as well as negative.

[$\because$ sin x is positive in Ist and IInd quadrants and negative in IIIrd quadrant]

$\Rightarrow$ $f(x)$ is neither increasing nor decreasing on $\left(0, \frac{\pi}{2}\right)$.

(*iv*) Let $f(x) = \tan x$

$\Rightarrow$ $f'(x) = \sec^2 x$

Since, $x \in \left(0, \frac{\pi}{2}\right) \Rightarrow 0 < x < \frac{\pi}{2}$ [$\because$ sec x is positive in Ist quadrant]

$\Rightarrow$ $\sec x > 0$

$\Rightarrow$ $\sec^2 x > 0$

$\Rightarrow$ $f'(x) > 0$ for all $x \in \left(0, \frac{\pi}{2}\right)$

$\Rightarrow$ $f(x)$ is strictly increasing on $\left(0, \frac{\pi}{2}\right)$.

Example 21. *On which of the following intervals is the function $x^{100} + \sin x - 1$ is strictly increasing.*

(*i*) $(-1, 1)$ (*ii*) $(0, 1)$

(*iii*) $\left(0, \frac{\pi}{2}\right)$ (*iv*) $\left(\frac{\pi}{2}, \pi\right)$.

Solution. Let $f(x) = x^{100} + \sin x - 1$ ...(1)

$\Rightarrow$ $f'(x) = 100\, x^{99} + \cos x.$

(*i*) When $x \in (-1, 1)$

$\Rightarrow$ $-1 < x < 1$

$\Rightarrow$ $-1 < x^{99} < 1$

$\Rightarrow$ $-100 < 100\, x^{99} < 100$

But, $\cos x$ lies between 0 and 1 for $x \in (-1, 1)$ *i.e.*, $0 < \cos x < 1$ for $x \in (-1, 1)$,

$\therefore$ $f'(x)$ can be positive as well as negative for $x \in (-1, 1)$

$\therefore$ $f(x)$ is neither increasing nor decreasing on $(-1, 1)$.

(*ii*) When $x \in (0, 1)$

$\Rightarrow$ $0 < x < 1$

$\Rightarrow$ $x^{99} > 0$

$\Rightarrow$ $100\, x^{99} > 0$

Also, $\cos x > 0$ for $x \in (0, 1)$

$\Rightarrow$ $f'(x) = (100\, x^{99} + \cos x) > 0$

$\Rightarrow$ $f'(x) > 0$ for all $x \in (0, 1)$

$\Rightarrow$ $f(x)$ is increasing on $(0, 1)$

(*iii*) When $x \in \left(0, \frac{\pi}{2}\right)$

$\Rightarrow$ $0 < x < \frac{\pi}{2}$

$\Rightarrow \quad x^{99} > 0 \Rightarrow 100\,x^{99} > 0$

and $\cos x > 0$ for $x \in \left(0, \frac{\pi}{2}\right)$ [$\because$ $\cos x$ is positive in Ist quadrant]

$\Rightarrow f'(x) = (100x^{99} + \cos x) > 0$

$\Rightarrow f'(x) > 0$ for all $x \in \left(0, \frac{\pi}{2}\right)$.

$\Rightarrow f(x)$ is increasing on $\left(0, \frac{\pi}{2}\right)$.

(*iv*) When $x \in \left(\frac{\pi}{2}, \pi\right)$

$\Rightarrow \quad \frac{\pi}{2} < x < \pi \quad \Rightarrow x^{99} > 1$

$\Rightarrow 100\,x^{99} > 100$

Also, $\quad -1 < \cos x < 0$ for $\left(\frac{\pi}{2} < x < \pi\right)$

$\Rightarrow \quad 0 > \cos x > -1$ for $\left(\frac{\pi}{2} < x < \pi\right)$

$$\Rightarrow \quad f'(x) = 100\,x^{99} + \cos x > 100 - 1$$
$$= 100\,x^{99} + \cos x > 99 > 0$$

$\Rightarrow f'(x) > 0$ for all $x \in \left(\frac{\pi}{2}, \pi\right)$

$\therefore$ $f(x)$ is increasing on $\left(\frac{\pi}{2}, \pi\right)$.

EXERCISE FOR PRACTICE

1. Find the interval in which the function given by $f(x) = 8 + 36x + 3x^2 - 2x^3$ is increasing or decreasing.
2. Find the interval in which the function $f(x) = x^3 - 6x^2 + 9x + 15$ is increasing or decreasing.
3. Show that $f(x) = \; = e^{1/x}$; $x \neq 0$ is a decreasing function for all $x \neq 0$.
4. Show that $f(x) = \log(\sin x)$ is increasing on $\left(0, \frac{\pi}{2}\right)$ and decreasing on $\left(\frac{\pi}{2}, \pi\right)$.
5. Show that the function $f(x) = \sin\left(2x + \frac{\pi}{4}\right)$ is decreasing on $\left(\frac{3\pi}{8}, \frac{5\pi}{8}\right)$.
6. Prove that the function $f(x) = x^3 - 6x^2 + 12x - 18$ is increasing on R.
7. Show that $f(x) = x^2 - x \sin x$ is an increasing function on $\left(0, \frac{\pi}{2}\right)$.
8. Show that $f(x) = \tan^{-1} x - x$ is a decreasing function on R.

9. Find the intervals on which the function $f(x) = \log(1 + x) - \frac{2x}{2 + x}$ is strictly increasing or strictly decreasing.

10. Show that the function $f(x) = \tan x - 4x$ is strictly decreasing function on $\left(-\frac{\pi}{3}, \frac{\pi}{3}\right)$.

Answers

1. Increasing on $(-2, 3)$. Decreasing on $(-\infty, -2)$, $(3, \infty)$.
2. Increasing on $(-\infty, -3)$, $(1, \infty)$. Decreasing on $(-3, 1)$.
9. Strictly increasing on $(-1, \infty)$. Strictly decreasing on $(-\infty, -1)$.

4

Differentials, Errors and Approximations

4.1 INTRODUCTION

Till now, we have been considering $\frac{dy}{dx}$ as a single entity, which represents the derivative of y w.r.t. x. In this chapter, we give a meaning to the symbols dx and dy in such a way that the original meaning of the symbol $\frac{dy}{dx}$ coincides with the quotient when dy is divided by dx.

4.2 DIFFERENTIALS

Let $y = f(x)$ be a given function of x.

Then, we know that $\lim\limits_{\delta x \to 0} \frac{f(x + \delta x) - f(x)}{\delta x} = f'(x)$

$$\Rightarrow \quad \frac{f(x+\delta x) - f(x)}{\delta x} = f'(x) + \varepsilon$$

where $\varepsilon \to 0$ as $\delta x \to 0$.

$\Rightarrow \quad f(x + \delta x) - f(x) = \delta x\,(f'(x) + \varepsilon)$

$\Rightarrow \quad f(x + \delta x) - f(x) = f'(x) \,.\, \delta x + \varepsilon \,.\, \delta x$

$\Rightarrow \quad f(x + \delta x) - f(x) = f'(x).\, \delta x$ (approximately)

$\Rightarrow \quad \delta y = f'(x) \,.\, \delta x$ $[\because\ f(x + \delta x) - f(x) = \delta y]$

$\Rightarrow \quad \delta y = \frac{dy}{dx} \,.\, \delta x$ $\left[\because\ f'(x) = \frac{dy}{dx}\right]$

Thus, if δx is an error in x, then the corresponding error δy in y can be calculated.

The quantities δx and $f'(x)\, \delta x$ are respectively called the differential of the independent variable x and the differential of the dependent variable y. The differentials of the variables, x and y are denoted by dx and dy respectively.

$\therefore \quad dx = \delta x \quad \text{and} \quad dy = f'(x)\, \delta x$

Here, the differential, dy of y is a function of x and δx.

4.3 ABSOLUTE ERROR

The error δx in x is called the absolute error in x.

4.4 RELATIVE ERROR

If δx is an error in x, then $\frac{\delta x}{x}$ is called the relative error in x.

4.5 PERCENTAGE ERROR

If δx is an error in x, then, $\frac{\delta x}{x} \times 100$ is called the percentage error in x.

Note. The formula $\delta y = f'(x) \,.\, \delta x$ is very useful in computing the small changes (errors) in dependent variable corresponding to small changes (errors) in the independent variables.

SOME SOLVED EXAMPLES

Example 1. *Find the approximate values of the following by using differentials :*

(*i*) $\sqrt[3]{26}$ (*ii*) $\sqrt[3]{127}$

(*iii*) $\sqrt[3]{66}$ (*iv*) $\sqrt{51}$

(*v*) $\sqrt{25.2}$ (*vi*) $\sqrt{401}$.

Solution. (*i*) Let $y = x^{1/3}$ ...(1)

$\Rightarrow \quad y + \delta y = (x + \delta x)^{1/3}$

$\Rightarrow \quad y + \delta y = (26)^{1/3}$ ...(2)

$\Rightarrow \quad y + \delta y = (27 - 1)^{1/3}$

$\therefore \quad x = 27$ and $\delta x = -1$

Let $dx = \delta x = -1$

Now, Differentiating (1) w.r.t. x, we get

$$\frac{dy}{dx} = \frac{1}{3} x^{-2/3}$$

$$\frac{dy}{dx} = \frac{1}{3x^{2/3}} \qquad ...(3)$$

$$\therefore \quad \left.\frac{dy}{dx}\right]_{x=27} = \frac{1}{3(27)^{2/3}} = \frac{1}{3(3^3)^{2/3}} = \frac{1}{3(9)} = \frac{1}{27} \qquad ...(4)$$

Since, δy is approximately equal to dy.

$$\therefore \quad dy = \frac{dy}{dx} \,.\, dx$$

$$\Rightarrow \quad dy = \frac{1}{27} \,.\, (-1) = \frac{-1}{27} \qquad [\because \text{ By using (4)}]$$

$$\therefore \quad (26)^{1/3} = y + \delta y \qquad [\because \text{ By using (2)}]$$

$$= x^{1/3} + \delta y \qquad [\because \text{ By using (1)}]$$

$$= (27)^{1/3} - \frac{1}{27} = 3 - \frac{1}{27} = \frac{80}{27} = 2.962963 \text{ (approx.)}.$$

(*ii*) Let $y = x^{1/3}$ ...(1)

$\Rightarrow$ $y + \delta y = (x + \delta x)^{1/3}$

$\Rightarrow$ $y + \delta y = (127)^{1/3}$ ...(2)

$\Rightarrow$ $y + \delta y = (125 + 2)^{1/3}$

$\therefore$ $x = 125$ and $\delta x = +2$

Let $dx = \delta x = +2$

Now, Differentiating (*i*) w.r.t. x, we get

$$\frac{dy}{dx} = \frac{1}{3} x^{-2/3}$$

$$\Rightarrow \quad \frac{dy}{dx} = \frac{1}{3x^{2/3}} \qquad \text{...(3)}$$

$$\therefore \quad \left.\frac{dy}{dx}\right]_{x=125} = \frac{1}{3(125)^{2/3}} = \frac{1}{3(5^3)^{2/3}} = \frac{1}{3(25)} = \frac{1}{75} \qquad \text{...(4)}$$

Since, δy is approximately equal to dy.

$$\therefore \quad dy = \frac{dy}{dx} . dx$$

$$= \frac{1}{75} . (+2) = \frac{+2}{75} \qquad [\because \text{ By using (4)}]$$

$$\therefore \quad (127)^{1/3} = y + \delta y \qquad [\because \text{ By using (2)}]$$

$$= x^{1/3} + \delta y \qquad [\because \text{ By using (1)}]$$

$$= (125)^{1/3} + \frac{2}{75} = 5 + \frac{2}{75} = \frac{375+2}{75}$$

$$= \frac{377}{75} = 5.0266 \text{ (approx.)}.$$

(*iii*) Please try yourself. $\left[\textbf{Ans. } (66)^{1/3} = \frac{97}{24} = 4.04166 \text{ (approx.)}\right]$

(*iv*) Let $y = \sqrt{x}$ ...(1)

$\Rightarrow$ $y + \delta y = \sqrt{x + \delta x}$

$\Rightarrow$ $y + \delta y = \sqrt{51}$ ...(2)

$\Rightarrow$ $y + \delta y = \sqrt{49 + 2}$

$\therefore$ $x = 49$ and $\delta x = 2$

Let $dx = \delta x = 2$

Now, Differentiating (1) w.r.t. x, we get

$$\frac{dy}{dx} = \frac{1}{2\sqrt{x}} \qquad \text{...(3)}$$

$$\therefore \quad \left.\frac{dy}{dx}\right]_{x=49} = \frac{1}{2\sqrt{49}} = \frac{1}{2(7)} = \frac{1}{14} \qquad \text{...(4)}$$

Since, δy is approximately equal to dy.

$$\therefore \quad dy = \frac{dy}{dx}.dx = \frac{1}{14}(2) = \frac{1}{7} \quad [\because \text{ By using (4)}]$$

$$\therefore \quad \sqrt{51} = y + \delta y \quad [\because \text{ By using (2)}]$$

$$= \sqrt{x} + \delta y = \sqrt{49} + \frac{1}{7} \quad [\because \text{ By using (1)}]$$

$$= 7 + \frac{1}{7} = \frac{49+1}{7} = \frac{50}{7} = 7.14285 \text{ (approx.).}$$

(*v*) Let $\quad y = \sqrt{x} \quad$ ...(1)

$$\Rightarrow \quad y + \delta y = \sqrt{x + \delta x}$$

$$\Rightarrow \quad y + \delta y = \sqrt{25.2} \quad \text{...(2)}$$

$$\Rightarrow \quad y + \delta y = \sqrt{25 + 0.2}$$

$$\therefore \quad x = 25 \quad \text{and} \quad \delta x = 0.2$$

Let $\quad dx = \delta x = 0.2$

Now, Differentiating (1) w.r.t. x, we get

$$\frac{dy}{dx} = \frac{1}{2\sqrt{x}} \quad \text{...(3)}$$

$$\therefore \quad \left.\frac{dy}{dx}\right]_{x=25} = \frac{1}{2\sqrt{25}} = \frac{1}{2(5)} = \frac{1}{10} \quad \text{...(4)}$$

Since, δy is approximately equal to dy.

$$\therefore \quad dy = \frac{dy}{dx}.dx = \left(\frac{1}{10}\right)(0.2) = \frac{2}{100} = 0.02 \quad [\because \text{ By using (4)}]$$

$$\therefore \quad \sqrt{25.2} = y + \delta y \quad [\because \text{ By using (2)}]$$

$$= \sqrt{x} + \delta y \quad [\because \text{ By using (1)}]$$

$$= \sqrt{25} + 0.02 = 5 + 0.02 = 5.02.$$

(*vi*) Please try yourself. $\left[\textbf{Ans. } \sqrt{401} = \frac{801}{40} = 20.025.\right]$

Example 2. *Find the approximate values of the following by using differentials :*

(*i*) $\sqrt{0.0037}$ (*ii*) $\sqrt[3]{0.026}$

(*iii*) $\sqrt[3]{0.009}$ (*iv*) $(255)^{1/4}$

(*v*) $(82)^{1/4}$.

Solution. (*i*) Let $\quad y = \sqrt{x} \quad$ ...(1)

$$\Rightarrow \quad y + \delta y = \sqrt{x + \delta x}$$

$$\Rightarrow \quad y + \delta y = \sqrt{0.0037} \quad \text{...(2)}$$

$\Rightarrow \quad y + \delta y = \sqrt{0.0036 + 0.0001}$

$\therefore \quad x = 0.0036 \quad \text{and} \quad \delta x = 0.0001.$

Now, Differentiating (1) w.r.t. x, we get

$$\frac{dy}{dx} = \frac{1}{2\sqrt{x}} \quad \text{...(3)}$$

$$\therefore \quad \left.\frac{dy}{dx}\right]_{x=0.0036} = \frac{1}{2\sqrt{0.0036}} = \frac{1}{2(0.06)} = \frac{1}{0.12} \quad \text{...(4)}$$

Since, δy is approximately equal to dy.

$$\therefore \quad dy = \frac{dy}{dx} . dx = \frac{1}{0.12} \times (0.0001) = \frac{1}{1200} \quad [\because \text{ By using (4)}]$$

$$\therefore \quad \sqrt{0.0037} = y + \delta y \quad [\because \text{ By using (2)}]$$

$$= \sqrt{x} + \delta y \quad [\because \text{ By using (1)}]$$

$$\sqrt{0.0036} + \frac{1}{1200} = 0.06 + 0.00083 = 0.06083.$$

(*ii*) Let $\quad y = x^{1/3}$ ...(1)

$\Rightarrow \quad y + \delta y = (x + \delta x)^{1/3}$

$\Rightarrow \quad y + \delta y = (0.026)^{1/3}$ ...(2)

$\Rightarrow \quad y + \delta y = (0.027 - 0.001)^{1/3}$

$\therefore \quad x = 0.027 \quad \text{and} \quad \delta x = -0.001$

Now, Differentiating (1) w.r.t. x, we get

$$\Rightarrow \quad \frac{dy}{dx} = \frac{1}{3} x^{-2/3}$$

$$\Rightarrow \quad \frac{dy}{dx} = \frac{1}{3x^{2/3}} \quad \text{...(3)}$$

$$\therefore \quad \left.\frac{dy}{dx}\right]_{x=0.027} = \frac{1}{3(0.027)^{2/3}} = \frac{1}{3[(0.3)^3]^{2/3}}$$

$$= \frac{1}{3(0.3)^2} = \frac{1}{3(0.09)} = \frac{1}{0.27} \quad \text{...(4)}$$

Since, δy is approximately equal to dy.

$$\therefore \quad dy = \frac{dy}{dx} . dx = \frac{1}{0.27} \times (-0.001) = \frac{-1}{270} \quad [\because \text{ By using (4)}]$$

$$\therefore \quad (0.026)^{1/3} = y + \delta y \quad [\because \text{ By using (2)}]$$

$$= x^{1/3} + \delta y = (0.027)^{1/3} - \frac{1}{270} \quad [\because \text{ By using (1)}]$$

$$= 0.3 - \frac{1}{270} = \frac{3}{10} - \frac{1}{270} = \frac{81-1}{270}$$

$$= \frac{80}{270} = \frac{8}{27} = 0.2963 \text{ (approx.)}.$$

(*iii*) Please try yourself. $\left[\textbf{Ans.} \ (0.009)^{1/3} = 0.2083 \ \text{(approx.)}\right]$

(*iv*) Let $y = x^{1/4}$ ...(1)

$\Rightarrow$ $y + \delta y = (x + \delta x)^{1/4}$

$\Rightarrow$ $y + \delta y = (255)^{1/4}$ ...(2)

$\Rightarrow$ $y + \delta y = (256 - 1)^{1/4}$

$\therefore$ $x = 256$ and $\delta x = -1$

Let $dx = \delta x = -1$

Now, Differentiating (1) w.r.t. x, we get

$$\frac{dy}{dx} = \frac{1}{4} x^{-3/4} = \frac{1}{4x^{3/4}} \quad \text{...(3)}$$

$$\therefore \quad \left.\frac{dy}{dx}\right]_{x=256} = \frac{1}{4(256)^{3/4}} = \frac{1}{4(4^4)^{3/4}} = \frac{1}{4(4^3)} = \frac{1}{4(64)} = \frac{1}{256} \quad \text{...(4)}$$

Since, δy is approximately equal to dy.

$$\therefore \quad dy = \frac{dy}{dx} \cdot dx = \frac{1}{256} \times (-1) = \frac{-1}{256} \quad [\because \text{ By using (4)}]$$

$$\therefore \quad (255)^{1/4} = y + \delta y \quad [\because \text{ By using (2)}]$$

$$= x^{1/4} + \delta y \quad [\because \text{ By using (1)}]$$

$$= (256)^{1/4} - \frac{1}{256}$$

$$= \frac{4}{1} - \frac{1}{256} = 4 - 0.003906 = 3.9960. \text{ (approx.)}$$

(*v*) Please try yourself. [**Ans.** $(82)^{1/4} = 3.009$ (approx.)]

Example 3. *Find the approximate values of the following by using differentials :*

(*i*) $(15)^{1/4}$ (*ii*) $(80)^{1/4}$

(*iii*) $(0.999)^{1/10}$ (*iv*) $(0.007)^{1/3}$

(*v*) $(31.9)^{1/5}$.

Solution. (*i*) Let $y = x^{1/4}$ ...(1)

$\Rightarrow$ $y + \delta y = (x + \delta x)^{1/4}$

$\Rightarrow$ $y + \delta y = (15)^{1/4}$ ...(2)

$\Rightarrow$ $y + \delta y = (16 - 1)^{1/4}$

$\therefore$ $x = 16$ and $\delta x = -1$

Let $dx = \delta x = -1$

Now, Differentiating (1) w.r.t. x, we get

$$\frac{dy}{dx} = \frac{1}{4}\, x^{-3/4} = \frac{1}{4\, x^{3/4}} \quad \text{...(3)}$$

$$\therefore \quad \left.\frac{dy}{dx}\right]_{x=16} = \frac{1}{4(16)^{3/4}} = \frac{1}{4(2^4)^{3/4}} = \frac{1}{4(2^3)} = \frac{1}{4(8)} = \frac{1}{32} \quad \text{...(4)}$$

Since, δy is approximately equal to dy

$$\therefore \quad dy = \frac{dy}{dx}\,.\, dx = \frac{1}{32}(-1) = -\frac{1}{32} \quad [\because \text{ By using (4)}]$$

$$\therefore \quad (15)^{1/4} = y + \delta y \quad [\because \text{ By using (2)}]$$

$$= x^{1/4} + \delta y \quad [\because \text{ By using (1)}]$$

$$= (16)^{1/4} - \frac{1}{32}$$

$$= 2 - \frac{1}{32} = \frac{64-1}{32} = \frac{63}{32} = 1.96875 \text{ (approx.).}$$

(*ii*) Please try yourself. $\left[\textbf{Ans. } (80)^{1/4} = \frac{323}{108} = 2.9907 \text{ (approx.)}\right]$

(*iii*) Let $y = x^{1/10}$ ...(1)

$\Rightarrow \quad y + \delta y = (x + \delta x)^{1/10}$

$\Rightarrow \quad y + \delta y = (0.999)^{1/10}$ ..(2)

$\Rightarrow \quad y + \delta y = (1 - 0.001)^{1/10}$

$\therefore \quad x = 1$ and $\delta x = -0.001$

Let $dx = \delta x = -0.001$

Now, Differentiating (1) w.r.t. x, we get

$$\frac{dy}{dx} = \frac{1}{10}\, x^{-\frac{9}{10}} = \frac{1}{10x^{9/10}} \quad \text{...(3)}$$

$$\therefore \quad \left.\frac{dy}{dx}\right]_{x=1} = \frac{1}{10(1)^{9/10}} = \frac{1}{10} \quad \text{...(4)}$$

Since, δy is approximately equal to dy

$$\therefore \quad dy = \frac{dy}{dx}\,.\, dx = \frac{1}{10} \times (-0.001) = \frac{-1}{10000} \quad [\because \text{ By using (4)}]$$

$$\therefore \quad (0.999)^{1/10} = y + \delta y \quad [\because \text{ By using (2)}]$$

$$= x^{1/10} + \delta y \quad [\because \text{ By using (1)}]$$

$$= (1)^{1/10} - \frac{1}{10000} = 1 - 0.0001$$

$$= 0.9999 \text{ (approx.).}$$

(*iv*) Let $y = x^{1/3}$ ...(1)

$\Rightarrow$ $y + \delta y = (x + \delta x)^{1/3}$

$\Rightarrow$ $y + \delta y = (0.007)^{1/3}$ ...(2)

$\Rightarrow$ $y + \delta y = (0.008 - 0.001)^{1/3}$

$\therefore$ $x = 0.008$ and $\delta x = -0.001$

Let $dx = \delta x = -0.001$

Now, Differentiating (1) w.r.t. x, we get

$$\frac{dy}{dx} = \frac{1}{3} x^{-2/3} = \frac{1}{3x^{2/3}} \quad ...(3)$$

$$\therefore \quad \left.\frac{dy}{dx}\right]_{x=0.008} = \frac{1}{3(0.008)^{2/3}} = \frac{1}{3[(0.2)^3]^{2/3}} = \frac{1}{3(0.2)^2}$$

$$= \frac{1}{3(0.04)} = \frac{1}{0.12} \quad ...(4)$$

Since, dy is approximately equal to δy

$$\therefore \quad dy = \frac{dy}{dx} . dx = \left(\frac{1}{0.12}\right) \times (-0.001) = \frac{-1}{120} \quad [\because \text{ By using (4)}]$$

$$\therefore \quad (0.007)^{1/3} = y + \delta y \quad [\because \text{ By using (2)}]$$

$$= x^{1/3} + \delta y \quad [\because \text{ By using (1)}]$$

$$= (0.008)^{1/3} - \frac{1}{120}$$

$$= 0.2 - \frac{1}{120} = \frac{2}{10} - \frac{1}{120} = \frac{24 - 1}{120} = \frac{23}{120} = 0.19166 \text{ (approx.)}.$$

(*v*) Let $y = x^{1/5}$ ...(1)

$\Rightarrow$ $y + \delta y = (x + \delta x)^{1/5}$

$\Rightarrow$ $y + \delta y = (31.9)^{1/5}$ ...(2)

$\Rightarrow$ $y + \delta y = (32 - 0.1)^{1/5}$

$\therefore$ $x = 32$ and $\delta x = -0.1$

Let $dx = \delta x = -0.1$

Now, Differentiating (1) w.r.t. x, we get

$$\frac{dy}{dx} = \frac{1}{5} x^{-4/5} = \frac{1}{5x^{4/5}} \quad ...(3)$$

$$\therefore \quad \left.\frac{dy}{dx}\right]_{x=32} = \frac{1}{5(32)^{4/5}} = \frac{1}{5(2^5)^{4/5}} = \frac{1}{5(2^4)} = \frac{1}{5(16)} = \frac{1}{80} \quad ...(4)$$

Since, δy is approximately equal to dy

$$\therefore \quad dy = \frac{dy}{dx} . dx = \frac{1}{80}(-0.1) = \frac{-1}{800} \quad [\because \text{ By using (4)}]$$

$\therefore \quad (31.9)^{1/5} = y + \delta y$ [$\because$ By using (2)]

$= x^{1/5} + \delta y$ [$\because$ By using (1)]

$$= (32)^{1/5} - \frac{1}{800} = 2 - \frac{1}{800} = \frac{1600-1}{800}$$

$$= \frac{1599}{800} = 1.99875 \text{ (approx.).}$$

Example 4. *Find the approximately values of the following by using differentials :*

(*i*) *log 4.98 ; Given log 5 = 1.6094*

(*ii*) *log 9.01 ; Given log 3 = 1.0986*

(*iii*) *log 10.02 ; Given log 10 = 2.3026*

(*iv*) *log 4.01 ; Given log 4 = 1.3863*

(*v*) *$\log_{10} 10.1$; Given $\log_{10} e = 0.4343$.*

Solution. (*i*) Let $\quad y = \log x$ ...(1)

$\Rightarrow \quad y + \delta y = \log (x + \delta x)$

$\Rightarrow \quad y + \delta y = \log (4.98)$ ...(2)

$\Rightarrow \quad y + \delta y = \log (5 - 0.02)$

$\therefore \quad x = 5$ and $\delta x = -0.02$

Let $\quad dx = \delta x = -0.02$

Now, Differentiating (1) w.r.t. x, we get

$$\Rightarrow \quad \frac{dy}{dx} = \frac{1}{x} \quad \text{...(3)}$$

$$\therefore \quad \left.\frac{dy}{dx}\right]_{x=5} = \frac{1}{5} \quad \text{...(4)}$$

Since, δy is approximately equal to dy.

$$\therefore \quad dy = \frac{dy}{dx} . dx = \frac{1}{5} \times (-0.02) = \frac{-2}{500}$$ [$\because$ By using (4)]

$\therefore \quad \log 4.98 = y + \delta y$ [$\because$ By using (2)]

$= \log x + \delta y$ [$\because$ By using (1)]

$$= \log 5 - \frac{2}{500}$$

$= 1.6094 - 0.0040 = 1.6054.$ [$\because$ log 5 = 1.6094 (Given)]

(*ii*) Please try yourself.

[**Hint.** Proceed as in part (*i*)]. [**Ans.** 2.1983]

(*iii*) Let $\quad y = \log x$ ...(1)

$\Rightarrow \quad y + \delta y = \log (x + \delta x)$

$\Rightarrow \quad y + \delta y = \log (10.02)$ ...(2)

$\Rightarrow \quad y + \delta y = \log (10 + 0.02)$

$\therefore \quad x = 10 \quad \text{and} \quad \delta x = 0.02$

Let $\quad dx = \delta x = 0.02$

Now, Differentiating (1) w.r.t. x, we get

$$\frac{dy}{dx} = \frac{1}{x} \quad ...(3)$$

$$\therefore \quad \left.\frac{dy}{dx}\right]_{x=10} = \frac{1}{10} \quad ...(4)$$

Since, δy is approximately equal to dy

$$\therefore \quad dy = \frac{dy}{dx} \cdot dx = \frac{1}{10} \times (0.02) = \frac{2}{1000} = \frac{1}{500} \quad [\because \text{ By using (4)}]$$

$$\therefore \quad \log(10.02) = y + \delta y \quad [\because \text{ By using (2)}]$$

$$= \log x + \delta y \quad [\because \text{ By using (1)}]$$

$$= \log 10 + \frac{2}{1000}$$

$$= 2.3026 + 0.002 \quad [\because \log 10 = 2.3026 \text{ (Given)}]$$

$$= 2.3046.$$

(*iv*) Let $\quad y = \log x \quad ...(1)$

$\Rightarrow \quad y + \delta y = \log(x + \delta x)$

$\Rightarrow \quad y + \delta y = \log(4.01) \quad ...(2)$

$\Rightarrow \quad y + \delta y = \log(4 + 0.01)$

$\therefore \quad x = 4 \quad \text{and} \quad \delta x = 0.01$

Let $\quad dx = \delta x = 0.01$

Now, Differentiating (1) w.r.t. x, we get

$$\frac{dy}{dx} = \frac{1}{x} \quad ...(3)$$

$$\therefore \quad \left.\frac{dy}{dx}\right]_{x=4} = \frac{1}{4} \quad ...(4)$$

Since, δy is approximately equal to dy

$$\therefore \quad dy = \frac{dy}{dx} \cdot dx = \frac{1}{4} \times (0.01) = \frac{1}{400} \quad [\because \text{ By using (4)}]$$

$$\therefore \quad \log(4.01) = y + \delta y \quad [\because \text{ By using (2)}]$$

$$= \log x + \delta y \quad [\because \text{ By using (1)}]$$

$$= \log 4 + \frac{1}{400}$$

$$= 1.3863 + 0.0025 \quad [\because \log 4 = 1.3863 \text{ (Given)}]$$

$$= 1.3888.$$

(*v*) Let $\quad y = \log_{10} x \quad ...(1)$

$\Rightarrow \quad y + \delta y = \log_{10}(x + \delta x)$

$\Rightarrow \quad y + \delta y = \log_{10} 10.1$ ...(2)

$\Rightarrow \quad y + \delta y = \log_{10} (10 + 0.1)$

$\therefore \quad x = 10$ and $\delta x = 0.1$

Let $dx = \delta x = 0.1$

Now, Differentiating (1) w.r.t. x, we get

$$\frac{dy}{dx} = \frac{1}{x} \log_{10} e \quad ...(3)$$

$$\therefore \quad \left.\frac{dy}{dx}\right]_{x=10} = \frac{1}{10} \times (0.4343) \quad [\because \log_{10} e = 0.4343 \text{ (Given)}]$$

$$= 0.04343$$

Since, δy is approximately equal to dy.

$$\therefore \quad dy = \frac{dy}{dx} . dx = (0.04343)(0.1) = 0.004343 \quad [\because \text{ By using (4)}]$$

$$\therefore \quad \log_{10} (10.1) = y + \delta y \quad [\because \text{ By using (2)}]$$

$$= \log_{10} x + \delta y \quad [\because \text{ By using (1)}]$$

$$= \log_{10} 10 + 0.004343$$

$$= 1 + 0.004343 \quad [\because \log_{10} 10 = 1]$$

$$= 1.004343.$$

Example 5. *Find the approximate values of the following by using differentials :*

(i) tan 46° Given : 1° = 0.01745 radians.

(ii) cos 61° Given : sin 60° = 0.86603 and 1° = 0.01745 radians.

(iii) $\tan^{-1} (0.999)$

(iv) $10^{3.01}$ *Given : log 10 = 2.3026.*

Solution. (*i*) Let $y = \tan x$ ...(1)

$\Rightarrow \quad y + \delta y = \tan (x + \delta x)$

$\Rightarrow \quad y + \delta y = \tan 46°$ ...(2)

$\Rightarrow \quad y + \delta y = \tan (45° + 1°)$

$\therefore \quad x = 45°$ and $\delta x = 1°$

Let $dx = \delta x = 1°$

Now, Differentiating (1) w.r.t. x, we get

$$\Rightarrow \quad \frac{dy}{dx} = \sec^2 x \quad ...(3)$$

$$\therefore \quad \left.\frac{dy}{dx}\right]_{x=45°} = \sec^2 45° = (\sqrt{2})^2 = 2 \quad ...(4)$$

Since, δy is approximately equal to dy.

$$\therefore \quad dy = \frac{dy}{dx} . dx = (2)(1°) = 2(0.01745) \quad [\because \text{ By using (4)}]$$

$$= 0.0349 \quad [\because 1° = 0.01745 \text{ radians (Given)}]$$

$$\therefore \quad \tan 46° = y + \delta y \quad [\because \text{ By using (2)}]$$

$$= \tan x + \delta y \quad [\because \text{ By using (1)}]$$

$= \tan 45° + 0.0349 = 1 + 0.0349$
$= 1.0349$ (approx.).

(*ii*) Let $y = \cos x$ ...(1)

$\Rightarrow \quad y + \delta y = \cos (x + \delta x)$

$\Rightarrow \quad y + \delta y = \cos 61°$ ...(2)

$\Rightarrow \quad y + \delta y = \cos (60° + 1°)$

$\therefore \quad x = 60° \quad \text{and} \quad \delta x = 1°$

Let $dx = \delta x = 1°$

Now, Differentiating (1) w.r.t. x, we get

$$\frac{dy}{dx} = -\sin x. \qquad ...(3)$$

$$\therefore \quad \left.\frac{dy}{dx}\right]_{x = 60°} = -\sin 60°$$

$= -0.86603$ ...(4) [$\because$ $\sin 60° = 0.86603$ radian (Given)]

Since, δy is approximately equal to dy

$$\therefore \quad dy = \frac{dy}{dx} . dx = (-0.86603)\ 1° \qquad [\because \text{ By using (4)}]$$

$= (-0.86603)(0.01745)$ [$\because$ $1° = 0.01745$ radian (Given)]

$= -0.015112$

$\therefore \quad \cos 61° = y + \delta y$ [$\because$ By using (2)]

$= \cos x + \delta y$ [$\because$ By using (1)]

$$= \cos 60° - 0.015112 = \frac{1}{2} - 0.015112 = 0.5 - 0.015112$$

$= 0.484888$ (approx.).

(*iii*) Let $y = \tan^{-1} x$ ...(1)

$\Rightarrow \quad y + \delta y = \tan^{-1} (x + \delta x)$

$\Rightarrow \quad y + \delta y = \tan^{-1} (0.999)$ ...(2)

$\Rightarrow \quad y + \delta y = \tan^{-1} (1 - 0.001)$

$\therefore \quad x = 1 \quad \text{and} \quad \delta x = -0.001$

Let $dx = \delta x = -0.001$

Now, Differentiating (1) w.r.t. x, we get

$$\frac{dy}{dx} = \frac{1}{1+x^2} \qquad ...(3)$$

$$\therefore \quad \left.\frac{dy}{dx}\right]_{x = 1} = \frac{1}{1+(1)^2} = \frac{1}{1+1} = \frac{1}{2} = 0.5 \qquad ...(4)$$

Since, δy is approximately equal to dy

$$\therefore \quad dy = \frac{dy}{dx} . dx = (0.5)(-0.001) = -0.0005 \qquad [\because \text{ By using (4)}]$$

$$\therefore \quad \tan^{-1}(0.999) = y + \delta y \qquad [\because \text{ By using (2)}]$$

$$= \tan^{-1} x + \delta y \qquad [\because \text{ By using (1)}]$$

$$= \tan^{-1} 1 - 0.0005$$

$$= \frac{\pi}{4} - 0.0005 = \frac{22}{7} \times \frac{1}{4} - 0.0005 = \frac{22}{28} - 0.0005$$

$$= 0.7857 - 0.0005$$

$$= 0.7852 \text{ (approx.).}$$

(*iv*) Let $y = 10^x$ ...(1)

$\Rightarrow \quad y + \delta y = 10^{x + \delta x}$ ...(2)

$\Rightarrow \quad y + \delta y = 10^{3.01}$

$\Rightarrow \quad y + \delta y = 10^{3 + 0.01}$

$\therefore \quad x = 3$ and $\delta x = 0.01$

Let $dx = \delta x = 0.01$

Now, Differentiating (1) w.r.t. x, we get

$$\frac{dy}{dx} = 10^x \,.\, \log 10 \qquad \text{...(3)}$$

$$\therefore \quad \left.\frac{dy}{dx}\right]_{x=3} = 10^3 \log 10$$

$$= 1000 \times 2.3026 \qquad [\because \log 10 = 2.3026 \text{ (Given)}]$$

$$= 2302.6 \qquad \text{...(4)}$$

Since, δy is approximately equal to dy.

$$\therefore \quad dy = \frac{dy}{dx} \,.\, dx = (2302.6)(0.01) = 23.026 \qquad [\because \text{ By using (4)}]$$

$$\therefore \quad 10^{3.01} = y + \delta y \qquad [\because \text{ By using (2)}]$$

$$= 10^x + \delta y \qquad [\because \text{ By using (1)}]$$

$$= 10^3 + 23.026 = 1000 + 23.026$$

$$= 1023.026 \text{ (approx.).}$$

Example 6. *If $y = x^2 + 3x + 7$ and if x changes from 5 to 5.01, find the actual change and approximate change in y.*

Solution. We have

$$y = x^2 + 3x + 7 \qquad \text{...(1)}$$

It is given that x changes from 5 to 5.01

$\therefore$ Let $x = 5$ and $x + \delta x = 5.01$

$\Rightarrow \quad \delta x = 5.01 - 5 = 0.01.$

Let $dx = \delta x = 0.01$

Now, Differentiating (1) w.r.t. x, we get

$$\frac{dy}{dx} = 2x + 3 \qquad \text{...(2)}$$

$$\therefore \quad \left.\frac{dy}{dx}\right]_{x=5} = 2(5) + 3 = 10 + 3 = 13 \qquad \text{...(3)}$$

Since, δy is approximately equal to dy.

$\therefore \quad dy = \frac{dy}{dx} \cdot dx = (13)(0.01)$ [$\because$ By using (3)]

$= 0.13$ (approx.)

From (1), we have $\quad y = x^2 + 3x + 7$

$\Rightarrow \quad y + \delta y = (x + \delta x)^2 + 3(x + \delta x) + 7$

$\therefore$ Actual change in $y = \delta y = (y + \delta y) - y$

$$= [(x + \delta x)^2 + 3(x + \delta x) + 7] - [x^2 + 3x + 7]$$
$$= x^2 + (\delta x)^2 + 2x\delta x + 3x + 3\delta x + 7 - x^2 - 3x - 7$$
$$= (\delta x)^2 + 2x\delta x + 3\delta x$$
$$= (0.01)^2 + 2(5)(0.01) + 3(0.01) \quad [\because \; x = 5, \delta x = 0.01]$$
$$= 0.0001 + 0.1 + 0.03 = 0.1301.$$

Example 7. *If the radius of a circle increases from 5 cm to 5.1 cm, find the increase in the area.*

Solution. Let r be the radius of the circle.

$\therefore \quad$ Area of the circle $= A = \pi r^2$ ...(1)

It is given that : radius r changes from 5 cm to 5.1 cm.

$\therefore$ Let $\quad r = 5$ cm and $r + \delta r = 5.1$

$\Rightarrow \quad \delta r = 5.1 - 5 = 0.1$

Let $\quad dr = \delta r = 0.1$

Now, Differentiating (1) w.r.t. r, we get

$$\frac{dA}{dr} = 2\pi r \quad \text{...(2)}$$

$$\therefore \quad \left.\frac{dA}{dr}\right]_{r=5} = 2\pi(5) = 10\pi \quad \text{...(3)}$$

Since, δy is approximately equal to dy.

$$\therefore \quad dy = \frac{dy}{dx} \cdot dx$$

$$\Rightarrow \quad dA = \frac{dA}{dr} \cdot dr \quad [\because \text{ By using (3)}]$$

$$= (10\pi)(0.1) = \pi \text{ cm}^2.$$

As change in area is positive.

$\therefore$ Approx. increase in the area is π cm^2.

Example 8. *A circular metal plate expands under heating so that its radius increases by 2%. Find the approximate increase in the area of the plate if the radius of the plate before heating is 10 cm.*

Solution. Let r be the radius of the circular mettalic plate.

$\therefore$ Area of the mettalic plate $= A = \pi r^2$ ...(1)

Let $dr = \delta r$ be the change in radius of the plate.

It is given that : $\frac{dr}{r} \times 100 = 2$

$\therefore$ When $r = 10$ cm

$$\therefore \quad \frac{dr}{10} \times 100 = 2$$

$$\Rightarrow \quad dr = \frac{2}{10} = 0.2$$

Now, Differentiating (1) w.r.t. r, we get

$$\frac{dA}{dr} = 2\pi r \qquad \text{...(2)}$$

$$\therefore \quad \left.\frac{dA}{dr}\right]_{r=10} = 2\pi(10) = 20\pi \qquad \text{...(3)}$$

Since, δy is approximately equal to dy.

$$\therefore \quad dy = \frac{dy}{dx} . dx$$

$$\Rightarrow \quad dA = \frac{dA}{dr} . dr \qquad [\because \text{ By using (3)}]$$

$$= (20\pi)(0.2) = 4\pi \text{ cm}^2.$$

$\therefore$ Approximate increase in area = 4π cm^2.

Example 9. *If the error committed in measuring the radius of a circle be 0.01%, find the corresponding error in calculating the area.*

Solution. Let r be the radius of the circle.

$\therefore$ Area of the circle $= A = \pi r^2$ ...(1)

Let $dr = \delta r$ be the change (error) in radius.

It is given that $\frac{dr}{r} \times 100 = 0.01$

$$\Rightarrow \quad dr = \frac{0.01\, r}{100} = 0.0001\, r$$

Now, Differentiating (1) w.r.t. r, we get

$$\frac{dA}{dr} = 2\pi r \qquad \text{...(2)}$$

Since, δy is approximately equal to dy.

$$\therefore \quad dy = \frac{dy}{dx} . dx$$

$$\Rightarrow \quad dA = \frac{dA}{dx} . dx = (2\pi r)(0.0001\, r)$$

$$= (0.0001)(2\pi r^2) \qquad \text{...(3)}$$

$\therefore$ Percentage change (error) in Area $= \frac{dA}{A} \times 100$ $\quad [\because$ By using (1) and (3)]

$$= \frac{(0.0001)(2\pi r^2)}{\pi r^2} \times 100 = 0.02\%.$$

Example 10. *Find the percentage error in calculating the volume of a cubical box if an error of 1% is made in measuring the length of the edges of the cube.*

Solution. Let x be the length of an edge of the cube.

$\therefore$ Volume of the cube = V = x^3 ...(1)

Let $dx = \delta x$ be the change (error) in the length of the edge.

It is given that $\frac{dx}{x} \times 100 = 1$

$$\Rightarrow \quad dx = \frac{1x}{100} = \frac{x}{100}$$

Now, Differentiating (1) w.r.t. x, we get

$$\frac{dV}{dx} = 3x^2 \quad ...(2)$$

Since, δy is approximately equal to dy.

$$\therefore \quad dy = \frac{dy}{dx} . dx$$

$$dV = \frac{dV}{dx} . dx = (3x^2)\left(\frac{x}{100}\right) = \frac{3x^3}{100} \quad ...(3)$$

Percentage change (error) in volume = $\frac{dV}{V} \times 100$ [$\because$ By using (1) and (3)]

$$= \frac{3x^3}{100 \times x^3} \times 100 = 3\%.$$

Example 11. *The radius of a sphere is measured as 4 cm with an error possibly as large as 0.01 cm. Find approximations to the greatest possible error, relative error and the percentage error in the computed volume.*

Solution. Let r be the radius of the sphere.

$\therefore$ Volume of the sphere = V = $\frac{4}{3}\pi r^3$ ...(1)

Let $dr = \delta r$ be the change in radius of the sphere.

It is given that : Error in measuring r is 0.01 cm.

$\therefore$ Let $r = 4$ cm, $\delta r = \pm 0.01$ cm.

Let $dr = \delta r = \pm 0.01$.

Now, Differentiating (1) w.r.t. r, we get

$$\frac{dV}{dr} = \frac{4}{3}\pi (3r^2) = 4\pi r^2 \quad ...(2)$$

$$\therefore \quad \left.\frac{dV}{dr}\right]_{r=4} = 4\pi(4)^2 = 4\pi(16) = 64\pi \quad ...(3)$$

Since, δy is approximately equal to dy.

$$\therefore \quad dy = \frac{dy}{dx} . dx$$

$\Rightarrow \qquad dV = \frac{dV}{dr} . dr = (64\pi)(\pm 0.01)$

$= \pm 0.64\ \pi$ (approx.) ...(4)

Relative error in volume $= \frac{dV}{V} = \frac{\pm 0.64\ \pi}{\frac{4}{3}\pi r^3}$ [$\because$ By using (1) and (4)]

$$= \pm \frac{3 \times 0.64\ \pi}{4\pi\, r^3} = \pm \frac{3 \times 0.64}{4(4)^3} = \pm \frac{1.92}{256} = \pm 0.0075$$

$\therefore$ Percentage error in volume

$$= \frac{dV}{V} \times 100 = \pm (0.0075) \times 100 = \pm 0.75\%.$$

Example 12. *The radius of a sphere shrinks from 10 cm to 9.99 cm. Find the approximate decrease in volume.*

Solution. Let r be the radius of the sphere.

$\therefore$ Volume of the sphere $= V = \frac{4}{3}\pi r^3$ ...(1)

It is given that r changes from 10 cm to 9.99 cm.

$\therefore$ Let $\qquad r = 10$ cm and $r + \delta r = 9.99$ cm

$\Rightarrow \qquad \delta r = 9.99 - 10 = -0.01$

Let $\qquad dr = \delta r = -0.01$

Now, Differentiating (1) w.r.t. r, we get

$$\frac{dV}{dr} = \frac{4}{3}\pi(3r^2) = 4\pi r^2 \qquad ...(2)$$

$$\therefore \quad \left.\frac{dV}{dr}\right]_{r=10} = 4\pi(10)^2 = 400\pi \qquad ...(3)$$

Since, δy is approximately equal to dy.

$$\therefore \quad dy = \frac{dy}{dx} . dx$$

$\Rightarrow \qquad dV = \frac{dV}{dr} . dr = (400\ \pi)(-0.01)$ [$\because$ By using (3)]

$= -4\pi$ (approx.).

Example 13. *If $y = \sin x$ and if x changes from $\frac{\pi}{2}$ to $\frac{22}{14}$, what is the approximate change in y.*

Solution. We have $\quad y = \sin x$ ...(1)

It is given that ; x changes from $\frac{\pi}{2}$ to $\frac{22}{14}$

$\therefore$ Let $\qquad x = \frac{\pi}{2}$ and $x + \delta x = \frac{22}{14}$

$$\Rightarrow \qquad \delta x = \frac{22}{14} - \frac{\pi}{2}.$$

Let $$dx = \delta x = \left(\frac{22}{14} - \frac{\pi}{2}\right).$$

Now, Differentiating (1) w.r.t. x, we get

$$\frac{dy}{dx} = \cos x \qquad ...(2)$$

$$\therefore \qquad \left.\frac{dy}{dx}\right]_{x=\frac{\pi}{2}} = \cos\frac{\pi}{2} = 0. \qquad ...(3)$$

Since, δy is approximately equal to dy.

$$\therefore \qquad dy = \frac{dy}{dx}\,.\,dx$$

$$= (0)\left(\frac{22}{14} - \frac{\pi}{2}\right) = 0. \qquad [\because \text{ By using (3)}]$$

Example 14. *If in a triangle ABC, the side c and angle C remains constant, while the remaining elements are changed slightly, using differentials show that :*

$$\frac{da}{\cos A} + \frac{db}{\cos B} = 0.$$

Solution. It is given that : In Δ ABC, the side c and angle C remains constant.

$$\Rightarrow \qquad \frac{c}{\sin C} = k \text{ (constant)}$$

$$\Rightarrow \qquad \frac{a}{\sin A} = \frac{b}{\sin B} = k \qquad \left[\because \frac{a}{\sin A} = \frac{b}{\sin B} = \frac{c}{\sin C}\right]$$

$$\Rightarrow \qquad a = k\sin A \quad \text{and} \quad b = k\sin B$$

$$\Rightarrow \qquad \frac{da}{dA} = k\cos A \text{ and } \frac{db}{dB} = k\cos B$$

$$\Rightarrow \qquad da = \left(\frac{da}{dA}\right).\,dA \quad \text{and} \quad db = \left(\frac{db}{dB}\right).\,dB$$

$$= (k\cos A)\,dA \quad \text{and} \quad db = (k\cos B)\,dB$$

$$\Rightarrow \qquad \frac{da}{\cos A} = k.dA \quad \text{and} \quad \frac{db}{\cos B} = k.dB.$$

$$\Rightarrow \qquad \frac{da}{\cos A} + \frac{db}{\cos B} = k.dA + k.dB$$

$$= k(dA + dB) = k.d(A + B) \qquad \begin{bmatrix} \because \text{ In } \Delta\text{ ABC}; A + B + C = \pi \\ \Rightarrow A + B = (\pi - c) = \text{constant} \\ \Rightarrow \qquad d(\pi - c) = 0 \end{bmatrix}$$

$$= k.d(\pi - c)$$
$$= k.0$$
$$= 0.$$

Example 15. *The area S of a triangle is calculated by measuring sides b, c and angle A. If there be an error δA in the measurement of A. Show that the relative error in area is*

$$\frac{\delta S}{S} = \cot A \,.\, \delta A.$$

Solution. In Δ ABC, It is given that

Area of triangle $\quad S = \frac{1}{2} bc \sin A \quad$ ...(1)

Let $dA = \delta A$ be the change in angle A.

Differentiating (1) w.r.t. A, we get

$$\frac{dS}{dA} = \frac{1}{2} bc \cos A$$

$$\therefore \quad dS = \left(\frac{dS}{dA}\right).dA$$

$$\Rightarrow \quad dS = \frac{1}{2} bc \cos A \,.\, dA \qquad ...(2)$$

Now, Relative error in area $S = \frac{dS}{S}$ $\quad$ [∵ By using (1) and (2)]

$$= \frac{\frac{1}{2} bc \cos A \,.\, dA}{\frac{1}{2} bc \sin A} = \cot A \,.\, dA$$

Example 16. *If a triangle ABC, inscribed in a fixed circle, be slightly varied in such a way as to have its vertices always on the circle, then show that :*

$$\frac{da}{\cos A} + \frac{db}{\cos B} + \frac{dc}{\cos C} = 0$$

Solution. As we know that by Sine Formula

$$\frac{a}{\sin A} = \frac{b}{\sin B} = \frac{c}{\sin C} = \text{constant} = k \text{ (say).}$$

$$\Rightarrow \quad \frac{a}{\sin A} = \frac{b}{\sin B} = \frac{c}{\sin C} = k$$

$$\Rightarrow \quad a = k \sin A$$

$$\Rightarrow \quad b = k \sin B \quad \text{and} \quad c = k \sin C$$

$$\Rightarrow \quad \frac{da}{dA} = k \cos A, \frac{db}{dB} = k \cos B \quad \text{and} \quad \frac{dc}{dC} = k \cos C$$

$$\therefore \quad da = \left(\frac{da}{dA}\right).dA = k \cos A \,.\, dA$$

$$\Rightarrow \quad \frac{da}{\cos A} = k.dA \quad ...(1)$$

$$db = \left(\frac{db}{dB}\right).dB$$

$$\Rightarrow \quad db = k \cos B \,.\, dB$$

$$\Rightarrow \quad \frac{db}{\cos B} = k.dB \quad ...(2)$$

and

$$dc = \left(\frac{dc}{dC}\right).dC$$

$$\Rightarrow \quad dc = k \cos C.\, dC$$

$$\Rightarrow \quad \frac{dc}{\cos C} = k.dC \quad ...(3)$$

Now, Adding equations (1), (2) and (3), we get

$$\frac{da}{\cos A} + \frac{db}{\cos B} + \frac{dc}{\cos C} = k.dA + k.dB + k.dC$$

$$= k(dA + dB + dC) = k.d(A + B + C)$$

$$= kd(\pi) = k.0$$

$$= 0. \qquad \left[\begin{array}{ll}\because & \text{In } \Delta\, ABC, A + B + C = \pi = \text{constant.} \\ \Rightarrow & d(A + B + C) = d(\pi) = 0\end{array}\right]$$

Example 17. *Find the percentage error in the area of an ellipse, when an error of 1% is made in measuring the major and minor axes.*

Solution. Let a and b be the semi-major and semi-minor axes of the ellipse.

$\therefore$ Area of the ellipse = A = $\pi\, ab$ ...(1)

Taking logarithms on both sides, we get

$$\Rightarrow \quad \log A = \log (\pi\, ab)$$

$$\Rightarrow \quad \log A = \log \pi + \log a + \log b \qquad [\because \ \log(uvw) = \log u + \log v + \log w]$$

Differentiating both sides, we have

$$\frac{1}{A}.\, dA = 0 + \frac{1}{a}.\, da + \frac{1}{b}.\, db$$

$$\Rightarrow \quad \frac{dA}{A} = \frac{da}{a} + \frac{db}{b}$$

$$\Rightarrow \quad \frac{dA}{A} \times 100 = \frac{da}{a} \times 100 + \frac{db}{b} \times 100 \quad ...(2) \ \text{[Multiplying both sides by 100]}$$

Now, It is given that the percentage error in measuring the major and minor axes is 1%.

i.e.,

$$\frac{da}{a} \times 100 = 1 \quad \text{and} \quad \frac{db}{b} \times 100 = 1.$$

$\therefore$ Equation (2) becomes

$$\frac{dA}{A} \times 100 = 1 + 1 = 2$$

$\therefore$ Percentage error in Area = 2%.

Example 18. *The pressure p and the volume V of a gas are connected by the relation $pV^{1/4} = a$ (constant). Find the percentage increase in pressure corresponding to diminution of $\frac{1}{2}$% in volume.*

Solution. We have

$$pV^{1/4} = a \quad ...(1)$$

It is given that : percentage decrease in volume *i.e.*,

$$\frac{dV}{V} \times 100 = \frac{-1}{2} \quad ...(2) \quad \text{[–ve sign is due to decrease in volume]}$$

From (1), we have

$$pV^{1/4} = a$$

Taking logarithms on both sides, we get

$$\Rightarrow \log (pV^{1/4}) = \log a$$

$$\Rightarrow \log p + \log V^{1/4} = \log a \quad [\because \log (m.n) = \log m + \log n]$$

$$\Rightarrow \log p + \frac{1}{4} \log V = \log a \quad [\because \log n^m = m \log n]$$

Differentiating both sides, we get

$$\Rightarrow \frac{1}{p} . dp + \frac{1}{4} . \frac{1}{V} dV = 0$$

$$\Rightarrow \frac{dp}{p} + \frac{1}{4} \frac{dV}{V} = 0 \quad ...(3)$$

$$\Rightarrow \frac{dp}{p} \times 100 + \frac{1}{4} \frac{dV}{V} \times 100 = 0 \quad \text{[Multiplying throughout by 100]}$$

$$\Rightarrow \frac{dp}{p} \times 100 + \frac{1}{4}\left(\frac{-1}{2}\right) = 0 \quad [\because \text{By using (2)}]$$

$$\Rightarrow \frac{dp}{p} \times 100 = \frac{1}{8}$$

$\therefore$ Percentage increase in pressure = $\frac{1}{8}$%.

Example 19. *The time T of oscillation of a simple pendulum of length l is given by $T = 2\pi\sqrt{\frac{l}{g}}$.*

Find the percentage error in T corresponding to an error of 2% in the value of l.

Solution. It is given that percentage error in the length l of the pendulum is 2%

i.e.,
$$\frac{dl}{l} \times 100 = 2 \quad ...(1)$$

Also,
$$T = 2\pi \sqrt{\frac{l}{g}}$$

$$\Rightarrow \qquad T = 2\pi \left(\frac{l}{g}\right)^{1/2}$$

Taking logarithms on both sides, we have

$$\log T = \log \left[2\pi \left(\frac{l}{g}\right)^{1/2}\right]$$

$$= \log 2\pi + \log \left(\frac{l}{g}\right)^{1/2} \qquad [\because \ \log (m.n) = \log m + \log n]$$

$$= \log 2\pi + \frac{1}{2} \log \left(\frac{l}{g}\right) \qquad [\because \ \log n^m = m \log n]$$

$$\Rightarrow \qquad \log T = \log 2\pi + \frac{1}{2} [\log l - \log g]$$

Differentiating both sides w.r.t. l, we get

$$\frac{1}{T} dT = 0 + \frac{1}{2}\left[\frac{1}{l} . dl - 0\right] \qquad [\because \ g \text{ is constant}]$$

$$\Rightarrow \qquad \frac{dT}{T} = \frac{1}{2} \frac{dl}{l}$$

$$\Rightarrow \qquad \frac{dT}{T} \times 100 = \frac{1}{2} \frac{dl}{l} \times 100 \qquad [\text{Multiplying both sides by } 100]$$

$$\Rightarrow \qquad \frac{dT}{T} \times 100 = \frac{1}{2} (2) = 1 \qquad [\because \ \text{By using (1)}]$$

$\therefore$ Percentage error in time T of oscillation = 1%.

EXERCISE FOR PRACTICE

1. Use differentials to calculate the approximate value of log 9.01. [Given : log 3 = 1.0986]
2. Find the percentage error in calculating the volume of a cubical box if an error of 1% is made in measuring the length of edge of the cube.
3. If the radius of a sphere increases by 0.01%. Find the percentage increase in the volume.
4. If $y = x^2 + 4$ and x changes from 2 to 2.1, find the approximate change in y.
5. A circular metal plate expands under heating so that its radius increases by 2%. Find the approximate increase in the area of the plate, if the radius of the plate before heating is 10 cm.
6. If x and y are positive variables connected by the implicit equation $xy^2 = 25$. If x is measured as 1 ± 0.001. Find the value of y.
7. The radius of a sphere is 3 cm. If an error of 0.03 cm is made in measuring the radius of the sphere. Find the error and the percentage error in measuring its surface area.

Answers

1. 2.1983 **2.** 3% **3.** 0.03

4. 8.4 **5.** 4π cm^2 **6.** 5 ∓ 0.0025

7. $\pm 0.72\pi$, $\pm 2\%$.

5

Rolle's and Lagrange's Theorems

5.1 INTRODUCTION

In this chapter, we shall study Rolle's Theorem and Lagrange's Theorem. These theorems are the direct applications of derivatives. Rolle's theorem is a special case of mean value theorem. These theorems would also help us to find a point on a curve where the tangent is horizontal or parallel to some chord of the curve.

5.2 ROLLE'S THEOREM

Statement. Let $f(x)$ be a real valued function defined on a closed interval $[a, b]$ such that :

(*i*) It is continuous on the closed interval $[a, b]$,

(*ii*) It is differentiable on the open interval (a, b), and

(*iii*) $f(a) = f(b)$.

Then, there exists at least one point C in the open interval (a, b) such that $f'(c) = 0$.

5.3 GEOMETRICAL INTERPRETATION OF ROLLE'S THEOREM

Let AB be the graph of the function $y = f(x)$ such that A and B are the points of the graph corresponding to $x = a$ and $x = b$ respectively.

(*i*) It is given that $f(x)$ is continuous on the interval $a \leq x \leq b$.

$\Rightarrow$ Graph of $f(x)$ is a continuous curve without any break between the points

$$x = a \quad \text{and} \quad x = b.$$

(*ii*) Also, it is given that, $f(x)$ is derivable on $a < x < b$.

$\Rightarrow$ Graph of $f(x)$ has a unique tangent at each point in the open interval (a, b).

(*iii*) Further, given that $f(a) = f(b)$.

$\Rightarrow$ Ordinates at $x = a$ and $x = b$ are equal. *i.e.*, the ordinates of A and B are equal so that the points A and B are equidistant from the x-axis and on the same side of it.

Graphs of functions satisfying all the above the conditions are as shown in the figures. From the figures, it is evident that there is at least one point P on the curve between A and B, the tangents at which is parallel to x-axis.

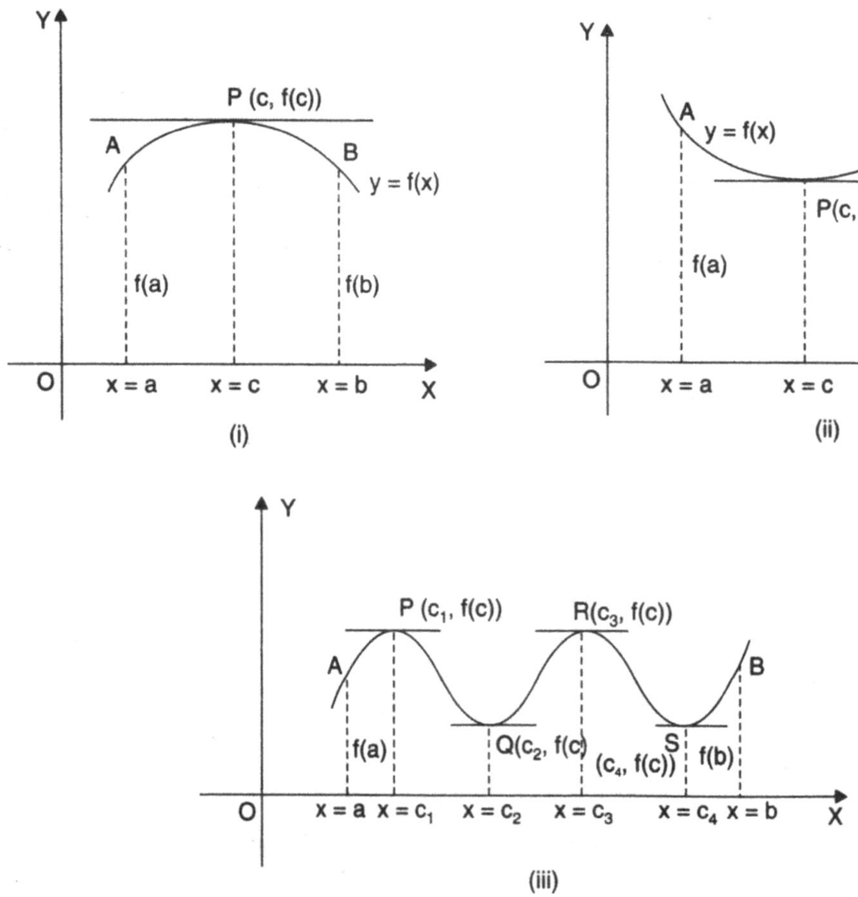

Since, the tangent at P is parallel to x-axis, therefore slope of tangent at P is zero.

If c be the abscissa of P, then $f'(c) = 0$

where $a < c < b$.

Hence, if all the three conditions are satisfied, then Rolle's Theorem confirms the existence of at least one point $c \in (a, b)$, where the tangent is parallel to x-axis.

Note. The converse of Rolle's Theorem is not true. *i.e.*, $f'(x)$ may be zero at a point in (a, b) without satisfying all the three conditions of Rolle's Theorem. *i.e.*, The three conditions of Rolle's Theorem are sufficient but not necessary for $f'(x) = 0$ for some $x \in (a, b)$.

5.3.1 Failure of Rolle's Theorem. Rolle's Theorem fails to hold good for a function which does not satisfy all the three conditions of the theorem. The theorem is not applicable to functions which fails to satisfy even one of the three conditions.

Remember.

(*i*) Every polynomial function is continuous and differentiable on R.

(*ii*) The functions sin x, cos x and e^x are continuous and differentiable on R.

(*iii*) Logarithmic function is continuous and differentiable in its domain.

(*iv*) Every differentiable function is continuous, but the converse is not true.

(*v*) If f and g are continuous (differentiable) functions on $[a, b]$, then the functions $f + g$, $f - g$, fg are also continuous (differentiable) on $[a, b]$. $\frac{f}{g}$ is also continuous (differentiable) on $[a, b]$ provided $g(x) \neq 0$ on $[a, b]$.

(*vi*) The function $|x|$ is not differentiable at $x = 0$.

(*vii*) The function $\tan x$ is not continuous at $x = \pm\frac{\pi}{2}, \pm\frac{3\pi}{2}, \pm\frac{5\pi}{2}$ .....

(*viii*) If $\lim_{x \to k} f'(x)$ is not real and finite, then $f(x)$ is not derivable at $x = k$.

(*ix*) A discontinuous function may not be differentiable.

(*x*) If the derivative of a function has finite and unique value on an interval, then the function is derivable on that interval.

5.4 LAGRANGE'S MEAN VALUE THEOREM

Statement. *Let $f(x)$ be a function defined on a closed interval $[a, b]$ such that*

(*i*) *It is continuous on the closed interval $[a, b]$.*

(*ii*) *It is differentiable on the open interval (a, b).*

Then, there exists at least one $c \in (a, b)$

such that $$\frac{f(b) - f(a)}{b - a} = f'(c).$$

Proof. Let us consider a function $F(x) = f(x) + Ax$, ...(1)

where A is a constant to be chosen such that

$$F(a) = F(b). \quad ...(2)$$

Now, $$F(a) = f(a) + Aa$$

and $$F(b) = f(b) + Ab$$

$\therefore$ $$F(a) = F(b)$$

$\Rightarrow$ $$f(a) + Aa = f(b) + Ab$$

$\Rightarrow$ $$A(a - b) = f(b) - f(a)$$

$\Rightarrow$ $$A = \frac{f(b) - f(a)}{a - b} \quad ...(3)$$

Since,

(*i*) $f(x)$ is continuous on $[a, b]$ and Ax, being a polynomial function, is continuous on $[a, b]$. Therefore, $F(x)$ being the sum of two continuous functions $f(x)$ and Ax, is continuous on $[a, b]$.

(*ii*) As $f(x)$ is differentiable on (a, b) and Ax, being a polynomial function is differentiable on (a, b). Therefore, $F(x)$ is differentiable, being the sum of two differentiable functions $f(x)$ and Ax.

(*iii*) $F(a) = F(b)$ [$\because$ By equation (2)]

$\therefore$ $F(x)$ satisfies all the three conditions of Rolle's Theorem.

$\therefore$ There exists at least one $c \in (a, b)$ such that

$$F'(c) = 0 \quad \text{...(4)}$$

Now, differentiating (1) w.r.t. x, we get

$$F'(x) = f'(x) + A$$

Put $x = c$ $\Rightarrow$ $F'(c) = f'(c) + A$

$\Rightarrow$ $0 = f'(c) + A$ [$\because$ By using (4)]

$\Rightarrow$ $f'(c) = -A$ ...(5)

$\therefore$ From equations (3) and (5), we get

$$f'(c) = -\frac{f(b) - f(a)}{a - b}$$

$$\Rightarrow \quad f'(c) = \frac{f(b) - f(a)}{b - a}$$

which proves the theorem.

5.5 GEOMETRICAL INTERPRETATION OF LAGRANGE'S MEAN VALUE THEOREM

Let A and B be the points on the graph of the function $y = f(x)$ corresponding to

$$x = a \quad \text{and} \quad x = b.$$

Then, co-ordinates of A and B are $(a, f(a))$ and $(b, f(b))$ respectively.

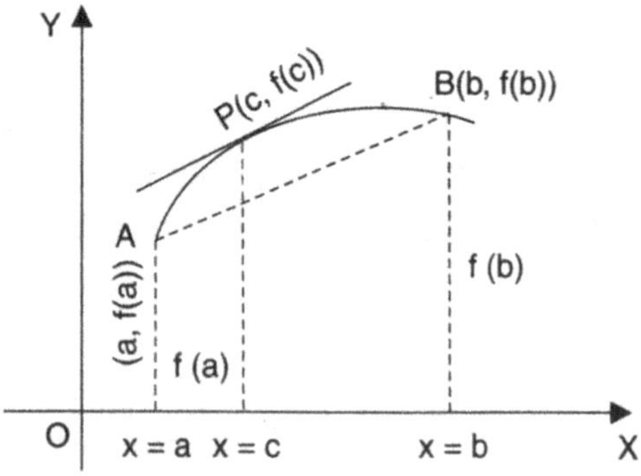

$\therefore$ Slope of chord $AB = \dfrac{f(b) - f(a)}{b - a}$

$$\left[\because \text{ slope} = \frac{y_2 - y_1}{x_2 - x_1}\right]$$

Since, (*i*) $f(x)$ is continuous on the interval $[a, b]$.

$\Rightarrow$ Graph of the function is continuous from A to B.

(*ii*) $f(x)$ is differentiable on the interval (a, b).

$\Rightarrow$ At every point between A and B, the curve has a unique tangent.

$\therefore$ From the figure, it is evident that, there is at least one point P between A and B, the tangent at which is parallel to the chord AB.

If c be the abscissa of P, then the slope of tangent at P is $f'(c)$.

$$\Rightarrow \quad f'(c) = \frac{f(b) - f(a)}{b - a}$$

$$\left[\begin{array}{l}\because \text{ For two parallel lines, having slopes } m_1 \text{ and } m_2. \\ m_1 = m_2 \text{ (i.e., their slopes are equal.)}\end{array}\right]$$

SOME SOLVED EXAMPLES

Example 1. *Verify Rolle's theorem for the following functions :*

(*i*) $x^2 - 8x + 12$ *on* $[2, 6]$ (*ii*) $x^2 - 4x + 3$ *on* $[1, 3]$

(*iii*) $x^2 + x - 6$ *on* $[-3, 2]$ (*iv*) $x^2 - 2x - 3$ *on* $[-1, 3]$.

Solution. (*i*) Let $f(x) = x^2 - 8x + 12$ ...(1)

$f(x)$ being a polynomial function is continuous and derivable for all values of x. Therefore,

(*a*) $f(x)$ is continuous on $[2, 6]$

(*b*) $f(x)$ is derivable on $(2, 6)$

(*c*) $f(2) = (2)^2 - 8(2) + 12$ and $f(6) = (6)^2 - 8(6) + 12$

$= 4 - 16 + 12$ $= 36 - 48 + 12$

$= 0$ $= 0$

$\therefore \quad f(2) = f(6)$.

Thus, $f(x)$ satisfies all the three conditions of Rolle's theorem.

$\therefore$ There exists at least one $c \in (2, 6)$ such that

$$f'(c) = 0$$

Now, differentiating (1) w.r.t. x, we get

$$f'(x) = 2x - 8$$

Put $x = c$, $\Rightarrow$ $f'(c) = 2c - 8$

$\Rightarrow$ $2c - 8 = 0$ $[\because f'(c) = 0]$

$\Rightarrow$ $2c = 8 \Rightarrow c = 4 \in (2, 6)$

Hence, Rolle's theorem is verified.

(*ii*) Let $f(x) = x^2 - 4x + 3$ ...(1)

$f(x)$ being a polynomial function is continuous and derivable for all values of x. Therefore,

(*a*) $f(x)$ is continuous on $[1, 3]$

(*b*) $f(x)$ is derivable on $(1, 3)$

(*c*) $f(1) = (1)^2 - 4(1) + 3$ and $f(3) = (3)^2 - 4(3) + 3$

$= 1 - 4 + 3$ $= 9 - 12 + 3$

$= 0$ $= 0$

$\therefore \quad f(1) = f(3)$

Thus, $f(x)$ satisfies all the three conditions of Rolle's theorem.

$\therefore$ There exists at least one $c \in (1, 3)$ such that

$$f'(c) = 0$$

Now, differentiating (1) w.r.t. x, we get

$$f'(x) = 2x - 4$$

Put $x = c$ $\Rightarrow$ $f'(c) = 2c - 4$

$\Rightarrow$ $2c - 4 = 0$ $[\because f'(c) = 0]$

$\Rightarrow \qquad 2c = 4 \Rightarrow c = 2 \in (1, 3)$

Hence, Rolle's theorem is verified.

(*iii*) Let $\qquad f(x) = x^2 + x - 6 \qquad$ ...(1)

$f(x)$ being a polynomial function is continuous and derivable for all values of x. Therefore,

(*a*) $f(x)$ is continuous on $[-3, 2]$

(*b*) $f(x)$ is derivable on $(-3, 2)$

(*c*) $f(-3) = (-3)^2 + (-3) - 6 = 9 - 3 - 6 = 0$ and $f(2) = (2)^2 + 2 - 6 = 4 + 2 - 6 = 0$

$\therefore \qquad f(-3) = f(2)$

Thus, $f(x)$ satisfies all the three conditions of Rolle's theorem.

$\therefore$ There exists at least one $c \in (-3, 2)$ such that

$$f'(c) = 0$$

Now, differentiating (1) w.r.t. x, we get

$$f'(x) = 2x + 1$$

Put $x = c$, $\Rightarrow \qquad f'(c) = 2c + 1$

$\Rightarrow \qquad 2c + 1 = 0 \qquad [\because f'(c) = 0]$

$$\Rightarrow \qquad 2c = -1 \Rightarrow c = \frac{-1}{2} \in (-3, 2)$$

Hence, Rolle's theorem is verified.

(*iv*) Let $\qquad f(x) = x^2 - 2x - 3 \qquad$ ...(1)

Please try yourself. **[Ans.** $c = 1$.]

Example 2. *Verify Rolle's theorem for the following functions :*

(*i*) $x(x-3)^2$ *on* $[0, 3]$ (*ii*) $\dfrac{8x^2}{3} - 2x$ *on* $\left[0, \dfrac{3}{4}\right]$

(*iii*) $x^2 - x - 6$ *on* $[-2, 3]$ (*iv*) $(x^2 - 1)(x - 2)$ *on* $[-1, 2]$

(*v*) $(x - 2)(x - 3)(x - 4)$ *on* $[2, 4]$.

Solution. (*i*) Let $\qquad f(x) = x(x - 3)^2$

$\Rightarrow \qquad f(x) = x(x^2 - 6x + 9)$

$\Rightarrow \qquad f(x) = x^3 - 6x^2 + 9 \qquad$ ...(1)

$f(x)$ being a polynomial function is continuous and derivable for all values of x. Therefore,

(*a*) $f(x)$ is continuous on $[0, 3]$

(*b*) $f(x)$ is derivable on $(0, 3)$

(*c*) $f(0) = 0$ and $f(3) = (3)^3 - 6(3)^2 + 9(3) = 27 - 54 + 27 = 0$

$\therefore \qquad f(0) = f(3)$

Thus, $f(x)$ satisfies all the three conditions of Rolle's theorem.

$\therefore$ There exists at least one $c \in (0, 3)$ such that

$$f'(c) = 0$$

Now, differentiating (1) w.r.t. x, we get

$$f'(x) = 3x^2 - 12x + 9$$

Put $x = c$, $\Rightarrow$ $f'(c) = 3c^2 - 12c + 9$

$\Rightarrow$ $3c^2 - 12c + 9 = 0$ $[\because\ f'(c) = 0]$

$\Rightarrow$ $c^2 - 4c + 3 = 0$

$\Rightarrow$ $(c - 1)(c - 3) = 0$

$\Rightarrow$ $c = 1, 3$ $[\because\ 3 \notin (1, 3)]$

Hence, Rolle's theorem is verified.

(*ii*) Let $f(x) = \dfrac{8x^2}{3} - 2x$ ...(1)

$f(x)$ being a polynomial function is continuous and derivable for all values of x. Therefore,

(*a*) $f(x)$ is continuous on $\left[0, \dfrac{3}{4}\right]$

(*b*) $f(x)$ is derivable on $\left(0, \dfrac{3}{4}\right)$

(*c*) $f(0) = 0$ and $f\left(\dfrac{3}{4}\right) = \dfrac{8}{3}\left(\dfrac{3}{4}\right)^2 - 2\left(\dfrac{3}{4}\right) = \dfrac{8}{3}\left(\dfrac{9}{16}\right) - \dfrac{3}{2} = \dfrac{3}{2} - \dfrac{3}{2} = 0$

$\therefore$ $f(0) = f\left(\dfrac{3}{4}\right)$

Thus, $f(x)$ satisfies all the three conditions of Rolle's theorem.

$\therefore$ There exists at least one $c \in \left(0, \dfrac{3}{4}\right)$ such that

$$f'(c) = 0$$

Now, differentiating (1) w.r.t. x, we get

$$f'(x) = \frac{16}{3}x - 2$$

Put $x = c$ $\Rightarrow$ $f'(c) = \dfrac{16}{3}c - 2$

$\Rightarrow$ $\dfrac{16}{3}c - 2 = 0$ $[\because\ f'(c) = 0]$

$\Rightarrow$ $\dfrac{16}{3}c = 2 \Rightarrow c = \dfrac{3}{8} \in \left(0, \dfrac{3}{4}\right)$

Hence, Rolle's theorem is verified.

(*iii*) Let $f(x) = x^2 - x - 6$ on $[-2, 3]$

Please try yourself. $\left[\textbf{Ans. } c = \dfrac{1}{2} \in (-2, 3)\right]$

(iv) Let $f(x) = (x^2 - 1)(x - 2)$

$\Rightarrow$ $f(x) = x^3 - 2x^2 - x + 2$ ...(1)

$f(x)$ being a polynomial function is continuous and derivable for all values of x. Therefore,

(a) $f(x)$ is continuous on $[-1, 2]$

(b) $f(x)$ is derivable on $(-1, 2)$

(c) $f(-1) = (-1)^3 - 2(-1)^2 - (-1) + 2$ and $f(2) = (2)^3 - 2(2)^2 - 2 + 2$

$= -1 - 2 + 1 + 2$ $\quad$ $= 8 - 8 - 2 + 2$

$= 0$ $\quad$ $= 0$

$\therefore$ $f(-1) = f(2)$

Thus, $f(x)$ satisfies all the three conditions of Rolle's theorem.

$\therefore$ There exists at least one $c \in (-1, 2)$ such that

$$f'(c) = 0$$

Now, differentiating (1) w.r.t. x, we get

$$f'(x) = 3x^2 - 4x - 1$$

Put $x = c$ $\Rightarrow$ $f'(c) = 3c^2 - 4c - 1$

$\Rightarrow$ $3c^2 - 4c - 1 = 0$ $\quad [\because f'(c) = 0]$

$$\Rightarrow \quad c = \frac{4 \pm \sqrt{16 - 4(3)(-1)}}{2(3)}$$

$$\left[\because \text{ For a quadratic equation } ax^2 + bx + c = 0 \Rightarrow x = \frac{-b \pm \sqrt{b^2 - 4ac}}{2a}\right]$$

$$= \frac{4 \pm \sqrt{28}}{6} = \frac{4 \pm 2\sqrt{7}}{6} = \frac{2(2 \pm \sqrt{7})}{6}$$

$$\Rightarrow \quad c = \frac{2 \pm \sqrt{7}}{3} \in (-1, 2)$$

Hence, Rolle's theorem is verified.

(v) Let $f(x) = (x - 2)(x - 3)(x - 4)$

$\Rightarrow$ $f(x) = (x - 2)(x^2 - 7x + 12)$

$\Rightarrow$ $f(x) = x^3 - 9x^2 + 26x - 24$ ...(1)

Further, proceed yourself as in part (v). $\left[\textbf{Ans. } c = \dfrac{9 \pm \sqrt{3}}{3} \in (2, 4).\right]$

Example 3. *Verify Rolle's theorem for the following functions :*

(i) $\sqrt{16 - x^2}$ *on* $[-4, 4]$ $\quad$ (ii) $\sqrt{1 - x^2}$ *on* $[-1, 1]$

(iii) $\sqrt{4 - x^2}$ *on* $[-2, 2]$.

Solution. (*i*) Let $f(x) = \sqrt{16 - x^2}$ ...(1)

Since, $f(x)$ has a unique and finite value for each x in the closed interval $[-4, 4]$.

(*a*) $\therefore$ $f(x)$ is continuous on $[-4, 4]$.

Now, differentiating (1) w.r.t. x, we get

$$f'(x) = \frac{1}{2\sqrt{16 - x^2}} \cdot \frac{d}{dx}(16 - x^2) = \frac{1}{2\sqrt{16 - x^2}}(-2x)$$

$$\Rightarrow \quad f'(x) = \frac{-x}{\sqrt{16 - x^2}} \quad \text{...(2)}$$

which exists uniquely on $(-4, 4)$.

(*b*) $f(x)$ is derivable on $(-4, 4)$

(*c*) $f(-4) = \sqrt{16 - 16}$ and $f(4) = \sqrt{16 - 16}$

$= 0 \qquad = 0$

$\therefore \quad f(-4) = f(4)$

Thus, $f(x)$ satisfies all the three conditions of Rolle's theorem.

$\therefore$ There exists at least one $c \in (-4, 4)$ such that

$f'(c) = 0$

Put $x = c$ in equation (2), we get

$$f'(c) = \frac{-c}{\sqrt{16 - c^2}}$$

$$\Rightarrow \quad \frac{-c}{\sqrt{16 - c^2}} = 0 \qquad [\because f'(c) = 0]$$

$$\Rightarrow \quad c = 0 \in (-4, 4)$$

Hence, Rolle's theorem is verified.

(*ii*) Let $f(x) = \sqrt{1 - x^2}$ ...(1)

Since, $f(x)$ has a unique and finite value for each x in the closed interval $[-1, 1]$.

(*a*) $\therefore$ $f(x)$ is continuous on $[-1, 1]$.

Now, differentiating (1) w.r.t. x, we get

$$f'(x) = \frac{1}{2\sqrt{1 - x^2}} \cdot \frac{d}{dx}(1 - x^2) = \frac{1}{2\sqrt{1 - x^2}}(-2x)$$

$$\Rightarrow \quad f'(x) = \frac{-x}{\sqrt{1 - x^2}} \quad \text{...(2)}$$

which exists uniquely on $(-1, 1)$.

(*b*) $f(x)$ is derivable on $(-1, 1)$

(*c*) $f(-1) = \sqrt{1 - 1}$ and $f(1) = \sqrt{1 - 1}$

$= 0 \qquad = 0$

$\therefore \quad f(-1) = f(1)$

Thus, $f(x)$ satisfies all the three conditions of Rolle's theorem.

$\therefore$ There exists at least one $c \in (-1, 1)$ such that

$$f'(c) = 0$$

Put $x = c$ in equation (2), we get

$$f'(c) = \frac{-c}{\sqrt{1-c^2}}$$

$$\Rightarrow \quad \frac{-c}{\sqrt{1-c^2}} = 0 \qquad [\because \; f'(c) = 0]$$

$$\Rightarrow \quad c = 0 \in (-1, 1)$$

Hence, Rolle's theorem is verified.

(*iii*) Let $\quad f(x) = \sqrt{4-x^2} \qquad$ ...(1)

Please try yourself. **[Ans.** $c = 0 \in (-2, 2)$.]

Example 4. *Verify Rolle's theorem for the following functions :*

(*i*) $|x^2 - 4|$ *on* $[-1, 1]$ (*ii*) $[x]$ *on* $[-1, 1]$

(*iii*) $\dfrac{x^2 - 3x - 4}{x - 5}$ *on* $[-1, 4]$ (*iv*) $\dfrac{x(x-2)}{x-1}$ *on* $[0, 2]$.

Solution. (*i*) Let $\quad f(x) = |x^2 - 4|$

Now, $\quad x^2 - 4 < 0$ if $x^2 < 4$ *i.e.,* if $x \in (-2, 2)$

$\Rightarrow \quad x^2 - 4 < 0$ on $[-1, 1] \qquad [\because \; [-1, 1]$ lies in $(-2, 2)]$

$\therefore \quad f(x) = -(x^2 - 4)$

$\Rightarrow \quad f(x) = -x^2 + 4 \qquad$ ...(1)

$f(x)$ being a polynomial function is continuous and derivable for all values of x. Therefore,

(*a*) $f(x)$ is continuous on $[-1, 1]$

(*b*) $f(x)$ is derivable on $(-1, 1)$

(*c*) $f(-2) = -(-2)^2 + 4$ and $f(2) = -(2)^2 + 4$

$= -4 + 4 \qquad = -4 + 4$

$= 0 \qquad = 0$

$\therefore \quad f(-2) = f(2)$

Thus, $f(x)$ satisfies all the conditions of Rolle's theorem.

$\therefore$ There exists at least one $c \in (-1, 1)$ such that

$$f'(c) = 0$$

Now, differentiating (1) w.r.t. x, we get

$$f'(x) = -2x$$

Put $x = c \;\Rightarrow\; f'(c) = -2c$

$$\Rightarrow \quad -2c = 0 \qquad [\because \; f'(c) = 0]$$

$\Rightarrow \qquad c = 0 \in (-1, 1)$

Hence, Rolle's theorem is verified.

(*ii*) Let $\qquad f(x) = [x] \qquad \text{...(1)}$

When $x = 0$

$$\text{L.H.L.} = \lim_{x \to 0^-} f(x) = \lim_{x \to 0^-} [x]$$

$$= \lim_{h \to 0} [0 - h] = -1$$

$$\text{R.H.L.} = \lim_{x \to 0^+} f(x) = \lim_{x \to 0^+} [x] = \lim_{h \to 0} [0 + h] = 0$$

$\therefore \qquad \text{L.H.L.} \neq \text{R.H.L.}$

$\therefore$ $f(x)$ is not continuous on $[-1, 1]$.

Thus, the condition of continuity at each point on $[-1, 1]$ is not satisfied.

Hence, Rolle's theorem is not applicable.

(*iii*) Let $\qquad f(x) = \dfrac{x^2 - 3x - 4}{x - 5} \qquad \text{...(1)}$

$f(x)$ being a rational function has a unique and finite value on each $x \in [-1, 4]$, except at $x = 5$.

(*a*) $\therefore$ $f(x)$ is continuous on $[-1, 4]$

Now, differentiating equation (1) w.r.t. x, we get

$$f'(x) = \frac{(x-5) \cdot \dfrac{d}{dx}(x^2 - 3x - 4) - (x^2 - 3x - 4) \cdot \dfrac{d}{dx}(x - 5)}{(x-5)^2}$$

$$= \frac{(x-5)(2x-3) - (x^2 - 3x - 4) \cdot 1}{(x-5)^2}$$

$$= \frac{2x^2 - 13x + 15 - x^2 + 3x + 4}{(x-5)^2}$$

$$= \frac{x^2 - 10x + 19}{(x-5)^2} \qquad \text{...(2)}$$

which exists uniquely on $(-1, 4)$.

(*b*) $\therefore$ $f(x)$ is derivable on $(-1, 4)$.

(*c*) $\qquad f(-1) = \dfrac{(-1)^2 - 3(-1) - 4}{-1 - 5}$ and $f(4) = \dfrac{(4)^2 - 3(4) - 4}{4 - 5}$

$= \dfrac{1 + 3 - 4}{-6} \qquad\qquad = \dfrac{16 - 12 - 4}{-1}$

$= 0 \qquad\qquad = 0$

$\therefore \qquad f(-1) = f(4)$

Thus, $f(x)$ satisfies all the three conditions of Rolle's theorem.

$\therefore$ There exists at least one $c \in (-1, 4)$ such that

$$f'(c) = 0$$

Put $x = c$ in equation (2), we get

$$f'(c) = \frac{c^2 - 10c + 19}{(c-5)^2}$$

$$\Rightarrow \quad \frac{c^2 - 10c + 19}{(c-5)^2} = 0 \qquad [\because f'(c) = 0]$$

$$\Rightarrow \quad c^2 - 10c + 19 = 0$$

$$\Rightarrow \quad c = \frac{10 \pm \sqrt{100 - 76}}{2}$$

$$\left[\begin{array}{l}\because \text{ For a quadratic equation } ax^2 + bx + c = 0 \\ \Rightarrow \ x = \dfrac{-b \pm \sqrt{b^2 - 4ac}}{2a}\end{array}\right]$$

$$\Rightarrow \quad c = \frac{10 \pm \sqrt{24}}{2} = \frac{10 \pm 2\sqrt{6}}{2}$$

$$\Rightarrow \quad c = 5 \pm \sqrt{6} \in (-1, 4)$$

Hence, Rolle's theorem is verified.

(*iv*) Let $\quad f(x) = \dfrac{x(x-2)}{x-1} \qquad ...(1)$

Since, the rational function $f(x)$ is not defined at $x = 1$.

$\therefore$ $f(x)$ is not continuous at $x = 1$.

Thus, the condition of continuity at each point on [0, 2] is not satisfied.

Hence, Rolle's theorem is not applicable.

Example 5. *Examine the applicability of Rolle's theorem for the following functions :*

(*i*) $|x|$ *on* $[-1, 1]$ (*ii*) $|x|$ *on* $[-2, 2]$

(*iii*) x^3 *on* $[-1, 1]$ (*iv*) $\sqrt{x}$ *on* $[-1, 1]$

(*v*) $x^{2/3}$ *on* $[-1, 1]$ (*vi*) $|x - 1|$ *on* $[0, 2]$

(*vii*) $\dfrac{x^2 + 5}{x - 2}$ *on* $[0, 4]$.

Solution. (*i*) Let $\quad f(x) = |x| \qquad ...(1)$

$$\Rightarrow \quad f(x) = \begin{Bmatrix} -x; & x < 0 \\ x; & x \geq 0 \end{Bmatrix}$$

$f(x)$ being a polynomial function is continuous for all values of x, except possibly at $x = 0$.

$\therefore$ When $x = 0$

$$\text{L.H.L.} = \lim_{x \to 0^-} f(x) = \lim_{x \to 0^-} (-x)$$

$$= \lim_{h \to 0} -(0 - h) = \lim_{h \to 0} h = 0$$

$$\text{R.H.L.} = \lim_{x \to 0^+} f(x) = \lim_{x \to 0^+} x$$

$$= \lim_{h \to 0} (0 + h) = \lim_{h \to 0} h = 0$$

$\therefore$ L.H.L. = R.H.L.

$\therefore$ $f(x)$ is continuous at $x = 0$.

(*a*) $f(x)$ is continuous on $[-1, 1]$

Since, $$f(x) = \begin{cases} -x; & -1 \le x \le 0 \\ x; & 0 \le x \le 1 \end{cases}$$

$$f'(x) = \begin{cases} -1; & -1 \le x \le 0 \\ 1; & 0 \le x \le 1 \end{cases}$$

$\therefore$ $L f'(0) \neq R f'(0)$

$\therefore$ $f(x)$ is not derivable at $x = 0$.

(*b*) $f(x)$ is not derivable on $(-1, 1)$

Thus, the condition of derivability at each point of the open interval $(-1, 1)$ is not satisfied. Hence, Rolle's theorem is not applicable.

(*ii*) Let $f(x) = |x|$ ...(1)

Please try yourself. [**Ans.** Rolle's theorem is not applicable.]

[**Hint.** Proceed as in part (*i*).]

(*iii*) Let $f(x) = x^3$ ...(1)

$f(x)$ being a polynomial function is continuous and derivable for all values of x. Therefore,

(*a*) $f(x)$ is continuous on $[-1, 1]$

(*b*) $f(x)$ is derivable on $(-1, 1)$

(*c*) $f(-1) = (-1)^3 = -1$ and $f(1) = (1)^3 = 1$

$\therefore$ $f(-1) \neq f(1)$

Hence, Rolle's theorem is not applicable.

(*iv*) Let $f(x) = \sqrt{x}$ ...(1)

Since, $f(x)$ is not defined on $[-1, 1]$.

$\therefore$ $f(x)$ is not continuous on $[-1, 1]$.

Hence, Rolle's theorem is not applicable.

(*v*) Let $f(x) = x^{2/3}$ ...(1)

(*a*) $f(x)$ is continuous on $[-1, 1]$.

Differentiating equation (1) w.r.t. x, we get

$$f'(x) = \frac{2}{3} x^{\frac{2}{3} - 1} = \frac{2}{3} x^{-1/3} = \frac{2}{3x^{1/3}}$$

$\Rightarrow$ $f'(x)$ is not defined at $x = 0$.

$\therefore$ $f(x)$ is not derivable on $(-1, 1)$.

Hence, Rolle's theorem is not applicable.

(*vi*) Let $f(x) = |x - 1|$ ...(1)

$\Rightarrow$ $f(x) = \begin{cases} 1 - x; & x < 1 \\ x - 1; & x \geq 1 \end{cases}$

$\Rightarrow$ $f'(x) = \begin{cases} -1; & x < 1 \\ 1; & x \geq 1 \end{cases}$

$\Rightarrow$ L.H.D. at $x = 1$ is -1 and R.H.D.at $x = 1$ is 1

$\therefore$ L.H.D. $\neq$ R.H.D. at $x = 1$

$\therefore$ $f(x)$ is not derivable in (0, 2).

Hence, Rolle's theorem is not applicable.

(*vii*) Let $f(x) = \dfrac{x^2 + 5}{x - 2}$ ...(1)

Since, the function $f(x)$ is not defined at $x = 2$.

$\therefore$ $f(x)$ is not continuous at $x = 2 \in [0, 4]$.

Hence, Rolle's theorem is not applicable.

Example 6. *Examine the applicability of Rolle's theorem for the following functions :*

(*i*) $(x - 1)^{2/5}$ *on [0, 3]* (*ii*) $3 + (x - 1)^{2/3}$ *on [0, 3]*

(*iii*) $f(x) = \begin{cases} 2x^2 + 6; & 0 \leq x \leq 2 \\ 5x + 4; & 2 < x \leq 7 \end{cases}$ *on [1, 4]* (*iv*) $f(x) = \tan x$ *on* $[0, \pi]$

(*v*) $f(x) = \begin{cases} x^2 + 1; & 0 \leq x \leq 1 \\ 3 - x; & 1 < x \leq 2 \end{cases}$.

Solution. (*i*) Let $f(x) = (x - 1)^{2/5}$ ...(1)

(*a*) $f(x)$ is continuous on [0, 3]

Differentiating equation (1) w.r.t. x, we get

$$f'(x) = \frac{2}{5}(x - 1)^{\frac{2}{5} - 1}$$

$$= \frac{2}{5}(x - 1)^{-\frac{3}{5}} = \frac{2}{5(x - 1)^{3/5}}$$

$\therefore$ $f'(x)$ is not defined at $x = 1$

$\therefore$ $f(x)$ is not derivable at $x = 1$

$\Rightarrow$ $f(x)$ is not derivable on (0, 3).

$\therefore$ Rolle's theorem is not applicable.

(*ii*) Let $f(x) = 3 + (x - 1)^{2/3}$ ...(1)

(*a*) $f(x)$ is continuous on [0, 3].

Differentiating equation (1) w.r.t. x, we get

$$f'(x) = 0 + \frac{2}{3}(x - 1)^{\frac{2}{3} - 1}$$

$$= \frac{2}{3}(x - 1)^{-\frac{1}{3}} = \frac{2}{3(x - 1)^{1/3}}$$

$\therefore$ $f'(x)$ is not defined at $x = 1$

$\therefore$ $f(x)$ is not derivable at $x = 1$.

$\Rightarrow$ $f(x)$ is not derivable on (0, 3).

Hence, Rolle's theorem is not applicable.

(*iii*) Let
$$f(x) = \begin{cases} 2x^2 + 6; & 0 \le x \le 2 \\ 5x + 4; & 2 < x < 7 \end{cases}$$

$f(x)$ being a polynomial function is continuous and derivable for all values of x. Therefore,

(*a*) $f(x)$ is continuous on [1, 4], except possibly at $x = 2$

$\therefore$ Continuity at $x = 2$

$$\text{L.H.L.} \lim_{x \to 2^-} f(x) = \lim_{x \to 2^-} (2x^2 + 6)$$
$$= 2(2)^2 + 6 = 2(4) + 6 = 8 + 6 = 14.$$
$$\text{R.H.L.} \lim_{x \to 2^+} f(x) = \lim_{x \to 2^+} (5x + 4)$$
$$= 5(2) + 4 = 10 + 4 = 14.$$

$\Rightarrow$ L.H.L = R.H.L.

Also, $f(2) = 2(2)^2 + 6 = 2(4) + 6 = 8 + 6 = 14$.

$\therefore$ $\lim_{x \to 2} f(x) = f(2)$

$\Rightarrow$ $f(x)$ is continuous at $x = 2$

$\therefore$ $f(x)$ is continuous on [1, 4].

(*b*) $f(x)$ is derivable on (1, 4), except possibly at $x = 2$.

$\therefore$ Derivability at $x = 2$:

$$\text{L.H.D.} \lim_{x \to 2^-} \frac{f(x) - f(2)}{x - 2} = \lim_{x \to 2^-} \frac{(2x^2 + 6) - (2(2)^2 + 6)}{x - 2}$$
$$= \lim_{x \to 2^-} \frac{2x^2 + 6 - 8 - 6}{x - 2} = \lim_{x \to 2^-} \frac{2x^2 - 8}{x - 2}$$
$$= \lim_{x \to 2^-} \frac{2(x^2 - 4)}{x - 2} = \lim_{x \to 2^-} \frac{2(x - 2)(x + 2)}{(x - 2)}$$
$$= \lim_{x \to 2^-} 2(x + 2) = 2(2 + 2) = 8.$$

$$\text{R.H.D.} \lim_{x \to 2^+} \frac{f(x) - f(2)}{x - 2} = \lim_{x \to 2^+} \frac{(5x + 4) - (5(2) + 4)}{x - 2}$$
$$= \lim_{x \to 2^+} \frac{5x + 4 - 14}{x - 2} = \lim_{x \to 2^+} \frac{5x - 10}{x - 2} = \lim_{x \to 2^+} \frac{5(x - 2)}{(x - 2)}$$
$$= \lim_{x \to 2^+} 5(1) = 5.$$

$\therefore$ L.H.D. $\neq$ R.H.D.

$\therefore$ $f(x)$ is not derivable at $x = 2$.

$\Rightarrow$ $f(x)$ is not derivable on (1, 4).

Hence, Rolle's theorem is not applicable.

(*iv*) Let $f(x) = \tan x$ ...(1)

Since $f(x)$ does not have a definite value at

$$x = \frac{\pi}{2} \in [0, \pi].$$

$\therefore$ $f(x)$ is not continuous at $x = \frac{\pi}{2} \in \left[0, \frac{\pi}{2}\right]$.

Hence, Rolle's theorem is not applicable.

(*v*) Let $f(x) = \begin{cases} x^2 + 1; & 0 \le x \le 1 \\ 3 - x; & 1 < x \le 2 \end{cases}$

$f(x)$ being a polynomial function is continuous and derivable for all values of x, except possibly at $x = 1$.

$\therefore$ Continuity at $x = 1$:

$$\text{L.H.L. } \lim_{x \to 1^-} f(x) = \lim_{x \to 1^-} (x^2 + 1) = (1)^2 + 1 = 2.$$

$$\text{R.H.L. } \lim_{x \to 1^+} f(x) = \lim_{x \to 1^+} (3 - x) = 3 - 1 = 2$$

$\therefore$ L.H.L = R.H.L.

Also, $f(1) = (1)^2 + 1 = 2$

$\therefore$ $\lim_{x \to 1} f(x) = f(1) = 2$

$\therefore$ $f(x)$ is continuous at $x = 1$.

Derivability at $x = 1$:

$$\text{L.H.D. } \lim_{x \to 1^-} \frac{f(x) - f(1)}{x - 1} = \lim_{x \to 1^-} \frac{(x^2 + 1) - (1 + 1)}{(x - 1)}$$

$$= \lim_{x \to 1^-} \frac{x^2 + 1 - 2}{x - 1} = \lim_{x \to 1^-} \frac{x^2 - 1}{x - 1} = \lim_{x \to 1^-} \frac{(x - 1)(x + 1)}{(x - 1)}$$

$$= \lim_{x \to 1^-} (x + 1) = 1 + 1 = 2.$$

$$\text{R.H.D. } \lim_{x \to 1^+} \frac{f(x) - f(1)}{x - 1} = \lim_{x \to 1^+} \frac{(3 - x) - (3 - 1)}{(x - 1)}$$

$$= \lim_{x \to 1^+} \frac{3 - x - 2}{x - 1} = \lim_{x \to 1^+} \frac{-x + 1}{x - 1} = \lim_{x \to 1^+} \frac{-(x - 1)}{x - 1}$$

$$= -1$$

$\therefore$ L.H.D. $\neq$ R.H.D.

$\therefore$ $f(x)$ is not differentiable at $x = 1$.

Hence, Rolle's theorem is not applicable.

Example 7. *Verify Rolle's theorem for the function*

$$f(x) = 2x^3 + x^2 - 4x - 2.$$

Solution. We have, $f(x) = 2x^3 + x^2 - 4x - 2$ ...(1)

$f(x)$ being a polynomial function is continuous and derivable for all values of x.

Since, the interval for the verification of Rolle's theorem is not given.

$\therefore$ Let $f(x) = 0$

$\Rightarrow$ $2x^3 + x^2 - 4x - 2 = 0$

$\Rightarrow$ $x^2 (2x + 1) - 2(2x + 1) = 0$

$\Rightarrow$ $(2x + 1)(x^2 - 2) = 0$

$\Rightarrow$ $x = -\dfrac{1}{2}$ or $x^2 = 2$

$\Rightarrow$ $x = -\dfrac{1}{2}$ or $x = \pm\sqrt{2}$

$\therefore$ We consider the interval $[-\sqrt{2}, \sqrt{2}]$.

(*i*) $f(x)$ is continuous on $[-\sqrt{2}, \sqrt{2}]$

(*ii*) $f(x)$ is derivable on $(-\sqrt{2}, \sqrt{2})$

(*iii*) $f(-\sqrt{2}) = 2(-\sqrt{2})^3 + (-\sqrt{2})^2 - 4(-\sqrt{2}) - 2$

$= 2(-2\sqrt{2}) + 2 + 4\sqrt{2} - 2$

$= -4\sqrt{2} + 2 + 4\sqrt{2} - 2 = 0$

and $f(\sqrt{2}) = 2(\sqrt{2})^3 + (\sqrt{2})^2 - 4(\sqrt{2}) - 2$

$= 2(2\sqrt{2}) + 2 - 4\sqrt{2} - 2$

$= 4\sqrt{2} + 2 - 4\sqrt{2} - 2 = 0$

$\therefore$ $f(-\sqrt{2}) = f(\sqrt{2})$

Thus, $f(x)$ satisfies all the three conditions of Rolle's theorem.

$\therefore$ There exists at least one $c \in (-\sqrt{2}, \sqrt{2})$ such that

$$f'(c) = 0$$

Now, differentiating (1) w.r.t. x, we get

$$f(x) = 6x^2 + 2x - 4$$

Put $x = c$ $\Rightarrow$ $f'(c) = 6c^2 + 2c - 4$

$\Rightarrow$ $6c^2 + 2c - 4 = 0$ $[\because f'(c) = 0]$

$\Rightarrow$ $3c^2 + c - 2 = 0$

$\Rightarrow$ $(3c - 2)(c + 1) = 0$

$\Rightarrow$ $c = \dfrac{2}{3}, -1 \in (-\sqrt{2}, \sqrt{2})$.

Hence, Rolle's theorem is verified.

Example 8. (*i*) *It is given that for the function $f(x) = x^3 - 6x^2 + ax + b$ on [1, 3], Rolle's theorem holds with $c = 2 + \frac{1}{\sqrt{3}}$. Find the values of a and b.*

(ii) It is given that for the function $f(x) = x^3 + bx^2 + ax$ on [1, 3], Rolle's theorem holds with $c = 2 + \frac{1}{\sqrt{3}}$. Find the values of a and b.

Solution. (*i*) We have,

$$f(x) = x^3 - 6x^2 + ax + b \qquad ...(1)$$

$f(x)$ being a polynomial function is continuous and derivable for all values of x. Therefore,

(*a*) $f(x)$ is continuous on [1, 3]

(*b*) $f(x)$ is derivable on (1, 3)

(*c*) $f(1) = (1)^3 - 6(1)^2 + a(1) + b$ and $f(3) = (3)^3 - 6(3)^2 + a(3) + b$

$= 1 - 6 + a + b$ $\qquad$ $= 27 - 54 + 3a + b$

$\Rightarrow f(1) = a + b - 5$ $\qquad$ $\Rightarrow f(3) = 3a + b - 27$

Since, Rolle's theorem holds for $f(x)$ on [1, 3].

$$\therefore \quad f(1) = f(3)$$

$$\Rightarrow \quad a + b - 5 = 3a + b - 27$$

$$\Rightarrow \quad a - 3a = -27 + 5$$

$$\Rightarrow \quad -2a = -22$$

$$\Rightarrow \quad a = 11$$

Now, differentiating equation (1) w.r.t. x, we get

$$f'(x) = 3x^2 - 12x + a \qquad [\because \ a = 11]$$

$$\Rightarrow \quad f'(x) = 3x^2 - 12x + 11 \qquad ...(2)$$

It is given that, Rolle's theorem holds with $c = \left(2 + \frac{1}{\sqrt{3}}\right)$

$$\Rightarrow \quad f'(c) = 0$$

Put $x = c$ in equation (2), we get

$$f'(c) = 3c^2 - 12c + 11$$

$$= 3\left(2 + \frac{1}{\sqrt{3}}\right)^2 - 12\left(2 + \frac{1}{\sqrt{3}}\right) + 11$$

$$= 3\left(4 + \frac{1}{3} + \frac{4}{\sqrt{3}}\right) - 24 - \frac{12}{\sqrt{3}} + 11 \qquad [\because \ \text{This is independent of } b]$$

$$= 12 + 1 + 4\sqrt{3} - 24 - 4\sqrt{3} + 11 = 0$$

$\therefore$ The value of a is 11 and b is arbitrary.

(*ii*) We have,

$$f(x) = x^3 + bx^2 + ax \qquad ...(1)$$

$f(x)$ being a polynomial function is continuous and derivable for all values of x. Therefore, (*a*) $f(x)$ is continuous on [1, 3]

(*b*) $f(x)$ is derivable on (1, 3)

(*c*) $f(1) = (1)^3 + b(1)^2 + a(1)$ and $f(3) = (3)^3 + b(3)^2 + a(3)$

$= 1 + b + a$ $\qquad$ $= 27 + 9b + 3a$

$\Rightarrow f(1) = a + b + 1$ $\qquad$ $\Rightarrow f(3) = 3a + 9b + 27$

Since, Rolle's theorem holds for $f(x)$ on [1, 3].

$$\therefore \quad f(1) = f(3)$$

$$\Rightarrow \quad a + b + 1 = 3a + 9b + 27$$

$$\Rightarrow \quad 2a + 8b + 26 = 0$$

$$\Rightarrow \quad a + 4b + 13 = 0 \qquad ...(2)$$

Now, differentiating equation (1) w.r.t. x, we get

$$f'(x) = 3x^2 + 2bx + a \qquad ...(3)$$

It is given that, Rolle's theorem holds with $c = \left(2 + \frac{1}{\sqrt{3}}\right)$

$$\therefore \quad f'(c) = 0$$

Put $x = c$ in equation (3), we get

$$f'(c) = 3c^2 + 2bc + a$$

$$\Rightarrow \quad 3\left(2 + \frac{1}{\sqrt{3}}\right)^2 + 2b\left(2 + \frac{1}{\sqrt{3}}\right) + a = 0 \qquad [\because f'(c) = 0]$$

$$\Rightarrow \quad 3\left(4 + \frac{1}{3} + \frac{4}{\sqrt{3}}\right) + 4b + \frac{2b}{\sqrt{3}} + a = 0$$

$$\Rightarrow \quad 12 + 1 + 4\sqrt{3} + 4b + \frac{2b}{\sqrt{3}} + a = 0$$

$$\Rightarrow \quad a + 4b + 13 + 4\sqrt{3} + \frac{2b}{\sqrt{3}} = 0 \qquad [\because \text{ By using (2)}]$$

$$\Rightarrow \quad 0 + 4\sqrt{3} + \frac{2b}{\sqrt{3}} = 0$$

$$\Rightarrow \quad \frac{2b}{\sqrt{3}} = -4\sqrt{3} \quad \Rightarrow \quad b = -6.$$

$\therefore$ From equation (2), we have

$$a + 4(-6) + 13 = 0 \qquad [\because b = -6]$$

$$\Rightarrow \quad a - 24 + 13 = 0$$

$$\Rightarrow \quad a - 11 = 0$$

$$\Rightarrow \quad a = 11$$

Hence, $\quad a = 11$ and $b = -6$.

Example 9. *Show that $f(x) = (x - a)^m (x - b)^n$, m, n being positive integers, satisfy Rolle's theorem on [a, b].*

Solution. We have,

$$f(x) = (x - a)^m (x - b)^n \quad ...(1)$$

Since, m, n are (+ve) integers. On expanding $(x - a)^m$ and $(x - b)^n$ by binomial theorem and then taking the product, we find that $f(x)$ is a polynomial of degree $(m + n)$.

$\therefore$ $f(x)$ being a polynomial function is continuous and derivable for all values of x. Therefore,

(*i*) $f(x)$ is continuous on $[a, b]$

(*ii*) $f(x)$ is derivable on (a, b)

(*iii*) $f(a) = (a - a)^m (a - b)^n = 0$ and $f(b) = (b - a)^m (b - b)^n = 0$

$\therefore$ $f(a) = f(b)$.

Thus, $f(x)$ satisfies all the three conditions of Rolle's theorem.

$\therefore$ There exists at least one $c \in (a, b)$ such that

$$f'(c) = 0$$

Now, differentiating (1) w.r.t. x, we get

$$\begin{aligned} f'(x) &= (x-a)^m \cdot \frac{d}{dx}(x-b)^n + (x-b)^n \cdot \frac{d}{dx}(x-a)^m \\ &= (x-a)^m \cdot n(x-b)^{n-1} + (x-b)^n \cdot m(x-a)^{m-1} \\ &= (x-a)^{m-1}(x-b)^{n-1}[n(x-a) + m(x-b)] \\ &= (x-a)^{m-1}(x-b)^{n-1}[nx - an + mx - bm] \\ &= (x-a)^{m-1}(x-b)^{n-1}[(m+n)x - (an+bm)] \end{aligned}$$

Put $x = c$

$\Rightarrow$ $f'(c) = (c-a)^{m-1}(c-b)^{n-1}[(m+n)c - (an+bm)]$

$\Rightarrow$ $(c-a)^{m-1}(c-b)^{n-1}[(m+n)c - (an+bm)] = 0$ $\quad [\because f'(c) = 0]$

$\Rightarrow$ $(m+n)c - (an+bm) = 0$ $\quad [\because c \neq a, c \neq b]$

$\Rightarrow$ $(m+n)c = an + bm$

$\Rightarrow$ $c = \dfrac{an+bm}{m+n} \in (a, b).$ $\quad \left[\because c = \left(\dfrac{an+bm}{m+n}\right) \text{ divides } (a, b) \text{ internally in the ratio } m : n\right]$

Hence, Rolle's theorem is verified.

Example 10. *Verify Rolle's theorem for the following functions :*

(*i*) $f(x) = x(x + 3) \cdot e^{-x/2}$ *on* $[-3, 0]$

(*ii*) $f(x) = e^{1-x^2}$ *on* $[-1, 1]$.

Solution. (*i*) We have,

$$f(x) = x(x + 3)e^{-x/2} \quad ...(1)$$

$f(x)$, being the product of three continuous functions is also continuous for all values of x. Therefore,

(*a*) $f(x)$ is continuous on $[-3, 0]$

(*b*) $f(x)$ is derivable on $(-3, 0)$

(c) $f(-3) = (-3)[-3+3]\, e^{-\left(-\frac{3}{2}\right)}$ and $f(0) = 0\,(0+3)\, e^0$

$= 0$ $\qquad\qquad$ $= 0$

$\therefore \qquad f(-3) = f(0)$

Thus, $f(x)$ satisfies all the three conditions of Rolle's theorem.

$\therefore$ There exists at least one $c \in (-3, 0)$ such that

$$f'(c) = 0$$

Now, differentiating equation (1) w.r.t. x, we get

$$f'(x) = (x^2 + 3x) \,.\, e^{-x/2} \,.\, \frac{d}{dx}\left(\frac{-x}{2}\right) + e^{-x/2}(2x+3)$$

$$= (x^2 + 3x) \,.\, e^{-x/2} \,.\, \left(\frac{-1}{2}\right) + e^{-x/2}(2x+3)$$

$$= e^{-x/2}\left[-\frac{x^2+3x}{2} + \frac{2x+3}{1}\right] = e^{-x/2}\left[\frac{-x^2-3x+4x+6}{2}\right]$$

$$= \frac{-x^2+x+6}{2} \,.\, e^{-x/2} \qquad \text{...(2)}$$

Put $x = c$, $\Rightarrow \quad f'(c) = \dfrac{-c^2+c+6}{2} \,.\, e^{-c/2} \qquad [\because f'(c) = 0]$

$\Rightarrow \quad \dfrac{-c^2+c+6}{2}\, e^{-c/2} = 0$

$\Rightarrow \quad -c^2 + c + 6 = 0$

$\Rightarrow \quad c^2 - c - 6 = 0$

$\Rightarrow \quad (c-3)(c+2) = 0$

$\Rightarrow \quad c = -2, 3$

$\Rightarrow \quad c = -2 \in (-3, 0)$

Hence, Rolle's theorem is verified.

(*ii*) We have,

$$f(x) = e^{1-x^2} \qquad \text{....(1)}$$

$f(x)$ being an exponential function is continuous and derivable for all values of x. Therefore,

(*a*) $f(x)$ is continuous on $[-1, 1]$

(*b*) $f(x)$ is derivable on $(-1, 1)$

(*c*) $f(-1) = e^{1-(-1)^2}$ and $f(1) = e^{1-(1)^2}$

$= e^{1-1} = e^0$ $\qquad\qquad$ $= e^{1-1} = e^0$

$= 1$ $\qquad\qquad$ $= 1$

$\therefore \qquad f(-1) = f(1)$

Thus, $f(x)$ satisfies all the three conditions of Rolle's theorem.

$\therefore$ There exists at least one $c \in (-1, 1)$ such that

$$f'(c) = 0$$

Now, differentiating equation (1) w.r.t. x, we get

$$f'(x) = e^{1-x^2} \cdot \frac{d}{dx}(1 - x^2) = e^{1-x^2} \cdot (0 - 2x)$$

$\Rightarrow$ $$f'(x) = -2xe^{1-x^2}$$

Put $x = c$

$\Rightarrow$ $$f'(c) = -2ce^{1-c^2} \qquad [\therefore \ f'(c) = 0]$$

$\Rightarrow$ $$-2ce^{1-c^2} = 0$$

$\Rightarrow$ $$c = 0 \qquad [\because \ e^{1-c^2} \neq 0]$$

$\Rightarrow$ $$c = 0 \in (-1, 1)$$

Hence, Rolle's theorem is verified.

Example 11. *Verify Rolle's theorem for the following functions :*

(*i*) *cos* $2x$ *on* $\left[\frac{-\pi}{4}, \frac{\pi}{4}\right]$ (*ii*) *sin* $2x$ *on* $\left[0, \frac{\pi}{2}\right]$

(*iii*) *sin* $2x - 2$ *sin* x *on* $[0, \pi]$ (*iv*) *sin* x – *sin* $2x$ *on* $[0, \pi]$.

Solution. (*i*) Let $f(x) = \cos 2x$ ...(1)

Since, cosine function is continuous for all values of x. Therefore,

(*a*) $f(x)$ is continuous on $\left[\frac{-\pi}{4}, \frac{\pi}{4}\right]$

Differentiating equation (1) w.r.t. x, we get

$$f'(x) = -2 \sin 2x \qquad ...(2)$$

$\therefore$ $f'(x)$ exists uniquely on $\left(\frac{-\pi}{4}, \frac{\pi}{4}\right)$.

(*b*) $\therefore$ $f(x)$ is derivable on $\left(\frac{-\pi}{4}, \frac{\pi}{4}\right)$.

(*c*) $f\left(\frac{-\pi}{4}\right) = \cos 2\left(\frac{-\pi}{4}\right)$ and $f\left(\frac{\pi}{4}\right) = \cos 2\left(\frac{\pi}{4}\right)$

$= \cos\left(-\frac{\pi}{2}\right)$ $\qquad = \cos\frac{\pi}{2}$ $\qquad [\because \ \cos(-\theta) = \cos\theta]$

$= \cos\left(\frac{\pi}{2}\right)$ $\qquad = 0$

$= 0$

$\Rightarrow$ $$f\left(\frac{-\pi}{4}\right) = f\left(\frac{\pi}{4}\right).$$

Thus, $f(x)$ satisfies all the conditions of Rolle's theorem.

$\therefore$ There exists at least one $c \in \left(\frac{-\pi}{4}, \frac{\pi}{4}\right)$ such that

$$f'(c) = 0$$

Put $x = c$ in equation (2), we get

$$f'(c) = -2 \sin 2c$$

$$\Rightarrow \quad -2 \sin 2c = 0 \qquad [\because f'(c) = 0]$$

$$\Rightarrow \quad \sin 2c = 0$$

$$\Rightarrow \quad c = 0 \in \left(\frac{-\pi}{4}, \frac{\pi}{4}\right).$$

Hence, Rolle's theorem is verified.

(*ii*) Let $\quad f(x) = \sin 2x \qquad \ldots(1)$

Since, sine function is continuous for all values of x. Therefore,

(*a*) $f(x)$ is continuous on $\left[0, \frac{\pi}{2}\right]$

Differentiating equation (1) w.r.t. x, we get

$$f'(x) = 2 \cos 2x \qquad \ldots(2)$$

$\therefore$ $f'(x)$ exists uniquely on $\left(0, \frac{\pi}{2}\right)$

(*b*) $f(x)$ is derivable on $\left(0, \frac{\pi}{2}\right)$.

(*c*) $f(0) = \sin 2(0)$ and $f\left(\frac{\pi}{2}\right) = \sin 2\left(\frac{\pi}{2}\right)$

$= \sin 0 \qquad = \sin \pi$

$= 0 \qquad = 0$

$$\therefore \quad f(0) = f\left(\frac{\pi}{2}\right)$$

Thus, $f(x)$ satisfies all the conditions of Rolle's theorem.

$\therefore$ There exists at least one $c \in \left(0, \frac{\pi}{2}\right)$ such that

$$f'(c) = 0$$

Put $x = c$ in equation (2), we get

$$f'(c) = 2 \cos 2c$$

$$\Rightarrow \quad 2 \cos 2c = 0 \qquad [\because f'(c) = 0]$$

$$\Rightarrow \quad \cos 2c = 0$$

$$\Rightarrow \quad 2c = \frac{\pi}{2}$$

$$\Rightarrow \quad c = \frac{\pi}{4} \in \left(0, \frac{\pi}{2}\right)$$

Hence, Rolle's theorem is verified.

(*iii*) Let $f(x) = \sin 2x - 2 \sin x$ ...(1)

Since, sine function is continuous for all values of x.

$\therefore$ $\sin 2x$ and $2 \sin x$ are both continuous functions for all values of x.

$f(x)$ being the difference of two continuous functions is also continuous for all values of x. Therefore,

(*a*) $f(x)$ is continuous on $[0, \pi]$

Differentiating equation (1) w.r.t. x, we get

$$f'(x) = 2 \cos 2x - 2 \cos x \quad ...(2)$$

$\therefore$ $f'(x)$ exists uniquely on $(0, \pi)$

(*b*) $\therefore$ $f(x)$ is derivable on $(0, \pi)$.

(*c*) $f(0) = \sin 0 - 2 \sin 0$ and $f(\pi) = \sin 2\pi - 2 \sin \pi$

$= 0 - 0$ $\qquad$ $= 0 - 0$

$= 0$ $\qquad$ $= 0$

$$\therefore \quad f(0) = f(\pi)$$

Thus, $f(x)$ satisfies all the conditions of Rolle's theorem.

$\therefore$ There exists at least one $c \in (0, \pi)$ such that

$$f'(c) = 0$$

Put $x = c$ in equation (2), we get

$$f'(c) = 2 \cos 2c - 2 \cos c$$

$$\Rightarrow \quad 2 \cos 2c - 2 \cos c = 0 \qquad [\because f'(c) = 0]$$

$$\Rightarrow \quad \cos 2c - \cos c = 0$$

$$\Rightarrow \quad 2 \cos^2 c - 1 - \cos c = 0$$

$$\Rightarrow \quad 2 \cos^2 c - \cos c - 1 = 0 \qquad [\Rightarrow \cos 2A = 2 \cos^2 A - 1]$$

$$\Rightarrow \quad \cos c = \frac{1 \pm \sqrt{1+8}}{4} = \frac{1 \pm \sqrt{9}}{4} = \frac{1 \pm 3}{4} \qquad \left[\begin{array}{l} \because \text{For quadratic equation } ax^2 + bx + c = 0 \\ \Rightarrow x = \dfrac{-b \pm \sqrt{b^2 - 4ac}}{2a} \end{array}\right]$$

$$\Rightarrow \quad \cos c = 1, -\frac{1}{2}$$

$$\Rightarrow \quad \cos c = 1 \quad \text{or} \quad \cos c = -\frac{1}{2}$$

$$\Rightarrow \quad \cos c = 1 = \cos 0 \quad \text{or} \quad \cos c = \frac{-1}{2} = \cos \frac{2\pi}{3}$$

$$\Rightarrow \quad c = 0 \qquad \Rightarrow c = \frac{2\pi}{3}$$

$$\therefore \quad c = \frac{2\pi}{3} \in (0, \pi)$$

Hence, Rolle's theorem is verified.

(*iv*) Let $f(x) = \sin x - \sin 2x$ ...(1)

Since, sine function is continuous for all values of x.

$\therefore$ $\sin x$ and $\sin 2x$ are both continuous functions for all values of x.

$f(x)$ being the difference of two continuous functions is also continuous for all values of x.

Therefore,

(*a*) $f(x)$ is continuous on $[0, \pi]$

Differentiating equation (1) w.r.t. x, we get

$$f'(x) = \cos x - 2\cos 2x \quad ...(2)$$

$\therefore$ $f'(x)$ exists uniquely on $(0, \pi)$

(*b*) $f(x)$ is derivable on $(0, \pi)$

(*c*) $f(0) = \sin 0° - \sin 0°$ and $f(\pi) = \sin \pi - \sin 2\pi$

$= 0 - 0$ $\qquad$ $= 0 - 0$

$= 0$ $\qquad$ $= 0$

$\therefore$ $f(0) = f(\pi)$

Thus, $f(x)$ satisfies all the conditions of Rolle's theorem.

$\therefore$ There exists at least one $c \in (0, \pi)$ such that

$$f'(c) = 0$$

Put $x = c$ in equation (2), we get

$$f'(c) = \cos c - 2\cos 2c$$

$\Rightarrow$ $\cos c - 2\cos 2c = 0$ $\qquad [\because f'(c) = 0]$

$[\because \cos 2A = 2\cos^2 A - 1]$

$\Rightarrow$ $\cos c - 2(2\cos^2 c - 1) = 0$

$\Rightarrow$ $\cos c - 4\cos^2 c + 2 = 0$

$\Rightarrow$ $4\cos^2 c - \cos c - 2 = 0$

$\Rightarrow$ $\cos c = \dfrac{1 \pm \sqrt{1 + 32}}{8}$ $\qquad \left[\begin{array}{l}\because \text{ For a quadratic equation} \\ ax^2 + bx + c = 0 \\ \Rightarrow x = \dfrac{-b \pm \sqrt{b^2 - 4ac}}{2a}\end{array}\right]$

$\Rightarrow$ $\cos c = \dfrac{1 \pm \sqrt{33}}{8}$

$\Rightarrow$ $\cos c = \dfrac{1 + \sqrt{33}}{8}$, $\qquad \cos c = \dfrac{1 - \sqrt{33}}{8}$

$\Rightarrow$ $\cos c = \dfrac{1 + 5.7445}{8}$, $\qquad \cos c = \dfrac{1 - 5.7445}{8}$

$\Rightarrow$ $\cos c = \dfrac{6.7445}{8}$, $\qquad \cos c = \dfrac{-4.7445}{8}$

$\Rightarrow$ $\cos c = 0.8431$, $\qquad \cos c = -0.5931$

$\Rightarrow$ $\cos c = \cos 32° 32'$, $\qquad -\cos c = 0.5931$

$\qquad$, $\cos(180° - c) = \cos 53° 37'$

$\Rightarrow \quad c = 32° \, 32' \in (0, \pi) \quad , \quad 180° - c = 53° \, 37'$

$\quad , \quad c = 126° \, 33' \in (0, \pi)$

Hence, Rolle's theorem is verified.

Example 12. *Verify Rolle's theorem for the following functions :*

(*i*) $\cos x + \sin x$ on $[0, 2\pi]$ (*ii*) $\sin x + \cos x - 1$ on $\left[0, \frac{\pi}{2}\right]$

(*iii*) $\sin^2 x$ on $[0, \pi]$ (*iv*) $\sin^4 x + \cos^4 x$ on $\left[0, \frac{\pi}{2}\right]$

(*v*) $e^x \cos x$ on $\left[-\frac{\pi}{2}, \frac{\pi}{2}\right]$ (*vi*) $\frac{\sin x}{e^x}$ on $[0, \pi]$.

Solution. (*i*) Let $f(x) = \cos x + \sin x$ ...(1)

Since, cosine function and sine function are continuous for all values of x.

$\therefore$ $f(x)$ being the sum of two continuous functions is also continuous for all values of x.

Therefore,

(*a*) $f(x)$ is continuous on $[0, 2\pi]$

Differentiating equation (1) w.r.t. x, we get

$$f'(x) = -\sin x + \cos x \quad ...(2)$$

$\therefore$ $f'(x)$ exists uniquely on $(0, 2\pi)$.

(*b*) $\therefore$ $f(x)$ is derivable on $(0, 2\pi)$.

(*c*) $f(0) = \cos 0° + \sin 0°$ and $f(2\pi) = \cos 2\pi + \sin 2\pi$

$= 1 + 0 \qquad = 1 + 0$

$= 1 \qquad = 1$

$$\therefore \quad f(0) = f(2\pi)$$

Thus, $f(x)$ satisfies all the conditions of Rolle's theorem.

$\therefore$ There exists at least one $c \in (0, 2\pi)$ such that

$$f'(c) = 0$$

Put $x = c$ in (2), we get

$$f'(c) = -\sin c + \cos c$$

$\Rightarrow \quad -\sin c + \cos c = 0 \qquad [\because f'(c) = 0]$

$\Rightarrow \quad \sin c = \cos c$

$\Rightarrow \quad \tan c = 1$

$\Rightarrow \quad c = \frac{\pi}{4}, \frac{5\pi}{4} \in (0, 2\pi)$

Hence, Rolle's theorem is verified.

(*ii*) Let $f(x) = \sin x + \cos x - 1$ ...(1)

Please try yourself.

[**Hint.** Proceed as in part (*i*)] $\left[\textbf{Ans. } c = \frac{\pi}{4} \in \left(0, \frac{\pi}{2}\right)\right]$

(*iii*) Let $f(x) = \sin^2 x$ ...(1)

Since, sine function is continuous for all values of x. Therefore,

(a) $f(x)$ is continuous on $[0, \pi]$

Differentiating equation (1) w.r.t. x, we get

$$f'(x) = 2 \sin x \cos x$$

$$\Rightarrow \quad f'(x) = \sin 2x \qquad [\because \; 2 \sin A \cos A = \sin 2A]$$

$\therefore$ $f'(x)$ exists uniquely on $(0, \pi)$

(b) $f(x)$ is derivable on $(0, \pi)$

(c) $f(0) = \sin^2 0° = 0$ and $f(\pi) = \sin^2 \pi = 0$

$$\therefore \quad f(0) = f(\pi)$$

Thus, $f(x)$ satisfies all the conditions of Rolle's theorem.

$\therefore$ There exists at least one $c \in (0, \pi)$ such that

$$f'(c) = 0$$

Put $x = c$ in equation (2), we get

$$f'(c) = \sin 2c$$

$$\Rightarrow \quad \sin 2c = 0 \qquad [\because \; f'(c) = 0]$$

$$\Rightarrow \quad 2c = \pi \;\Rightarrow\; c = \frac{\pi}{2} \in (0, \pi)$$

Hence, Rolle's theorem is verified.

(iv) Let $\quad f(x) = \sin^4 x + \cos^4 x \qquad ...(1)$

$$= \sin^4 x + \cos^4 x + 2 \sin^2 x \cos^2 x - 2 \sin^2 x \cos^2 x$$

[Add and subtract $2 \sin^2 x \cos^2 x$]

$$= (\sin^2 x + \cos^2 x)^2 - 2 \sin^2 x \cos^2 x$$

$$= 1 - 2 \sin^2 x \cos^2 x \qquad [\because \; \sin^2 A + \cos^2 A = 1]$$

$$= 1 - \frac{1}{2}(2 \sin x \cos x)^2$$

$$= 1 - \frac{1}{2}(\sin 2x)^2 \qquad [\because \; 2 \sin A \cos A = \sin 2A]$$

$$\Rightarrow \quad f(x) = 1 - \frac{1}{2} \sin^2 2x \qquad ...(2)$$

Since, constant function and sine function are continuous for all values of x.

$\therefore$ $f(x)$ being the difference of two continuous functions is also a continuous function for all values of x. Therefore,

(a) $f(x)$ is continuous on $\left[0, \frac{\pi}{2}\right]$.

Differentiating equation (2), w.r.t. x, we get

$$f'(x) = 0 - \frac{1}{2} \,.\, 2 \sin 2x \,.\, \cos 2x \,.\, 2 = -2 \sin 2x \cos 2x$$

$$\Rightarrow \quad f'(x) = -\sin 4x \qquad ...(3)$$

$\therefore$ $f'(x)$ exists uniquely on $\left(0, \frac{\pi}{2}\right)$.

(b) $\therefore$ $f(x)$ is derivable on $\left(0, \frac{\pi}{2}\right)$.

(c) $f(0) = 1 - \frac{1}{2}\sin^2 0°$ and $f\left(\frac{\pi}{2}\right) = 1 - \frac{1}{2}\sin^2 2\left(\frac{\pi}{2}\right)$

$= 1 - \frac{1}{2}(0)$ $\qquad$ $= 1 - \frac{1}{2}\sin^2 \pi$

$= 1 - 0 = 1$ $\qquad$ $= 1 - 0 = 1$

$\therefore \quad f(0) = f\left(\frac{\pi}{2}\right)$

Thus, $f(x)$ satisfies all the conditions of Rolle's theorem.

$\therefore$ There exists at least one $c \in \left(0, \frac{\pi}{2}\right)$ such that

$$f'(c) = 0$$

Put $x = c$ in equation (3), we get

$$f'(c) = -\sin 4c \qquad [\because f'(c) = 0]$$

$\Rightarrow \quad -\sin 4c = 0$

$\Rightarrow \quad \sin 4c = 0$

$\Rightarrow \quad 4c = 0, \pi, 2\pi \,.......$

$\Rightarrow \quad c = 0, \frac{\pi}{4}, \frac{\pi}{2}, ...$

$\Rightarrow \quad c = \frac{\pi}{4} \in \left(0, \frac{\pi}{2}\right)$

Hence, Rolle's theorem is verified.

(v) Let $\quad f(x) = e^x \cos x \qquad ...(1)$

Since, exponential function and cosine function are continuous for all values of x.

$\therefore$ $f(x)$ being the product of two continuous functions is also continuous for all values of x. Therefore,

(a) $f(x)$ is continuous on $\left[-\frac{\pi}{2}, \frac{\pi}{2}\right]$

Differentiating equation (1) w.r.t. x, we get

$$f'(x) = e^x . (-\sin x) + \cos x . e^x$$

$\Rightarrow \quad f'(x) = e^x . (\cos x - \sin x) \qquad ...(2)$

$\therefore$ $f'(x)$ exists uniquely on $\left(-\frac{\pi}{2}, \frac{\pi}{2}\right)$

(b) $\therefore$ $f(x)$ is derivable on $\left(-\frac{\pi}{2}, \frac{\pi}{2}\right)$.

(c) $f\left(-\frac{\pi}{2}\right) = e^{-\pi/2} . \cos\left(-\frac{\pi}{2}\right)$ and $f\left(\frac{\pi}{2}\right) = e^{\pi/2} . \cos\frac{\pi}{2}$

$= e^{-\pi/2} . \cos\frac{\pi}{2}$ $\qquad$ $= e^{\pi/2} . 0$ $\qquad [\because \cos(-\theta) = \cos\theta]$

$= e^{-\pi/2} . 0$ $\qquad$ $= 0$

$= 0$

$\therefore \qquad f\left(-\frac{\pi}{2}\right) = f\left(\frac{\pi}{2}\right)$

Thus, $f(x)$ satisfies all the conditions of Rolle's theorem.

$\therefore$ There exists at least one $c \in \left(-\frac{\pi}{2}, \frac{\pi}{2}\right)$ such that

$$f'(c) = 0$$

Put $x = c$ in equation (2), we get

$$f'(c) = e^c (\cos c - \sin c) \qquad [\because f'(c) = 0]$$

$\Rightarrow \qquad e^c (\cos c - \sin c) = 0$

$\Rightarrow \qquad \cos c - \sin c = 0$

$\Rightarrow \qquad \cos c = \sin c$

$\Rightarrow \qquad \tan c = 1$

$\Rightarrow \qquad c = \frac{\pi}{4} \in \left(-\frac{\pi}{2}, \frac{\pi}{2}\right)$

Hence, Rolle's theorem is verified.

(*vi*) Let $\qquad f(x) = \frac{\sin x}{e^x} \qquad ...(1)$

Since, sine function and exponential function are continuous for all values of x.

$\therefore$ $f(x)$ being the quotient of two continuous functions is also continuous for all values of x. Therefore,

(*a*) $f(x)$ is continuous on $[0, \pi]$.

Differentiating equation (1) w.r.t. x, we get

$$f'(x) = \frac{e^x . \cos x - \sin x . e^x}{(e^x)^2} = \frac{e^x(\cos x - \sin x)}{(e^x)^2}$$

$$\Rightarrow \qquad f'(x) = \frac{\cos x - \sin x}{e^x} \qquad ...(2)$$

$\therefore$ $f'(x)$ exists uniquely on $(0, \pi)$

(*b*) $\therefore$ $f(x)$ is derivable on $(0, \pi)$

(*c*) $f(0) = \frac{\sin 0^\circ}{e^0}$ and $f(\pi) = \frac{\sin \pi}{e^\pi}$

$= \frac{0}{1} \qquad\qquad = \frac{0}{e^\pi}$

$= 0 \qquad\qquad = 0$

$\therefore \qquad f(0) = f(\pi)$

Thus, $f(x)$ satisfies all the conditions of Rolle's theorem.

$\therefore$ There exists at least one $c \in (0, \pi)$ such that

$$f'(c) = 0$$

Put $x = c$ in equation (2), we get

$$f'(c) = \frac{\cos c - \sin c}{e^c} \qquad [\because f'(c) = 0]$$

$$\Rightarrow \quad \frac{\cos c - \sin c}{e^c} = 0$$

$$\Rightarrow \quad \cos c - \sin c = 0$$

$$\Rightarrow \quad \cos c = \sin c$$

$$\Rightarrow \quad \tan c = 1$$

$$\Rightarrow \quad c = \frac{\pi}{4} \in (0, \pi)$$

Hence, Rolle's theorem is verified.

Example 13. *Verify Rolle's theorem for the function $f(x) = \log(x^2 + 2) - \log 3$ on $[-1, 1]$.*

Solution. We have,

$$f(x) = \log(x^2 + 2) - \log 3 \qquad \left[\because \log m - \log n = \log\left(\frac{m}{n}\right)\right]$$

$$\Rightarrow \quad f(x) = \log\left(\frac{x^2 + 2}{3}\right) \qquad ...(1)$$

Since, logarithm function is continuous for all values of x. Therefore,

(*i*) $f(x)$ is continuous on $[-1, 1]$

Differentiating equation (1) w.r.t. x, we get

$$f'(x) = \frac{1}{\left(\frac{x^2+2}{3}\right)} \cdot \frac{d}{dx}\left(\frac{x^2+2}{3}\right) = \frac{3}{x^2+2}\left(\frac{2x+0}{3}\right)$$

$$\Rightarrow \quad f'(x) = \frac{2x}{x^2+2} \qquad ...(2)$$

$\therefore$ $f'(x)$ exists uniquely on $(-1, 1)$.

(*ii*) $\therefore$ $f(x)$ is derivable on $(-1, 1)$.

(*iii*) $f(-1) = \log\left(\frac{(-1)^2+2}{3}\right)$ and $f(1) = \log\left(\frac{(1)^2+2}{3}\right)$

$= \log\left(\frac{1+2}{3}\right)$ $\qquad$ $= \log\left(\frac{1+2}{3}\right)$

$= \log 1$ $\qquad$ $= \log 1$

$= 0$ $\qquad$ $= 0$

$$\therefore \quad f(-1) = f(1)$$

Thus, $f(x)$ satisfies all the conditions of Rolle's theorem.

$\therefore$ There exists at least one $c \in (-1, 1)$ such that

$$f'(c) = 0$$

Put $x = c$ in equation (2), we get

$$\Rightarrow \quad f'(c) = \frac{2c}{c^2+2}$$

$$\Rightarrow \quad \frac{2c}{c^2+2} = 0 \qquad [\because f'(c) = 0]$$

$\Rightarrow \quad 2c = 0 \quad \Rightarrow \quad c = 0 \in (-1, 1)$

Hence, Rolle's theorem is verified.

Example 14. *Verify Rolle's theorem for the function :*

$$f(x) = \log \left[\frac{x^2 + ab}{(a+b)x}\right] \text{ on } [a, b],$$

where a and b are positive real numbers.

Solution. We have,

$$f(x) = \log \left[\frac{x^2 + ab}{(a+b)x}\right]$$

$$= \log (x^2 + ab) - \log [(a + b)\, x] \qquad \left[\because \log\left(\frac{m}{n}\right) = \log m - \log n\right]$$

$$\Rightarrow \quad f(x) = \log (x^2 + ab) - \log (a + b) - \log x \qquad \text{...(1)} \; [\because \log (m \,.\, n) = \log m + \log n]$$

Since, logarithmic function is continuous for all values of x.

$\therefore$ $f(x)$ being the sum of continuous functions is also continuous for all values of x. Therefore,

(*i*) $f(x)$ is continuous on $[a, b]$.

Differentiating equation (1) w.r.t. x, we get

$$f'(x) = \frac{1}{(x^2 + ab)}(2x + 0) - 0 - \frac{1}{x}$$

$$= \frac{2x}{x^2 + ab} - \frac{1}{x} = \frac{2x^2 - x^2 - ab}{x(x^2 + ab)}$$

$$\Rightarrow \quad f'(x) = \frac{x^2 - ab}{x(x^2 + ab)} \qquad \text{...(2)}$$

$\therefore$ $f'(x)$ exists uniquely on $[a, b]$.

(*ii*) $\therefore$ $f(x)$ is derivable on (a, b).

(*iii*)
$$f(a) = \log (a^2 + ab) - \log (a + b) - \log a$$
$$= \log [a\, (a + b)] - \log (a + b) - \log a$$
$$= \log a + \log (a + b) - \log (a + b) - \log a = 0$$
$$[\because \log (m \,.\, n) = \log m + \log n]$$

and
$$f(b) = \log (b^2 + ab) - \log (a + b) - \log b$$
$$= \log [b\, (a + b)] - \log (a + b) - \log b$$
$$= \log b + \log (a + b) - \log (a + b) - \log b = 0$$

$\therefore \quad f(a) = f(b).$

Thus, $f(x)$ satisfies all the conditions of Rolle's theorem.

$\therefore$ There exists at least one $c \in (a, b)$ such that

$$f'(c) = 0$$

Put $x = c$ in equation (2), we get

$$f'(c) = \frac{c^2 - ab}{c(c^2 + ab)}$$

$\Rightarrow \quad \dfrac{c^2 - ab}{c(c^2 + ab)} = 0 \qquad [\because f'(c) = 0]$

$\Rightarrow \quad c^2 - ab = 0$

$\Rightarrow \quad c^2 = ab$

$\Rightarrow \quad c = \pm\sqrt{ab}$

$\Rightarrow \quad c = \sqrt{ab} \in (a, b)$

Hence, Rolle's theorem is verified.

Example 15. *At what points on the curve $y = \cos x - 1$, $x \in [0, 2\pi]$, is the tangent parallel to the x-axis.*

Solution. Let $\quad f(x) = \cos x - 1 \qquad ...(1)$

Since, cosine function and constant function are continuous for all values of x.

$\therefore$ $f(x)$ being the difference of two continuous functions is also continuous for all values of x. Therefore,

(*i*) $f(x)$ is continuous on $[0, 2\pi]$

Differentiating equation (1) w.r.t. x, we get

$f'(x) = -\sin x \qquad ...(2)$

$\therefore$ $f'(x)$ exists uniquely on $(0, 2\pi)$.

(*ii*) $\therefore$ $f(x)$ is derivable on $(0, 2\pi)$.

(*iii*) $f(0) = \cos 0° - 1$ and $f(2\pi) = \cos 2\pi - 1$

$= 1 - 1 \qquad\qquad = 1 - 1$

$= 0 \qquad\qquad = 0$

$\therefore \quad f(0) = f(2\pi)$

Thus, $f(x)$ satisfies all the conditions of Rolle's theorem.

$\therefore$ There exists at least one $c \in (0, 2\pi)$ such that

$$f'(c) = 0$$

Put $x = c$ in equation (2), we get

$$f'(c) = -\sin c$$

$\Rightarrow \quad -\sin c = 0 \qquad [\because f'(c) = 0]$

$\Rightarrow \quad \sin c = 0$

$\Rightarrow \quad c = \pi \in (0, 2\pi)$

$\therefore \quad f'(c) = 0$ at $c = \pi$

Also, when $x = \pi$

From (1), we have,

$$y = f(x) = \cos \pi - 1$$
$$= -1 - 1 = -2$$

$\therefore$ At point $(\pi, -2)$; $f'(x) = 0$ *i.e.,* $\dfrac{dy}{dx} = 0$

$\Rightarrow$ At point $(\pi, -2)$, slope of tangent is zero.

i.e., the tangent is parallel to x-axis.

Example 16. *Using Rolle's theorem, find the points on the curve $y = 16 - x^2$; $x \in [-1, 1]$, where the tangent is parallel to x-axis.*

Solution. Let $\quad y = f(x) = 16 - x^2$ ...(1)

$f(x)$ being a polynomial function is continuous and derivable for all values of x. Therefore,

(*i*) $f(x)$ is continuous on $[-1, 1]$

(*ii*) $f(x)$ is derivable on $(-1, 1)$

(*iii*) $f(-1) = 16 - (-1)^2 = 16 - 1 = 15$ and $f(1) = 16 - (1)^2 = 16 - 1 = 15$

$\therefore \quad f(-1) = f(1)$

Thus, $f(x)$ satisfies all the conditions of Rolle's theorem.

$\therefore$ There exists at least one $c \in (-1, 1)$ such that

$$f'(c) = 0$$

Now, differentiating equation (1), w.r.t. x, we get

$$f'(x) = -2x$$

Put $x = c \Rightarrow \quad f'(c) = -2c$

$\Rightarrow \quad -2c = 0 \quad [\because f'(c) = 0]$

$\Rightarrow \quad c = 0 \in (-1, 1)$

$\therefore \quad f'(c) = 0$ at $c = 0$

Also, when $x = 0$

From (1), we have

$$y = f(x) = 16 - 0 = 16$$

$\therefore$ At point $(0, 16)$, $f'(x) = 0$ *i.e.,* $\dfrac{dy}{dx} = 0$

$\Rightarrow$ At point $(0, 16)$, slope of tangent is zero. *i.e.,* the tangent is parallel to x-axis.

Example 17. *Verify Lagrange's mean value theorem for the following functions :*

(*i*) $x(x - 2)$ *on* $[1, 2]$ (*ii*) $2x^2 - 10x + 29$ *on* $[2, 7]$

(*iii*) $x + \dfrac{1}{x}$ *on* $[1, 3]$ (*iv*) $x(x - 1)(x - 2)$ *on* $\left[0, \dfrac{1}{2}\right]$

(*v*) $x^3 + x^2 - 6x$ *on* $[-1, 4]$.

Solution. (*i*) Let $\quad f(x) = x(x - 2)$

$\Rightarrow \quad f(x) = x^2 - 2x$ ...(1)

$f(x)$ being a polynomial function is continuous and derivable for all values of x. Therefore,

(*a*) $f(x)$ is continuous on $[1, 2]$

(*b*) $f(x)$ is derivable on $(1, 2)$

Thus, $f(x)$ satisfies both the conditions of Lagrange's mean value theorem.

∴ There exists at least one $c \in (1, 2)$ such that

$$f'(c) = \frac{f(b) - f(a)}{b - a} \qquad ...(2)$$

where $a = 1, b = 2$

Differentiating equation (1) w.r.t. x, we get

$$f'(x) = 2x - 2$$

Put $x = c \Rightarrow \quad f'(c) = 2c - 2 \qquad ...(3)$

∴ From equations (2) and (3), we have

$$2c - 2 = \frac{f(2) - f(1)}{2 - 1}$$

$$\Rightarrow \quad 2c - 2 = \frac{[(2)^2 - 2(2)] - [(1)^2 - 2(1)]}{1}$$

$$\Rightarrow \quad 2c - 2 = (4 - 4) - (1 - 2)$$

$$\Rightarrow \quad 2c - 2 = 0 - (-1)$$

$$\Rightarrow \quad 2c - 2 = 1 \Rightarrow 2c = 1 + 2$$

$$\Rightarrow \quad c = \frac{3}{2} = 1.5 \in (1, 2)$$

Hence, Lagrange's mean value theorem is verified.

(*ii*) Let $\quad f(x) = 2x^2 - 10x + 29 \qquad ...(1)$

$f(x)$ being a polynomial function is continuous and derivable for all values of x. Therefore,

(*a*) $f(x)$ is continuous on [2, 7]

(*b*) $f(x)$ is derivable on (2, 7).

Thus, $f(x)$ satisfies both the conditions of Lagrange's mean value theorem.

∴ There exists at least one $c \in (2, 7)$ such that

$$f'(c) = \frac{f(b) - f(a)}{b - a} \qquad ...(2)$$

where $a = 2, b = 7$

Differentiating equation (1) w.r.t. x, we get

$$f'(x) = 4x - 10$$

Put $x = c \Rightarrow \quad f'(c) = 4c - 10 \qquad ...(3)$

∴ From equations (2) and (3), we have

$$4c - 10 = \frac{f(7) - f(2)}{7 - 2}$$

$$\Rightarrow \quad 4c - 10 = \frac{[2(7)^2 - 10(7) + 29] - [2(2)^2 - 10(2) + 29]}{7 - 2}$$

$$\Rightarrow \quad 4c - 10 = \frac{2(49) - 70 + 29 - 2(4) + 20 - 29}{5}$$

$$\Rightarrow \qquad 4c - 10 = \frac{98 - 70 - 8 + 20}{5}$$

$$\Rightarrow \qquad 4c - 10 = \frac{40}{5}$$

$$\Rightarrow \qquad 4c - 10 = 8 \;\Rightarrow\; 4c = 8 + 10 \;\Rightarrow\; c = \frac{18}{4} = \frac{9}{2} = 4.5$$

$$\Rightarrow \qquad c = 4.5 \in (2, 7)$$

Hence, Lagrange's mean value theorem is verified.

(*iii*) Let $\qquad f(x) = x + \frac{1}{x}$ ...(1)

$$\Rightarrow \qquad f(x) = \frac{x^2 + 1}{x}$$

$f(x)$ being a quotient of two continuous functions is also continuous for all values of x. Therefore,

(*a*) $f(x)$ is continuous on [1, 3]

Differentiating (1) w.r.t. x, we get

$$f'(x) = 1 - \frac{1}{x^2} \qquad ...(2)$$

which is uniquely defined on (1, 3)

(*b*) $\therefore$ $f(x)$ is derivable on (1, 3).

Thus, $f(x)$ satisfies both the conditions of L.M.V. theorem.

$\therefore$ There exists at least one value of $c \in (1, 3)$

such that $\qquad f'(c) = \frac{f(b) - f(a)}{b - a}$ ...(3)

where $a = 1$, $b = 3$

Put $x = c$ in equation (2), we get

$$f'(c) = 1 - \frac{1}{c^2} \qquad ...(4)$$

$\therefore$ From equations (3) and (4), we have

$$1 - \frac{1}{c^2} = \frac{f(3) - f(1)}{3 - 1}$$

$$\Rightarrow \qquad 1 - \frac{1}{c^2} = \frac{\left(3 + \frac{1}{3}\right) - \left(1 + \frac{1}{1}\right)}{2}$$

$$\Rightarrow \qquad 1 - \frac{1}{c^2} = \frac{\frac{10}{3} - 2}{2}$$

$$\Rightarrow \qquad 1 - \frac{1}{c^2} = \frac{4}{3} \times \frac{1}{2} = \frac{2}{3}$$

$\Rightarrow \qquad \dfrac{1}{c^2} = 1 - \dfrac{2}{3}$

$\Rightarrow \qquad \dfrac{1}{c^2} = \dfrac{1}{3} \Rightarrow c^2 = 3 \Rightarrow c = \pm\sqrt{3}$

$\Rightarrow \qquad c = \sqrt{3} \in (1, 3)$

Hence, Lagrange's mean value theorem is verified.

(*iv*) Let $\qquad f(x) = x(x-1)(x-2)$

$\Rightarrow \qquad f(x) = (x^2 - x)(x-2)$

$\Rightarrow \qquad f(x) = x^3 - 3x^2 + 2x \qquad \ldots(1)$

$f(x)$ being a polynomial function is continuous and derivable for all values of x. Therefore,

(*a*) $f(x)$ is continuous on $\left[0, \dfrac{1}{2}\right]$

(*b*) $f(x)$ is derivable on $\left(0, \dfrac{1}{2}\right)$.

Thus, $f(x)$ satisfies both the conditions of L.M.V. theorem

$\therefore$ There exists at least one $c \in \left(0, \dfrac{1}{2}\right)$ such that

$$f'(c) = \frac{f(b) - f(a)}{b - a} \qquad \ldots(2)$$

where $a = 0, \quad b = \dfrac{1}{2}$.

Differentiating equation (1) w.r.t. x, we get

$$f'(x) = 3x^2 - 6x + 2$$

Put $x = c \Rightarrow \qquad f'(c) = 3c^2 - 6c + 2 \qquad \ldots(3)$

$\therefore$ From equations (2) and (3), we have

$$3c^2 - 6c + 2 = \frac{f\left(\frac{1}{2}\right) - f(0)}{\frac{1}{2} - 0}$$

$$\Rightarrow \qquad 3c^2 - 6c + 2 = \frac{\left[\left(\frac{1}{2}\right)^3 - 3\left(\frac{1}{2}\right)^2 + 2\left(\frac{1}{2}\right)\right] - [0]}{\frac{1}{2}}$$

$$\Rightarrow \qquad 3c^2 - 6c + 2 = 2\left[\frac{1}{8} - \frac{3}{4} + 1\right] = 2\left[\frac{1 - 6 + 8}{8}\right] = \frac{3}{4}$$

$$\Rightarrow \qquad 3c^2 - 6c + 2 = \frac{3}{4}$$

$\Rightarrow \quad 12c^2 - 24c + 8 = 3$

$\Rightarrow \quad 12c^2 - 24c + 5 = 0$

$$\left[\because \text{ For a quadratic equation } ax^2 + bx + c = 0 \Rightarrow x = \frac{-b \pm \sqrt{b^2 - 4ac}}{2a}\right]$$

$$\Rightarrow \quad c = \frac{24 \pm \sqrt{576 - 240}}{24}$$

$$\Rightarrow \quad c = \frac{24 \pm \sqrt{336}}{24} = \frac{24 \pm 18.33}{24}$$

$$\Rightarrow \quad c = \frac{42.33}{24}, \frac{5.67}{24}$$

$$\Rightarrow \quad c = 1.76, 0.23$$

$$\Rightarrow \quad c = 0.23 \in \left(0, \frac{1}{2}\right)$$

Hence, Lagrange's mean value theorem is verified.

(*v*) Let $\quad f(x) = x^3 + x^2 - 6x \qquad ...(1)$

Please try yourself. **[Ans.** $c = 2 \in (-1, 4)$**]**

Example 18. *Verify Lagrange's mean value theorem for the following functions :*

(*i*) $(x-3)(x-6)(x-9)$ *on* $[3, 5]$ (*ii*) $(x-4)(x-6)(x-8)$ *on* $[4, 10]$

(*iii*) $\sqrt{25 - x^2}$ *on* $[-3, 4]$ (*iv*) $(x-1)^{2/3}$ *on* $[1, 2]$.

Solution. (*i*) Let $\quad f(x) = (x-3)(x-6)(x-9)$

$$= (x^2 - 9x + 18)(x - 9)$$

$$\Rightarrow \quad f(x) = x^3 - 18x^2 + 99x - 162 \qquad ...(1)$$

$f(x)$ being a polynomial function is continuous and derivable for all values of x.

(*a*) $f(x)$ is continuous on $[3, 5]$

(*b*) $f(x)$ is derivable on $(3, 5)$.

Thus, $f(x)$ satisfies both the conditions of L.M.V. theorem.

$\therefore$ There exists at least one $c \in (3, 5)$ such that

$$f'(c) = \frac{f(b) - f(a)}{b - a}, \text{ where } \quad a = 3, b = 5 \qquad ...(2)$$

Differentiating equation (1) w.r.t. x, we get

$$f'(x) = 3x^2 - 36x + 99$$

Put $x = c \Rightarrow \quad f'(c) = 3c^2 - 36c + 99 \qquad ...(3)$

$\therefore$ From equations (2) and (3), we have

$$3c^2 - 36c + 99 = \frac{f(5) - f(3)}{5 - 3}$$

$$\Rightarrow \quad 3c^2 - 36c + 99 = \frac{[(5)^3 - 18(5)^2 + 99(5) - 162] - [(3)^3 - 18(3)^2 + 99(3) - 162]}{2}$$

$$\Rightarrow \quad 3c^2 - 36c + 99 = \frac{125 - 450 + 495 - 162 - 27 + 162 - 297 - 162}{2}$$

$$\Rightarrow \quad 3c^2 - 36c + 99 = \frac{8}{2}$$

$$\Rightarrow \quad 3c^2 - 36c + 99 = 4$$

$$\Rightarrow \quad 3c^2 - 36c + 95 = 0 \qquad \left[\begin{array}{l} \because \text{ For a quadratic equation } ax^2 + bx + c = 0 \\ \Rightarrow \quad x = \dfrac{-b \pm \sqrt{b^2 - 4ac}}{2a} \end{array}\right]$$

$$\Rightarrow \quad c = \frac{36 \pm \sqrt{1296 - 1140}}{6}$$

$$\Rightarrow \quad c = \frac{36 \pm \sqrt{156}}{6} = \frac{36 \pm 12.49}{6}$$

$$\Rightarrow \quad c = \frac{48.49}{6}, \frac{23.51}{6}$$

$$\Rightarrow \quad c = 8.08, 3.92$$

$$\Rightarrow \quad c = 3.92 \in (3, 5).$$

Hence, Lagrange's mean value theorem is verified.

(*ii*) Let $f(x) = (x - 4)(x - 6)(x - 8)$

$$= (x^2 - 10x + 24)(x - 8)$$

$$\Rightarrow \quad f(x) = x^3 - 18x^2 + 104x - 192$$

Proceed further as in part (*i*). [**Ans.** $c = 8 \in (4, 10)$]

(*iii*) Let $f(x) = \sqrt{25 - x^2}$ ...(1)

Since, $f(x)$ has a finite and unique value for all $x \in [-3, 4]$. Therefore,

(*a*) $f(x)$ is continuous as $[-3, 4]$

Differentiating equation (1) w.r.t. x, we get

$$f'(x) = \frac{1}{2\sqrt{25 - x^2}} \cdot (-2x)$$

$$f'(x) = \frac{-x}{\sqrt{25 - x^2}} \qquad ...(2)$$

which exists uniquely on $(-3, 4)$.

(*b*) $f(x)$ is derivable on $(-3, 4)$.

Thus, $f(x)$ satisfies both the conditions of L.M.V. theorem.

$\therefore$ There exists at least one $c \in (-3, 4)$ such that

$$f'(c) = \frac{f(b) - f(a)}{b - a} \qquad ...(3)$$

where $a = -3$, $b = 4$

Put $x = c$ in equation (2), we get

$$f'(c) = \frac{-c}{\sqrt{25 - c^2}} \qquad ...(4)$$

$\therefore$ From equations (3) and (4), we have

$$\frac{-c}{\sqrt{25-c^2}} = \frac{f(4) - f(-3)}{4 - (-3)}$$

$$\Rightarrow \quad \frac{-c}{\sqrt{25-c^2}} = \frac{\left[\sqrt{25-(4)^2}\right] - \left[\sqrt{25-(-3)^2}\right]}{4+3}$$

$$\Rightarrow \quad \frac{-c}{\sqrt{25-c^2}} = \frac{\sqrt{25-16} - \sqrt{25-9}}{7}$$

$$\Rightarrow \quad \frac{-c}{\sqrt{25-c^2}} = \frac{3-4}{7}$$

$$\Rightarrow \quad \frac{-c}{\sqrt{25-c^2}} = \frac{-1}{7}$$

$$\Rightarrow \quad \sqrt{25-c^2} = 7c$$

$$\Rightarrow \quad 25 - c^2 = 49c^2 \qquad \text{[Squaring both sides]}$$

$$\Rightarrow \quad 50c^2 = 25$$

$$\Rightarrow \quad c^2 = \frac{25}{50} = \frac{1}{2}$$

$$\Rightarrow \quad c = \pm\frac{1}{\sqrt{2}} \in (-3, 4).$$

Hence, Lagrange's mean value theorem is verified.

(*iv*) Let $\qquad f(x) = (x-1)^{2/3} \qquad ...(1)$

Since, $f(x)$ has a finite and unique value for all $x \in [1, 2]$. Therefore,

(*a*) $f(x)$ is continuous on $[1, 2]$

Differentiating equation (1) w.r.t. x, we get

$$f'(x) = \frac{2}{3}(x-1)^{-1/3} \Rightarrow f'(x) = \frac{2}{3(x-1)^{1/3}} \qquad ...(2)$$

which exists uniquely on (1, 2).

(*b*) $\therefore$ $f(x)$ is derivable on (1, 2).

Thus, $f(x)$ satisfies both the conditions of L.M.V. theorem.

$\therefore$ There exists at least one $c \in (1, 2)$ such that

$$f'(c) = \frac{f(b) - f(a)}{b - a} \qquad ...(3)$$

where $a = 1, \quad b = 2.$

Put $x = c$ in equation (2), we get

$$f'(c) = \frac{2}{3(c-1)^{1/3}} \qquad ...(4)$$

$\therefore$ From equations (3) and (4), we have

$$\frac{2}{3(c-1)^{1/3}} = \frac{f(2) - f(1)}{2 - 1}$$

$$\Rightarrow \quad \frac{2}{3(c-1)^{1/3}} = \frac{(2-1)^{2/3} - (1-1)^{2/3}}{1}$$

$$\Rightarrow \quad \frac{2}{3(c-1)^{1/3}} = \frac{1-0}{1}$$

$$\Rightarrow \quad 3(c-1)^{1/3} = 2$$

$$\Rightarrow \quad (c-1)^{1/3} = \frac{2}{3}$$

$$\Rightarrow \quad c - 1 = \left(\frac{2}{3}\right)^3 = \frac{8}{27} \qquad \text{[Cubing both sides]}$$

$$\Rightarrow \quad c = \frac{8}{27} + 1 = \frac{35}{27} = 1.29 \in (1, 2)$$

Hence, Lagrange's mean value theorem is verified.

Example 19. *Verify Lagrange's mean value theorem for the following functions :*

(*i*) $\dfrac{1}{x+2}$ *on [1, 4]* (*ii*) $\dfrac{1}{4x-1}$ *on [1, 4]*

(*iii*) $\dfrac{x^5 - 1}{x+6}$ *on [– 8, 0].*

Solution. (*i*) Let $f(x) = \dfrac{1}{x+2}$ ...(1)

Since, $f(x)$ has a finite and unique value for all $x \in [1, 4]$. Therefore,

(*a*) $f(x)$ is continuous on [1, 4].

Differentiating equation (1) w.r.t. x, we get

$$f'(x) = \frac{-1}{(x+2)^2} \qquad ...(2)$$

which exists uniquely on (1, 4).

(*b*) ∴ $f(x)$ is derivable on (1, 4).

Thus, $f(x)$ satisfies both the conditions of L.M.V. theorem.

∴ There exists at least one $c \in (1, 4)$ such that

$$f'(c) = \frac{f(b) - f(a)}{b - a} \qquad ...(3)$$

where $a = 1, b = 4$.

Put $x = c$ in equation (2), we get

$$f'(c) = \frac{-1}{(c+2)^2} \qquad ...(4)$$

∴ From equations (3) and (4), we have

$$\frac{-1}{(c+2)^2} = \frac{f(4) - f(1)}{4 - 1}$$

$$\Rightarrow \quad \frac{-1}{(c+2)^2} = \frac{\left(\frac{1}{4+2}\right) - \left(\frac{1}{1+2}\right)}{3} = \frac{\left(\frac{1}{6}\right) - \left(\frac{1}{3}\right)}{3}$$

$$= \frac{1}{3}\left[\frac{1}{6} - \frac{1}{3}\right] = \frac{1}{3}\left[\frac{1-2}{6}\right]$$

$$\Rightarrow \quad \frac{-1}{(c+2)^2} = \frac{-1}{18} \quad \Rightarrow \quad (c+2)^2 = 18$$

$$\Rightarrow \quad (c+2) = \pm\sqrt{18} \quad \Rightarrow \quad c + 2 = \pm 3\sqrt{2}$$

$$\Rightarrow \quad c = -2 \pm 3\sqrt{2} \quad \Rightarrow \quad c = -2 \pm 4.24$$

$$\Rightarrow \quad c = 2.24, -6.24 \quad \Rightarrow \quad c = 2.24 \in (1, 4)$$

Hence, Lagrange's mean value theorem is verified.

(*ii*) Let $$f(x) = \frac{1}{4x - 1} \qquad ...(1)$$

Since, $f(x)$ has a finite and unique value for all $x \in [1, 4]$. Therefore,

(*a*) $f(x)$ is continuous on [1, 4].

Differentiating equation (1) w.r.t. x, we get

$$f'(x) = \frac{-1}{(4x-1)^2} \cdot (4)$$

$$\Rightarrow \quad f'(x) = \frac{-4}{(4x-1)^2} \qquad ...(2)$$

which exists uniquely on (1, 4).

(*b*) ∴ $f(x)$ is derivable on (1, 4).

Thus, $f(x)$ satisfies both the conditions of L.M.V. theorem.

$\therefore$ There exists at least one $c \in (1, 4)$ such that

$$f'(c) = \frac{f(b) - f(a)}{b - a} \quad \text{...(3)}$$

where $a = 1, b = 4$.

Put $x = c$ in equation (2), we get

$$f'(c) = \frac{-4}{(4c - 1)^2} \quad \text{...(4)}$$

$\therefore$ From equations (3) and (4), we have

$$\frac{-4}{(4c-1)^2} = \frac{f(4) - f(1)}{4 - 1}$$

$$\Rightarrow \quad \frac{-4}{(4c-1)^2} = \frac{\left[\frac{1}{4(4) - 1}\right] - \left[\frac{1}{4(1) - 1}\right]}{3}$$

$$\Rightarrow \quad \frac{-4}{(4c-1)^2} = \frac{\left(\frac{1}{16 - 1}\right) - \left(\frac{1}{4 - 1}\right)}{3}$$

$$\Rightarrow \quad \frac{-4}{(4c-1)^2} = \frac{1}{3}\left[\frac{1}{15} - \frac{1}{3}\right] = \frac{1}{3}\left[\frac{1-5}{15}\right] = \frac{-4}{45}$$

$$\Rightarrow \quad \frac{-4}{(4c-1)^2} = \frac{-4}{45}$$

$$\Rightarrow \quad (4c - 1)^2 = 45$$

$$\Rightarrow \quad (4c - 1) = \pm\sqrt{45}$$

$$\Rightarrow \quad 4c = 1 \pm \sqrt{45}$$

$$\Rightarrow \quad c = \frac{1 \pm \sqrt{45}}{4} = \frac{1 \pm 6.71}{4}$$

$$\Rightarrow \quad c = 1.92, -1.42$$

$$\Rightarrow \quad c = 1.92 \in (1, 4)$$

Hence, Lagrange's mean value theorem is verified

(*iii*) Let
$$f(x) = \frac{x^5 - 1}{x + 6} \quad \text{...(1)}$$

$f(x)$ being a rational function is continuous on $R - \{-6\}$.

$\therefore$ $f(x)$ is not continuous on $[-8, 0]$.

Hence, Lagrange's mean value theorem is not applicable.

Example 20. *Examine the applicability of Lagrange's mean value theorem for the following functions :*

(i) $|x|$ *on* $[-1, 1]$ (ii) $x^{3/4}$ *on* $[0, 16]$

(iii) $x^{2/3}$ *on* $[-2, 2]$ (iv) $f(x) = \begin{cases} 1+3x \;;\; x \le 1 \\ 2x^2+7 \;;\; x > 1 \end{cases}$ *on* $[0, 3]$

(v) $f(x) = \begin{cases} 2+x^3 & ; x \le 1 \\ 3x & ; x > 1 \end{cases}$ *on* $[-1, 2]$.

Solution. (i) Let $f(x) = |x|$ ...(1)

$$\Rightarrow \quad f(x) = \begin{cases} -x & ; -1 \le x < 0 \\ x & ; 0 \le x \le 1 \end{cases}$$

$f(x)$ being a polynomial function is continuous for all values of x, except possibly at $x = 0$.

$\therefore$ When $x = 0$

$$\text{L.H.L.} = \lim_{x \to 0^-} f(x) = \lim_{x \to 0^-} (-x)$$

$$= \lim_{h \to 0} -(0-h) = \lim_{h \to 0} h = 0$$

$$\text{R.H.L.} = \lim_{x \to 0^+} f(x) = \lim_{x \to 0^+} x$$

$$= \lim_{h \to 0} (0+h) = \lim_{h \to 0} h = 0$$

$\therefore$ L.H.L. = R.H.L.

$\therefore$ $f(x)$ is continuous at $x = 0$.

$\Rightarrow$ $f(x)$ is continuous on $[-1, 1]$.

Since, $$f(x) = \begin{cases} -x & ; -1 \le x \le 0 \\ x & ; 0 \le x \le 1 \end{cases}$$

$$\Rightarrow \quad f'(x) = \begin{cases} -1 & ; -1 \le x \le 0 \\ 1 & ; 0 \le x \le 1 \end{cases}$$

$\therefore$ $Lf'(0) \ne Rf'(0)$.

$\therefore$ $f(x)$ is not derivable at $x = 0$.

$\Rightarrow$ $f(x)$ *is* not derivable on $(-1, 1)$.

Hence, Lagrange's mean value theorem is not applicable.

(ii) Let $f(x) = x^{3/4}$ ...(1)

Since, $f(x)$ has a finite and unique value for all $x \in [0, 16]$. Therefore,

(a) $f(x)$ is continuous on $[0, 16]$

Differentiating equation (1) w.r.t. x, we get

$$f'(x) = \frac{3}{4} x^{\frac{3}{4}-1} = \frac{3}{4} x^{-1/4}$$

$$\Rightarrow \quad f'(x) = \frac{3}{4(x^{1/4})} \quad ...(2)$$

which exists uniquely on $(0, 16)$.

(*b*) $\therefore$ $f(x)$ is derivable on (0, 16).

Thus, $f(x)$ satisfies both the conditions of L.M.V. theorem.

$\therefore$ There exists at least one $c \in (0, 16)$ such that

$$f'(c) = \frac{f(b) - f(a)}{b - a} \qquad ...(3)$$

where $a = 0$, $b = 16$.

Put $x = c$ in equation (2), we get

$$f'(c) = \frac{3}{4\,c^{1/4}} \qquad ...(4)$$

$\therefore$ From equations (3) and (4), we have

$$\frac{3}{4\,c^{1/4}} = \frac{f(16) - f(0)}{16 - 0} = \frac{(16)^{3/4} - 0}{16 - 0} = \frac{8 - 0}{16}$$

$$\Rightarrow \qquad \frac{3}{c^{1/4}} = 2 \Rightarrow c^{1/4} = \frac{3}{2}$$

$$\Rightarrow \qquad c = \left(\frac{3}{2}\right)^4 = \frac{81}{16} = 5.06 \in (0, 16)$$

Hence, Lagrange's mean value theorem is verified.

(*iii*) Let $\qquad f(x) = x^{2/3} \qquad ...(1)$

Since, $f(x)$ has a finite and unique value for all $x \in [-2, 2]$. Therefore,

(*a*) $f(x)$ is continuous on $[-2, 2]$

Differentiating equation (1) w.r.t. x, we get

$$f'(x) = \frac{2}{3}\,x^{-1/3} = \frac{2}{3x^{1/3}} \qquad ...(2)$$

$\therefore$ $f'(x)$ is not defined at $x = 0$

$\therefore$ $f(x)$ is not derivable on $(-2, 2)$.

Hence, Lagrange's mean value theorem is not applicable.

(*iv*) We have, $\qquad f(x) = \begin{Bmatrix} 1 + 3x & ; & x \le 1 \\ 2x^2 + 7 & ; & x > 1 \end{Bmatrix} \qquad ...(1)$

$f(x)$ being a polynomial function is continuous for all values of x except possibly at $x = 1$.

$\therefore$ $f(x)$ is continuous on [0, 3] except possibly at $x = 1$.

$\therefore$ Continuity at $x = 1$:

$$\text{L.H.L.} = \lim_{x \to 1^-} f(x) = \lim_{x \to 1^-} (1 + 3x) = 1 + 3(1) = 4$$

$$\text{R.H.L.} = \lim_{x \to 1^+} f(x) = \lim_{x \to 1^+} (2x^2 + 7) = 2(1)^2 + 7 = 9$$

$\therefore$ $\qquad$ L.H.L. $\ne$ R.H.L. at $x = 1$

$\therefore$ $f(x)$ is not continuous at $x = 1$.

$\therefore$ $f(x)$ is not continuous on [0, 3].

Hence, Lagrange's mean value theorem is not applicable.

(*v*) We have, $$f(x) = \begin{cases} 2 + x^3 ; & x \le 1 \\ 3x \;\; ; & x > 1 \end{cases} \quad ...(1)$$

$f(x)$ being a polynomial function is continuous for all values of x except possibly at $x = 1$

$\therefore$ $f(x)$ is continuous on $[-1, 2]$ except possibly at $x = 1$

$\therefore$ Continuity at $x = 1$.

$$\text{L.H.L.} = \lim_{x \to 1^-} f(x) = \lim_{x \to 1^-} (2 + x^3)$$
$$= 2 + (1)^3 = 3.$$
$$\text{R.H.L.} = \lim_{x \to 1^+} f(x) = \lim_{x \to 1^+} 3x = 3(1) = 3$$

$\therefore$ L.H.L. = R.H.L. at $x = 1$

Also, $f(1) = 3$.

$\therefore$ $f(x)$ is continuous at $x = 1$.

(*a*) $f(x)$ is continuous on $[-1, 2]$.

Derivability at $x = 1$:

Differentiating equation (1) w.r.t. x, we get

$$f'(x) = \begin{cases} 3x^2 ; & x \le 1 \\ 3 \;\; ; & x > 1 \end{cases} \quad ...(2)$$

$\therefore$ L.H.D. = $Lf'(1) = 3(1)^2 = 3$

R.H.D. = $Rf'(1) = 3$

$\therefore$ L.H.D. = R.H.D. at $x = 1$.

$\therefore$ $f(x)$ is derivable at $x = 1$.

(*b*) $f(x)$ is derivable on $(-1, 2)$.

Thus, $f(x)$ satisfies both the conditions of L.M.V. theorem

$\therefore$ There exists at least one $c \in (-1, 2)$ such that

$$f'(c) = \frac{f(b) - f(a)}{b - a} \quad ...(3)$$

where $a = -1$, $b = 2$.

$$\Rightarrow \quad 3c^2 = \frac{f(2) - f(-1)}{2 - (-1)} \quad [\because \text{ By using (2) at } x = c]$$

$$\Rightarrow \quad 3c^2 = \frac{[3(2)] - [2 + (-1)^3]}{2 + 1}$$

$$\Rightarrow \quad 3c^2 = \frac{6 - 1}{3} \Rightarrow 3c^2 = \frac{5}{3} \Rightarrow c^2 = \frac{5}{9}$$

$$\Rightarrow \quad c = \pm \frac{\sqrt{5}}{3} = \pm \frac{2.23}{3}$$

$$\Rightarrow \quad c = \pm 0.74 \in (-1, 2)$$

Hence, Lagrange's mean value theorem is verified.

Example 21. *Verify Lagrange's mean value theorem for the following functions*

(*i*) *log x on [1, 2]* (*ii*) *log x on [1, 5]* (*iii*) *log x on [1, e].*

Solution. (*i*) Let $f(x) = \log x$ ...(1)

Since, logarithmic function is continuous and derivable for all values of $x > 0$. Therefore,

(*a*) $f(x)$ is continuous on $[1, 2]$

(*b*) $f(x)$ is derivable on $(1, 2)$.

Thus, $f(x)$ satisfies both the conditions of Lagrange's mean value theorem.

$\therefore$ There exists at least one $c \in (1, 2)$ such that

$$f'(c) = \frac{f(b) - f(a)}{b - a} \qquad \text{...(2)}$$

where $a = 1, b = 2$.

Differentiating equation (1) w.r.t. x, we get

$$f'(x) = \frac{1}{x}$$

Put $x = c \Rightarrow$
$$f'(c) = \frac{1}{c} \qquad \text{...(3)}$$

$\therefore$ From equations (2) and (3), we have

$$\frac{1}{c} = \frac{f(2) - f(1)}{2 - 1}$$

$$\Rightarrow \quad \frac{1}{c} = \frac{\log 2 - \log 1}{1} = \frac{\log 2 - 0}{1} = \log 2 \qquad [\because \log 1 = 0]$$

$$\Rightarrow \quad c = \frac{1}{\log 2}$$

$$\Rightarrow \quad c = \log_2 e = \frac{\log_{10} e}{\log_{10} 2} = \frac{0.4343}{0.3010}$$

$$\Rightarrow \quad c = 1.44 \in (1, 2).$$

Hence, Lagrange's mean value theorem is verified.

(*ii*) Let $f(x) = \log x$ ...(1)

Please try yourself.

[**Hint.** Proceed as in part (*i*)] [**Ans.** $c = 4, \log_5 e \in (1, 5)$]

(*iii*) Let $f(x) = \log x$ ...(1)

Since, logarithmic function is continuous and derivable for all values of $x > 0$. Therefore,

(*a*) $f(x)$ is continuous on $[1, e]$.

(*b*) $f(x)$ is derivable on $(1, e)$

Thus, $f(x)$ satisfies both the conditions of L.M.V. theorem.

$\therefore$ There exists at least one $c \in (1, e)$ such that

$$f'(c) = \frac{f(b) - f(a)}{b - a} \quad ...(2)$$

where $a = 1$, $b = e$,

Differentiating equation (1) w.r.t. x, we get

$$f'(x) = \frac{1}{x}$$

Put $x = c \Rightarrow \quad f'(c) = \frac{1}{c} \quad ...(3)$

$\therefore$ From equations (2) and (3), we have

$$\frac{1}{c} = \frac{\log e - \log 1}{e - 1} = \frac{1 - 0}{e - 1} = \frac{1}{e - 1} \quad \left\{ \because \begin{array}{l} \log e = 1 \\ \log 1 = 0 \end{array} \right\}$$

$$c = e - 1 \in (1, e)$$

Hence, Lagrange's mean value theorem is verified.

Example 22. *Verify Lagrange's mean value theorem for the following functions :*

(i) $x - 2 \sin x$ *on* $[-\pi, \pi]$ *(ii)* $\sin x - \sin 2x$ *on* $[0, \pi]$

(iii) $2 \sin x + \sin 2x$ *on* $[0, \pi]$ *(iv)* $\tan^{-1} x$ *on* $[0, 1]$

(v) e^x *on* $[0, 1]$.

Solution. (*i*) Let $f(x) = x - 2 \sin x \quad ...(1)$

Since, x and $2 \sin x$ are both continuous functions for all values of x.

$\therefore$ $f(x)$ being the difference of two continuous functions is also continuous for all values of x. Therefore,

(*a*) $f(x)$ is continuous on $[-\pi, \pi]$.

Differentiating equation (1) w.r.t. x, we get

$$f'(x) = 1 - 2 \cos x \quad ...(2)$$

which exists uniquely on $(-\pi, \pi)$.

(*b*) $\therefore$ $f(x)$ is derivable on $(-\pi, \pi)$.

Thus, $f(x)$ satisfies both the conditions of L.M.V. theorem.

$\therefore$ There exists at least one $c \in (-\pi, \pi)$ such that

$$f'(c) = \frac{f(b) - f(a)}{b - a} \quad ...(3)$$

where $a = -\pi$, $b = \pi$.

Put $x = c$ in equation (2), we get

$$f'(c) = 1 - 2 \cos c \quad ...(4)$$

$\therefore$ From equations (3) and (4), we have

$$1 - 2 \cos c = \frac{f(\pi) - f(-\pi)}{\pi - (-\pi)}$$

$$\Rightarrow \quad 1 - 2 \cos c = \frac{(\pi - 2 \sin \pi) - (-\pi - 2 \sin (-\pi))}{\pi + \pi} \quad [\because \sin(-\theta) = -\sin \theta)$$

$$\Rightarrow \quad 1 - 2 \cos c = \frac{(\pi - 0) - (-\pi + 2 \sin \pi)}{2\pi}$$

$$\Rightarrow \quad 1 - 2\cos c = \frac{\pi + \pi - 0}{2\pi} = \frac{2\pi}{2\pi} = 1$$

$$\Rightarrow \quad 1 - 2\cos c = 1$$

$$\Rightarrow \quad 2\cos c = 0 \quad \Rightarrow \quad \cos c = 0$$

$$\Rightarrow \quad c = \pm\frac{\pi}{2} \in (-\pi, \pi).$$

Hence, Lagrange's mean value theorem is verified.

(*ii*) Let $\quad f(x) = \sin x - \sin 2x \quad$ ...(1)

Since, $\sin x$ and $\sin 2x$ are both continuous functions for all values of x.

$\therefore$ $f(x)$ being the difference of two continuous functions is also continuous for all values of x. Therefore,

(*a*) $f(x)$ is continuous on $[0, \pi]$.

Differentiating equation (1) w.r.t. x, we get

$$f'(x) = \cos x - 2\cos 2x \quad \text{...(2)}$$

which exists uniquely on $(0, \pi)$.

(*b*) $\therefore$ $f(x)$ is derivable on $(0, \pi)$.

Thus, $f(x)$ satisfies both the conditions of L.M.V. theorem.

$\therefore$ There exists at least one $c \in (0, \pi)$ such that

$$f'(c) = \frac{f(b) - f(a)}{b - a} \quad \text{...(3)}$$

where $a = 0$, $b = \pi$.

Put $x = c$ in equation (2), we get

$$f'(c) = \cos c - 2\cos 2c \quad \text{...(4)}$$

$\therefore$ From equations (3) and (4), we have

$$\cos c - 2\cos 2c = \frac{f(\pi) - f(0)}{\pi - 0}$$

$$\Rightarrow \quad \cos c - 2\cos 2c = \frac{[\sin\pi - \sin 2\pi] - [\sin 0° - \sin 0°]}{\pi - 0}$$

$$\Rightarrow \quad \cos c - 2\cos 2c = \frac{(0-0)-(0-0)}{\pi} = \frac{0}{\pi}$$

$\Rightarrow \cos c - 2(2\cos^2 c - 1) = 0 \quad [\because \cos 2A = 2\cos^2 A - 1]$

$\Rightarrow \quad \cos c - 4\cos^2 c + 2 = 0$

$\Rightarrow \quad 4\cos^2 c - \cos c - 2 = 0 \quad \left[\because \text{For a quadratic equation } ax^2 + bx + c = 0 \Rightarrow x = \frac{-b \pm \sqrt{b^2 - 4ac}}{2a}\right]$

$$\Rightarrow \quad \cos c = \frac{1 \pm \sqrt{1 + 32}}{8}$$

$\Rightarrow \quad \cos c = \dfrac{1 \pm \sqrt{33}}{8} = \dfrac{1 \pm 5.7445}{8}$

$\Rightarrow \quad \cos c = \dfrac{6.7445}{8}, \dfrac{-4.7445}{8}$

$\Rightarrow \quad \cos c = 0.8431, 0.5931$

$\Rightarrow \quad \cos c = 0.8431 \quad \text{or} \quad \cos(\pi - c) = 0.5931$

$\Rightarrow \quad c = 32° \, 32' \quad \text{or} \quad \pi - c = 53.37 \qquad [\because \; \pi = 180°]$

$\Rightarrow \quad c = 126° \, 23'.$

$\Rightarrow \quad c = 32° \, 32', 126° \, 23' \in (0, \pi) \text{ or } (0, 180°)$

(*iii*) Let $\quad f(x) = 2 \sin x + \sin 2x \qquad ...(1)$

Please try yourself.

[**Hint.** Proceed as in part (*ii*) of the same example.] $\left[\textbf{Ans.} \quad c = \dfrac{\pi}{3} \in (0, \pi)\right]$

(*iv*) Let $\quad f(x) = \tan^{-1} x \qquad ...(1)$

Since, $f(x)$ has a finite and unique value on [0, 1]. Therefore,

(*a*) $f(x)$ is continuous on [0, 1].

Differentiating equation (1) w.r.t. x, we get

$$f'(x) = \frac{1}{1+x^2} \qquad ...(2)$$

which exists uniquely on (0, 1).

(*b*) $\therefore$ $f(x)$ is derivable on (0, 1).

Thus, $f(x)$ satisfies both the conditions of L.M.V. theorem.

$\therefore$ There exists at least one $c \in (0, 1)$ such that

$$f'(c) = \frac{f(b) - f(a)}{b - a} \qquad ...(3)$$

where $a = 0$, $b = 1$.

Put $x = c$ in equation (2), we get

$$f'(c) = \frac{1}{1+c^2} \qquad ...(4)$$

$\therefore$ From equations (3) and (4), we have

$$\frac{1}{1+c^2} = \frac{f(1) - f(0)}{1 - 0}$$

$$\Rightarrow \quad \frac{1}{1+c^2} = \frac{\tan^{-1} 1 - \tan^{-1} 0}{1} = \left(\frac{\pi}{4} - 0\right)$$

$$\Rightarrow \quad 1 + c^2 = \frac{4}{\pi} \quad \Rightarrow \quad c^2 = \frac{4}{\pi} - 1 = \frac{4}{\frac{22}{7}} - 1 = \frac{28}{22} - 1$$

$$\Rightarrow \quad c^2 = \frac{28 - 22}{22} = \frac{6}{22} = 0.27$$

$\Rightarrow \qquad c = \pm \sqrt{0.27}$

$\Rightarrow \qquad c = \pm 0.52$

$\Rightarrow \qquad c = 0.52 \in (0, 1).$

Hence, Lagrange's mean value theorem is verified.

(*v*) Let $\qquad f(x) = e^x \qquad$ ...(1)

Since, exponential function is continuous for all values of x. Therefore,

(*a*) $f(x)$ is continuous on $[0, 1]$

Differentiating equation (1) w.r.t. x, we get

$$f'(x) = e^x \qquad ...(2)$$

which exists uniquely on $(0, 1)$.

(*c*) $\therefore$ $f(x)$ is derivable on $(0, 1)$.

Thus, $f(x)$ satisfies both the conditions of L.M.V. theorem.

$\therefore$ There exists at least one $c \in (0, 1)$ such that

$$f'(c) = \frac{f(b) - f(a)}{b - a} \qquad ...(3)$$

where $a = 0$, $b = 1$.

Put $x = c$ in equation (2), we get

$$f'(c) = e^c \qquad ...(4)$$

$\therefore$ From equations (3) and (4), we have

$$\Rightarrow \qquad e^c = \frac{f(1) - f(0)}{1 - 0} = \frac{e^1 - e^0}{1} = e - 1 \qquad [\because \ e^0 = 1)$$

$$\Rightarrow \qquad c = \log(e - 1) \in (0, 1)$$

$[\because \ 1 < e - 1 < e \ \Rightarrow \ \log 1 < \log(e - 1) < \log e \Rightarrow 0 < \log(e - 1) < 1$
$\Rightarrow \log(e - 1) \in (0, 1)]$

Hence, Lagrange's mean value theorem is verified.

Example 23. *Using Lagrange's mean value theorem, find a point on the curve* $y = \sqrt{x^2 - 4}$ *defined on the interval [2, 4], where the tangent is parallel to the chord joining the end points of the curve.*

Solution. Let $\qquad f(x) = \sqrt{x^2 - 4} \qquad$ ...(1)

Since, $f(x)$ has a finite and unique value for all $x \in [2, 4]$. Therefore,

(*i*) $f(x)$ is continuous on $[2, 4]$.

Differentiating equation (1) w.r.t. x, we get

$$f'(x) = \frac{1}{2\sqrt{x^2 - 4}} \cdot (2x)$$

$$\Rightarrow \qquad f'(x) = \frac{x}{\sqrt{x^2 - 4}} \qquad ...(2)$$

which exists uniquely on $(2, 4)$.

(*ii*) $f(x)$ is derivable on (2, 4) .

Thus, $f(x)$ satisfies both the conditions of L.M.V. theorem.

$\therefore$ There exists at least one $c \in (2, 4)$ such that

$$f'(c) = \frac{f(b) - f(a)}{b - a} \qquad ...(3)$$

where $a = 2, b = 4$.

Put $x = c$ in equation (2), we get

$$f'(c) = \frac{c}{\sqrt{c^2 - 4}} \qquad ...(4)$$

$\therefore$ From equations (3) and (4), we have

$$\frac{c}{\sqrt{c^2 - 4}} = \frac{f(4) - f(2)}{4 - 2}$$

$$\Rightarrow \quad \frac{c}{\sqrt{c^2 - 4}} = \frac{\sqrt{(4)^2 - 4} - \sqrt{(2)^2 - 4}}{2} = \frac{\sqrt{16 - 4} - \sqrt{4 - 4}}{2}$$

$$= \frac{\sqrt{12} - 0}{2} = \frac{2\sqrt{3}}{2} = \sqrt{3}$$

$$\Rightarrow \quad c = \sqrt{3}\sqrt{c^2 - 4}$$

$$\Rightarrow \quad c = \sqrt{3c^2 - 12} \qquad \text{[Squaring both sides]}$$

$$\Rightarrow \quad c^2 = 3c^2 - 12$$

$$\Rightarrow \quad 2c^2 = 12 \Rightarrow c^2 = 6 \Rightarrow c = \pm\sqrt{6}$$

$$\Rightarrow \quad c = \sqrt{6} \in (2, 4)$$

$$\therefore \quad f(c) = [f(x)]_{x=c} = \sqrt{(\sqrt{6})^2 - 4} = \sqrt{6 - 4} = \sqrt{2}$$

$\therefore$ $(c, f(c)) = \left(\sqrt{6}, \sqrt{2}\right)$ is a point on the curve $y = \sqrt{x^2 - 4}$ such that the tangent at which is parallel to the chord joining the end points of the curve.

Example 24. (*i*) *Find a point on the parabola* $y = (x - 3)^2$, *where the tangent is parallel to the chord joining (3, 0) and (4, 1).*

(*ii*) *On the curve* $y = x^2$, *find a point at which tangent is parallel to the chord joining (0, 0) and (1, 1).*

(*iii*) *Find a point on the curve* $y = x^3$, *where the tangent is parallel to the chord joining (1, 1) and (3, 27).*

Solution. (*i*) Let $f(x) = (x - 3)^2$ ...(1)

$f(x)$ being a polynomial function is continuous and derivable for all values of x. Therefore,

(*a*) $f(x)$ is continuous on [3, 4]

(*b*) $f(x)$ is derivable on (3, 4).

Thus, $f(x)$ satisfies both the conditions of L.M.V. theorem

$\therefore$ There exists at least one value of $c \in (3, 4)$ such that

$$f'(c) = \frac{f(b) - f(a)}{b - a} \qquad ...(2)$$

where $a = 3$, $b = 4$.

Differentiating equation (1), w.r.t. x, we get

$$f'(x) = 2(x - 3)$$

Put $x = c \Rightarrow \quad f'(c) = 2(c - 3) \qquad ...(3)$

$\therefore$ From equations (2) and (3), we have

$$2c - 6 = \frac{f(4) - f(3)}{4 - 3}$$

$$\Rightarrow \quad 2c - 6 = \frac{(4-3)^2 - (3-3)^2}{1}$$

$$\Rightarrow \quad 2c - 6 = \frac{1 - 0}{1}$$

$$\Rightarrow \quad 2c = 1 + 6 = 7$$

$$\Rightarrow \quad c = \frac{7}{2} = 3.5 \in (3, 4).$$

$$\therefore \quad f(c) = [f(x)]_{\text{at } x = c} = (x - 3)^2 = \left(\frac{7}{2} - 3\right)^2 = \left(\frac{7-6}{2}\right)^2 = \left(\frac{1}{2}\right)^2 = \frac{1}{4}$$

$$\therefore \quad (c, f(c)) = \left(\frac{7}{2}, \frac{1}{4}\right).$$

Now, from (2), we have

$$f'\left(\frac{7}{2}\right) = \frac{f(4) - f(3)}{4 - 3}.$$

$\therefore$ Slope of chord joining the points (3, 0) and (4, 1) is equal to the slope of tangent at the point $\left(\frac{7}{2}, \frac{1}{4}\right)$.

(*ii*) Let $\qquad f(x) = x^2 \qquad ...(1)$

(*a*) $f(x)$ being a polynomial function is continuous and derivable for all values of x. Therefore,

$f(x)$ is continuous on [0, 1].

(*b*) $f(x)$ is derivable on (0, 1).

Thus, $f(x)$ satisfies both the conditions of L.M.V. theorem

$\therefore$ There exists at least one $c \in (0, 1)$ such that

$$f'(c) = \frac{f(b) - f(a)}{b - a} \qquad ...(2)$$

where $a = 0$, $b = 1$.

Differentiating equation (1) w.r.t. x, we get

$$f'(x) = 2x$$

Put $x = c \Rightarrow f'(c) = 2c$ ...(3)

$\therefore$ From equations (2) and (3), we have

$$2c = \frac{f(1) - f(0)}{1 - 0} = \frac{(1)^2 - 0}{1 - 0} = \frac{1 - 0}{1}$$

$$\Rightarrow \qquad 2c = 1 \Rightarrow c = \frac{1}{2} \in (0, 1).$$

$$\therefore \qquad f(c) = [f(x)]_{\text{at } x = c} = x^2 = \left(\frac{1}{2}\right)^2 = \frac{1}{4}$$

$$\therefore \qquad (c, f(c)) = \left(\frac{1}{2}, \frac{1}{4}\right).$$

Now, from equation (2), we have

$$f'\left(\frac{1}{2}\right) = \frac{f(1) - f(0)}{1 - 0}$$

$\therefore$ Slope of the chord joining the points (0, 0) and (1, 1) is equal to the slope of tangent at the point $\left(\frac{1}{2}, \frac{1}{4}\right)$.

(iii) Let $f(x) = x^3$.

Please try yourself.

[**Hint.** Proceed as in part *(ii)* of the same example.]

$$\left[\textbf{Ans. } \left(\frac{\sqrt{39}}{3}, \frac{13\sqrt{39}}{9}\right) \textit{ or } \left(\sqrt{\frac{13}{3}}, \frac{13}{3}\sqrt{\frac{13}{3}}\right)\right]$$

Example 25. *If $f(x) = x\,(1 - \log x)$ and $0 < a < b$, then show that there exists at least one $c \in (a, b)$ such that*

$$(a - b) \log c = b(1 - \log b) - a\,(1 - \log a).$$

Solution. We have, $f(x) = x(1 - \log x)$ ...(1)

Since, x and $(1 - \log x)$ are both continuous functions for all values of x.

$\therefore$ $f(x)$ being the product of two continuous functions is also continuous for all values of x. Therefore,

(i) $f(x)$ is continuous on $[a, b]$.

Differentiating equation (1) w.r.t. x, we get

$$f'(x) = x \,.\, \frac{d}{dx}(1 - \log x) + (1 - \log x)\frac{d}{dx}(x)$$

$$= x \,.\, \left(0 - \frac{1}{x}\right) + (1 - \log x) \,.\, 1$$

$$= -1 + 1 - \log x$$

$$\Rightarrow \qquad f'(x) = -\log x \qquad ...(2)$$

which exists uniquely on (a, b).

(*ii*) $\therefore$ $f(x)$ is derivable in (a, b).

Thus, $f(x)$ satisfies both the conditions of L.M.V. theorem.

$\therefore$ There exists at least one $c \in (a, b)$ such that

$$f'(c) = \frac{f(b) - f(a)}{b - a} \qquad ...(3)$$

Put $x = c$ in equation (2), we get

$$f'(c) = -\log c \qquad ...(4)$$

$\therefore$ From equations (3) and (4), we have

$$-\log c = \frac{f(b) - f(a)}{b - a}$$

$$\Rightarrow \qquad -\log c = \frac{[b(1 - \log b)] - [a\,(1 - \log a)]}{b - a}$$

$$\Rightarrow \qquad -(b - a)\log c = b\,(1 - \log b) - a(1 - \log a)$$

$$\Rightarrow \qquad (a - b)\log c = b\,(1 - \log b) - a(1 - \log a).$$

Example 26. *Using Lagrange's mean value theorem, show that*

$$\left(\frac{b-a}{b}\right) < \log\left(\frac{b}{a}\right) < \left(\frac{b-a}{a}\right).$$

where $0 < a < b$.

Solution. Let $\qquad f(x) = \log x$ on $[a, b] \qquad ...(1)$

Since, Logarithmic function is continuous and derivable for all values of $x > 0$. Therefore,

(*i*) $f(x)$ is continuous on $[a, b]$

(*ii*) $f(x)$ is derivable on (a, b).

Thus, $f(x)$ satisfies both the conditions of L.M.V. theorem

$\therefore$ There exists at least one $c \in (a, b)$ such that

$$f'(c) = \frac{f(b) - f(a)}{b - a} \qquad ...(2)$$

Differentiating equation (1) w.r.t. x, we get

$$f'(x) = \frac{1}{x}$$

Put $x = c \Rightarrow \qquad f'(c) = \frac{1}{c} \qquad ...(3)$

$\therefore$ From equations (2) and (3), we have

$$\frac{1}{c} = \frac{f(b) - f(a)}{b - a}$$

$$\Rightarrow \qquad \frac{1}{c} = \frac{\log b - \log a}{b - a} \qquad \left[\because \log m - \log n = \log \frac{m}{n}\right]$$

$$\Rightarrow \qquad \frac{1}{c} = \frac{\log\left(\frac{b}{a}\right)}{b - a} \qquad \text{...(4)}$$

Since, $\quad c \in (a, b)$

$$\Rightarrow \qquad a < c < b$$

$$\Rightarrow \qquad \frac{1}{a} > \frac{1}{c} > \frac{1}{b}$$

$$\Rightarrow \qquad \frac{1}{b} < \frac{1}{c} < \frac{1}{a}$$

$$\Rightarrow \qquad \frac{1}{b} < \frac{\log\left(\frac{b}{a}\right)}{b - a} < \frac{1}{a} \qquad [\because \text{By using (4)}]$$

$$\Rightarrow \qquad \left(\frac{b-a}{b}\right) < \log\left(\frac{b}{a}\right) < \left(\frac{b-a}{a}\right). \qquad \text{[Multiplying throughout by } (b - a)]$$

Example 27. *Let f and g be differentiable on [0, 1] such that*

$$f(0) = 2,\ g(0) = 0,\ f(1) = 6 \text{ and } g(1) = 2,$$

show that there exists at least one $c \in (0, 1)$ such that $f'(c) = 2g'(c)$.

Solution. It is given that, f and g are differentiable functions on [0, 1].

Since, every differentiable function is continuous.

$\therefore$ f and g are also continuous functions on [0,1].

Thus, f and g satisfies both the conditions of L.M.V. theorem.

$\therefore$ There exists at least one $c \in (0, 1)$ such that

$$f'(c) = \frac{f(b) - f(a)}{b - a} \qquad \text{...(1)}$$

where $a = 0$, $b = 1$.

$$\Rightarrow \qquad f'(c) = \frac{f(1) - f(0)}{1 - 0}$$

$$\Rightarrow \qquad f'(c) = \frac{6 - 2}{1} = 4 \qquad [\because f(0) = 2 \text{ and } f(1) = 6 \text{ (Given)}]$$

$$\Rightarrow \qquad f'(c) = 4 \qquad \text{...(2)}$$

And, $$g'(c) = \frac{g(b) - g(a)}{b - a} \qquad \text{...(3)}$$

where $a = 0$, $b = 1$.

$$\Rightarrow \qquad g'(c) = \frac{g(1) - g(0)}{b - a}$$

$$\Rightarrow \qquad g'(c) = \frac{2-0}{1-0} = 2 \qquad [\because \; g(0) = 0 \text{ and } g(1) = 2 \text{ (Given)}]$$

$$\Rightarrow \qquad g'(c) = 2 \qquad ...(4)$$

$\therefore$ From equations (2) and (4), we have

$$f'(c) = 2\,g'(c).$$

Example 28. *Using Lagrange's mean value theorem, show that*

$$\tan x > x \text{ for all } x \in \left(0, \frac{\pi}{2}\right).$$

Solution. Let $\quad f(x) = \tan x$ on $[0, x] \subset \left(0, \frac{\pi}{2}\right)$ ...(1)

Clearly, $f(x)$ is continuous and derivable on $[0, x]$.

Thus, $f(x)$ satisfies both the conditions of L.M.V. theorem.

$\therefore$ There exists at least one $c \in (0, x)$ such that

$$f'(c) = \frac{f(b) - f(a)}{b - a} \qquad ...(2)$$

where $a = 0,\ b = x$.

Differentiating equation (1) w.r.t. x, we get

$$f'(x) = \sec^2 x$$

Put $x = c \Rightarrow \quad f'(c) = \sec^2 c$ ...(3)

$\therefore$ From equations (2) and (3), we have

$$\sec^2 c = \frac{f(x) - f(a)}{x - 0}$$

$$\Rightarrow \qquad \sec^2 c = \frac{\tan x - \tan 0}{x}$$

$$\Rightarrow \qquad \sec^2 c = \frac{\tan x - 0}{x}$$

$$\Rightarrow \qquad \frac{\tan x}{x} = \sec^2 c > 1 \text{ for all } x \in \left(0, \frac{\pi}{2}\right).$$

$$\Rightarrow \qquad \frac{\tan x}{x} > 1 \text{ for all } x \in \left(0, \frac{\pi}{2}\right)$$

$$\Rightarrow \qquad \tan x > x \text{ for all } x \in \left(0, \frac{\pi}{2}\right).$$

Hence proved.

EXERCISE FOR PRACTICE

Verify Rolle's theorem for the following functions : (Q. No. 1—6).

1. $x^2 - 5x + 6$ on $[2, 3]$
2. $(x^2 - 1)(x - 2)$ on $[-1, 2]$
3. $\sqrt{1-x^2}$ on $[-1, 1]$
4. $\sin x - 1$ on $\left[\frac{\pi}{2}, \frac{5\pi}{2}\right]$
5. $x(x-1)^2$ on $[0, 1]$
6. $\sin x + \cos x - 1$ on $\left[0, \frac{\pi}{2}\right]$
7. **Using Rolle's theorem, find the points on the curve $y = 16 - x^2$, $x \in [-1, 1]$, where the tangent is parallel to the x-axis.**
8. **Using Rolle's theorem, find the points on the curve $y = x^2 - 4x$, $x \in [0, 4]$, where the tangent is parallel to x-axis.**

Verify Langrange's mean value theorem for the following functions : (Q. No. 9—11)

9. $\log x$ on $[1, 5]$
10. $2x^2 - 3x + 1$ on $[1, 3]$
11. $\cos x$ on $\left[0, \frac{\pi}{2}\right]$
12. Find a point on the curve $y = x^3$, where the tangent is parallel to the chord joining the points (1, 1) and (3, 27).

Answers

1. $c = 2.5$
2. $c = -0.22, 1.55$
3. $c = 0$
4. $c = \frac{3\pi}{2}$
5. $c = \frac{1}{3} \in (0, 1)$
6. $c = \frac{\pi}{4} \in \left(0, \frac{\pi}{2}\right)$
7. $(0, 16)$
8. $(2, -4)$
9. $c = 4, \log_5 e \in (1, 5)$
10. $c = 2$
11. $c = \sin^{-1}(0.64)$
12. $\left(\sqrt{\frac{13}{3}}, \frac{13}{3}\sqrt{\frac{13}{3}}\right)$.

6

Maxima and Minima

6.1 INTRODUCTION

In the previous chapters, we have learnt about various applications of differentiation. In this chapter, we shall study the use of derivative in finding the maximum and minimum values of functions in its domain. The largest and smallest values of a function, if they exist, are called absolute maximum and the absolute minimum of the function. We will also discuss the applications of maxima and minima in solving some applied problems.

6.2 MAXIMUM VALUE

Let $f(x)$ be a real valued function defined on an interval I. Then $f(x)$, is said to have the maximum value in I, if there exists a point a in I such that :

$$f(x) \leq f(a) \text{ for all } x \in \text{I}.$$

Here, $f(a)$ is called the maximum value of $f(x)$ in the interval I and the point a is called the point of maximum of $f(x)$ in I.

6.3 MINIMUM VALUE

Let $f(x)$ be a real valued function defined on an interval I. Then $f(x)$ is said to have the minimum value in I, if there exists a point a in I such that :

$$f(x) \geq f(a) \text{ for all } x \in \text{I}.$$

Here, $f(a)$ is called the minimum value of $f(x)$ in the interval I and the point a is called the point of minimum of $f(x)$ in I.

6.4 LOCAL MAXIMUM VALUE

A function $f(x)$ is said to have a local maximum at $x = a$, if there is an open interval I containing a such that :

$$f(x) < f(a) \text{ for all } x \in \text{I}.$$

A local maximum value is also called a relative maximum value or simply a maximum value.

6.5 LOCAL MINIMUM VALUE

A function $f(x)$ is said to have a local minimum at $x = a$, if there is an open interval I containing a such that :

$$f(x) > f(a) \text{ for all } x \in \text{I}.$$

A local minimum value is also called a relative minimum value or simply a minimum value.

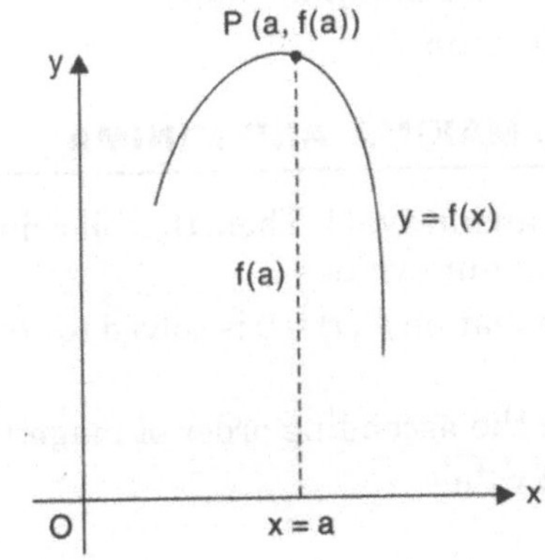

Fig. (*i*)
Point P is a point of local maximum and $f(a)$ is the local maximum value, at $x = a$.

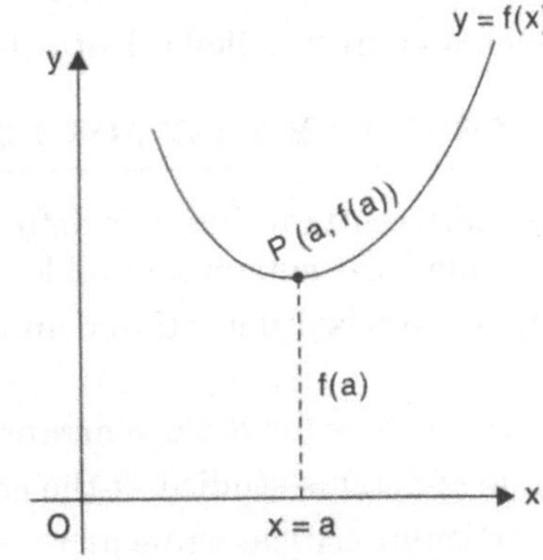

Fig. (*ii*)
Point P is a point of local minimum and $f(a)$ is the local minimum value, at $x = a$.

6.6 EXTREMUM VALUE

The maximum or minimum value of a function is called an extremum (or extreme) value of the function.

6.6.1 Theorem. If $f'(x)$ exists in an interval $[a, b]$ and $f(x)$ has a maximum or minimum value at $x = c \in (a, b)$. Then, $f'(c) = 0$.

Proof. The proof is beyond the scope of this book.

6.7 STATIONARY VALUE

The values of x for which $f'(x)$ is zero *i.e.*, $f'(x) = 0$ are called stationary points or turning points and the corresponding values of $f(x)$ are called stationary values or turning values of $f(x)$.

6.8 FIRST DERIVATIVE TEST FOR LOCAL MAXIMA AND MINIMA

Let $f(x)$ be a differentiable function defined on an interval I. Then, the following procedure is followed to find the local maximum and local minimum values :

1. The function $f(x)$ is differentiated and the equation $f'(x) = 0$ is solved to get the critical values for $f(x)$.

2. The critical values for $f(x)$ are arranged in the ascending order of magnitude.

3. The sign of $f'(x)$ is studied in the immediate neighbourhood of critical values.

4. (*i*) If for a particular critical value $x = a$, the sign of $f'(x)$ changes from +ve to –ve as x increases through $x = a$, then,

$f(a)$ is a **Local Maximum Value.**

(*ii*) If for a particular critical value $x = a$, the sign of $f'(x)$ changes from –ve to +ve as x increases through $x = a$, then :

$f(a)$ is a **Local Minimum Value.**

(*iii*) If for a particular critical value $x = a$, the sign of $f'(x)$ does not changes as x increases through $x = a$, then

$f(a)$ is neither a local maximum value nor a local minimum value.

In fact, such a point is called a **Point of Inflexion**.

6.9 SECOND DERIVATIVE TEST FOR LOCAL MAXIMA AND MINIMA

Let $f(x)$ be a differentiable function defined on an interval I. Then, the following procedure is followed to find the local maximum and local minimum values :

1. The function $f(x)$ is differentiated and the equation $f'(x) = 0$ is solved to get the critical values for $f(x)$.
2. The critical values for $f(x)$ are arranged in the ascending order of magnitude.
3. The value of $f''(x)$ is studied at the critical values.
4. For a particular critical value at $x = a$,

(*i*) If $f''(a) < 0$.

Then, $f(a)$ is a local maximum value.

(*ii*) If $f''(a) > 0$.

Then, $f(a)$ is a local minimum value.

(*iii*) If $f''(a) = 0$ or ∞.

Then, the test fails and the first derivative test is to be followed to study the nature of $f(a)$.

6.10 ABSOLUTE MAXIMUM VALUE

A function $f(x)$ is said to have the greatest value or absolute maximum value at a point $x = a$ in its domain, if $f(x) \leq f(a)$ for all x in the domain of $f(x)$.

The absolute maximum value is also called the maximum value and $x = a$ is the point of maximum.

6.11 ABSOLUTE MINIMUM VALUE

A function $f(x)$ is said to have the smallest value or absolute minimum value at a point $x = a$ in its domain, if $f(a) \leq f(x)$ for all x in the domain of $f(x)$.

The absolute minimum value is also called the minimum value and $x = a$ is the point of minimum.

6.12 MAXIMA AND MINIMA IN A CLOSED INTERVAL

Let $f(x)$ be a continuous function defined on $[a, b]$. Let the local minimum value of $f(x)$ be m and let the local maximum value of $f(x)$ be M.

Then, (*i*) The minimum value of $f(x)$ in $[a, b]$ is the smallest of m, $f(a)$ and $f(b)$.

(*ii*) The maximum value of $f(x)$ in $[a, b]$ is the greatest of M, $f(a)$ and $f(b)$.

Remarks :

1. The plurals of maximum and minimum are maxima and minima respectively.
2. The end-points of the interval cannot be the points of local extremum.
3. A local maximum (minimum) value may not be the absolute maximum (minimum) or greatest (smallest) value.

4. Every extreme point is a critical point but every critical point is not an extreme point.

5. The absolute maximum value of a function is never less than the absolute minimum value of the function. But it may be possible that a local maximum value is less than a local minimum value.

6. If $f(x)$ is a continuous function in its domain, then at least one maxima and one minima must lie between two equal values of x.

7. Maxima and Minima occur alternately, that is, between two maxima there is one minimum and *vice-versa*.

SOME SOLVED EXAMPLES

Example 1. *Show that the following functions do not have local maxima or minima :*

(i) $2x + 7$ (ii) $\dfrac{2x+3}{5x+7}$

(iii) $x^3 - 3x^2 + 5x + 11$ (iv) $x^3 + x^2 + x + 1$

(v) $x^3 - x^2 + 7x + 3$ (vi) e^{x+1}

(vii) $\cos x$; $0 < x < \pi$ (viii) $\log x$.

Solution. (i) Let $f(x) = 2x + 7$

$\Rightarrow$ $f'(x) = 2$

$\therefore$ For a local maximum or minimum, we have

$$f'(x) = 0$$

$\Rightarrow$ $2 = 0$,

which is not possible.

Hence, $f(x)$ does not have local maxima or minima.

(ii) Let $$f(x) = \frac{2x+3}{5x+7}$$

$$\Rightarrow \quad f'(x) = \frac{(5x+7)(2) - (2x+3)(5)}{(5x+7)^2}$$

$$= \frac{10x + 14 - 10x - 15}{(5x+7)^2} = \frac{-1}{(5x+7)^2}$$

$\therefore$ For a local maximum or minimum, we have

$$f'(x) = 0$$

$$\Rightarrow \quad \frac{-1}{(5x+7)^2} = 0,$$

which is not possible.

Hence, $f(x)$ does not have local maxima or minima.

(iii) Let $f(x) = x^3 - 3x^2 + 5x + 11$

$\Rightarrow$ $f'(x) = 3x^2 - 6x + 5$

$\therefore$ For a local maximum or minimum, we have

$$f'(x) = 0$$

$\Rightarrow$ $3x^2 - 6x + 5 = 0$

$$\Rightarrow \qquad x = \frac{6 \pm \sqrt{36-60}}{6} = \frac{6 \pm \sqrt{-24}}{6}$$

$$\Rightarrow \qquad x = \frac{6 \pm 2\sqrt{6}\, i}{6} \notin \text{R}.$$

i.e., this equation gives imaginary values of x.

Since, $f'(x) = 0$ does not give any real value of x. Hence, $f(x)$ does not have local maxima or minima.

(*iv*) Let $\qquad f(x) = x^3 + x^2 + x + 1$

$\Rightarrow \qquad f'(x) = 3x^2 + 2x + 1$

$\therefore$ For a maximum or minimum, we have

$$f'(x) = 0$$

$$\Rightarrow \qquad 3x^2 + 2x + 1 = 0$$

$$\Rightarrow \qquad x = \frac{-2 \pm \sqrt{4-12}}{6} = \frac{-2 \pm \sqrt{-8}}{6}$$

$$\Rightarrow \qquad x = \frac{-2 \pm 2\sqrt{2}\, i}{6} \notin \text{R}$$

i.e., this equation gives imaginary values of x.

Since, $f'(x) = 0$ does not give any real value of x.

Hence, $f(x)$ does not have local maxima or minima.

(*v*) Let $\qquad f(x) = x^3 - x^2 + 7x + 3$

Please try yourself.

[**Hint.** Proceed as part (*iv*).]

(*vi*) Let $\qquad f(x) = e^{x+1}$

$$f'(x) = e^{x+1} . 1 = e^{x+1}$$

$\therefore$ For a maximum or minimum, we have

$$f'(x) = 0$$

$$\Rightarrow \qquad e^{x+1} = 0,$$

which is not possible for any value of x.

Hence, $f(x)$ does not have local maxima or minima.

(*vii*) Let $\qquad f(x) = \cos x \,;\, 0 < x < \pi$

$\Rightarrow \qquad f'(x) = -\sin x$

$\therefore$ For a maximum or minimum, we have

$$f'(x) = 0$$

$$\Rightarrow \qquad -\sin x = 0,$$

which is not possible for any $x \in (0, \pi)$.

Hence, $f(x)$ does not have local maxima or minima on $(0, \pi)$.

(*viii*) Let $\qquad f(x) = \log x$

$$\Rightarrow \qquad f'(x) = \frac{1}{x}$$

$\therefore$ For a maximum or minimum, we have

$$f'(x) = 0$$

$\Rightarrow$ $\frac{1}{x} = 0$, which is not possible.

Hence, $f(x)$ does not have local maxima or minima.

Example 2. *Find the local maximum and local minimum values of the following functions :*

(*i*) $2x^2 - 3x + 5$ (*ii*) $3x^2 - 4x + 2$

(*iii*) $x^3 - 6x^2 + 9x - 8$ (*iv*) $x^3 - 6x^2 + 12x - 8$

(*v*) $2x^3 - 15x^2 - 36x + 18$ (*vi*) $2x^3 - 21x^2 + 36x - 20$

(*vii*) $(x - 1)(x + 2)^2$.

Solution. (*i*) Let $f(x) = 2x^2 - 3x + 5$ ...(1)

$\Rightarrow$ $f'(x) = 4x - 3$ ...(2)

$\therefore$ For a local maximum or local minimum, we have

$$f'(x) = 0$$

$\Rightarrow$ $4x - 3 = 0 \Rightarrow 4x = 3$

$\Rightarrow$ $x = \frac{3}{4}$

$\therefore$ The critical value for $f(x)$ is

$$x = \frac{3}{4} = 0.75$$

When x is slightly less than $\frac{3}{4}$, say (0.7) [$\because$ By using (2)]

$\therefore$ $f'(0.7) = 4(0.7) - 3 = (0.28 - 3) = (-\text{ve})$

When x is slightly greater than $\frac{3}{4}$, say (0.8) [$\because$ By using (2)]

$\therefore$ $f'(0.8) = 4(0.8) - 3 = (3.2 - 3) = (+\text{ve})$

$\therefore$ $f'(x)$ changes sign from (–ve) to (+ve) as x increases through $\frac{3}{4}$.

$\therefore$ $x = \frac{3}{4}$ is a point of local minimum and the corresponding local minimum value is given by

$$f\left(\frac{3}{4}\right) = 2\left(\frac{3}{4}\right)^2 - 3\left(\frac{3}{4}\right) + 5 \quad [\because \text{ By using (1)}]$$

$$= 2\left(\frac{9}{16}\right) - \frac{9}{4} + 5 = \frac{9}{8} - \frac{9}{4} + 5 = \frac{9 - 18 + 40}{8} = \frac{31}{8}$$

$\therefore$ Local minimum value is $\frac{31}{8}$ at $x = \frac{3}{4}$.

Alternative method** :

Let $f(x) = 2x^2 - 3x + 5$ ...(1)

$\Rightarrow$ $f'(x) = 4x - 3$ ...(2)

and $f''(x) = 4$ ...(3)

$\therefore$ For a local maximum or minimum, we have

$$f'(x) = 0$$

$\Rightarrow$ $4x - 3 = 0 \Rightarrow 4x = 3 \Rightarrow x = \frac{3}{4}$ [$\because$ By using (2)]

$\therefore$ The critical value for $f(x)$ is $x = \frac{3}{4}$.

Now, $f''(x)$ at $x = \frac{3}{4}$

$\therefore$ $f''\left(\frac{3}{4}\right) = 4 > 0$ [$\because$ By using (3)]

$\therefore$ $x = \frac{3}{4}$ is a point of local minimum and the corresponding local minimum value is given by

$$f\left(\frac{3}{4}\right) = 2\left(\frac{3}{4}\right)^2 - 3\left(\frac{3}{4}\right) + 5 = \frac{9}{8} - \frac{9}{4} + 5 = \frac{31}{8}.$$ [$\because$ By using (1)]

$\therefore$ Local minimum value is $\frac{31}{8}$ at $x = \frac{3}{4}$.

(*ii*) Let $f(x) = 3x^2 - 4x + 2$ ...(1)

$\Rightarrow$ $f'(x) = 6x - 4$ ...(2)

and $f''(x) = 6$ ...(3)

$\therefore$ For a local maximum or minimum, we have

$$f'(x) = 0$$

$\Rightarrow$ $6x - 4 = 0 \Rightarrow 6x = 4 \Rightarrow x = \frac{4}{6} \Rightarrow x = \frac{2}{3}$ [$\because$ By using (2)]

$\therefore$ The critical value for $f(x)$ is $x = \frac{2}{3}$.

Now, $f''(x)$ at $x = \frac{2}{3} \Rightarrow f''\left(\frac{2}{3}\right) = 6 > 0$ [$\because$ By using (3)]

$\therefore$ $x = \frac{2}{3}$ is a point of local minimum and the corresponding local minimum value is given by

**[It can be easily observed that the second derivative method is comparitively easier to apply. So, the reader is advised that the first derivative method should be applied only when :

(*a*) It is asked to use in the question itself.

(*b*) It is difficult to compute the second derivative.

(*c*) The second derivative test fails. *i.e.*, when $f''(a) = 0$ or ∞.]

(*vii*) Let $f(x) = (x - 1)(x + 2)^2$

$\Rightarrow \quad f(x) = (x - 1)(x^2 + 4x + 4)$

$\Rightarrow \quad f(x) = x^3 + 3x^2 - 4 \qquad ...(1)$

$\Rightarrow \quad f'(x) = 3x^2 + 6x \qquad ...(2)$

and $\quad f''(x) = 6x + 6 \qquad ...(3)$

$\therefore$ For a local maximum or minimum, we have

$f'(x) = 0$

$\Rightarrow \quad 3x^2 + 6x = 0 \qquad [\because$ By using (2)]

$\Rightarrow \quad 3x(x + 2) = 0$

$\Rightarrow \quad x = 0, -2$

$\therefore$ The critical values for $f(x)$ in ascending order are $x = -2, 0$.

Now, $f''(x)$ at $x = -2$

$\therefore \quad f''(-2) = 6(-2) + 6 = -12 + 6 = -6 < 0 \qquad [\because$ By using (3)]

$\therefore$ $x = -2$ is a point of local maximum and the corresponding local maximum value is given by

$$f(-2) = (-2)^3 + 3(-2)^2 - 4 \qquad [\because \text{ By using (1)}]$$
$$= -8 + 12 - 4 = 0$$

Also, $f''(x)$ at $x = 0$

$\therefore \quad f''(0) = 6(0) + 6 = 0 + 6 = 6 > 0 \qquad [\because$ By using (3)]

$\therefore$ $x = 0$ is a point of local minimum and the corresponding local minimum value is given by

$$f(0) = 0 + 0 - 4 = -4 \qquad [\because \text{ By using (1)}]$$

$\therefore$ Local minimum value is -4 at $x = 0$ and

Local maximum value is 0 at $x = -2$.

Example 3. *Find the points of local maxima and local minima, if any, of the following functions. Also find the local maximum and local minimum values.*

(*i*) $x^4 - 62x^2 + 120x + 9$ (*ii*) $3x^4 - 2x^3 - 6x^2 + 6x + 1$

(*iii*) $\frac{-3}{4}x^4 + 2x^3 + \frac{9}{2}x^2 + 100$ (*iv*) $\frac{-3}{4}x^4 - 8x^3 + \frac{45}{2}x^2 + 105$

(*v*) $6x^5 - 45x^4 + 80x^3 + 6$.

Solution. (*i*) Let $f(x) = x^4 - 62x^2 + 120x + 9 \qquad ...(1)$

$f'(x) = 4x^3 - 124x + 120 \qquad ...(2)$

and $\quad f''(x) = 12x^2 - 124 \qquad ...(3)$

$\therefore$ For a local maximum or minimum, we have

$f'(x) = 0$

$\Rightarrow \quad 4x^3 - 124x + 120 = 0 \qquad [\because$ By using (2)]

$\Rightarrow \quad 4(x^3 - 31x + 30) = 0$

$\Rightarrow \quad x^3 - 31x + 30 = 0$

$\Rightarrow \quad (x - 1)(x^2 + x - 30) = 0$

$\Rightarrow \quad (x - 1)(x + 6)(x - 5) = 0$

$\Rightarrow \quad x = 1, -6, 5$

$\therefore$ The critical values for $f(x)$ in ascending order are $x = -6, 1, 5$.

Now, $f''(x)$ at $x = -6$

$\Rightarrow \quad f''(-6) = 12(-6)^2 - 124 = 432 - 124 = 308 > 0 \quad [\because$ By using (3)]

$x = -6$ is a point of local minimum and the corresponding local minimum value is given by

$$f(-6) = (-6)^4 - 62(-6)^2 + 120(-6) + 9$$
$$= 1296 - 2232 - 720 + 9 = -1647$$

$f''(x)$ at $x = 1$, $f''(1) = 12(1)^2 - 124 = 12 - 124 = -112 < 0$

$\therefore$ $x = 1$ is a point of local maximum and the corresponding local maximum value is given by

$$f(1) = (1)^4 - 62(1)^2 + 120(1) + 9 \quad [\because \text{ By using (1)}]$$
$$= 1 - 62 + 120 + 9 = 68$$

$f''(x)$ at $x = 5$, $f''(5) = 12(5)^2 - 124 = 300 - 124 = 176 > 0$

$\therefore$ $x = 5$ is a point of local minimum and the corresponding local minimum value is given by

$$f(5) = (5)^4 - 62(5)^2 + 120(5) + 9$$
$$= 625 - 1550 + 600 + 9 = -316$$

$\therefore$ Local minimum value $= -1647$ at $x = -6$

Local maximum value $= 68$ at $x = 1$

Local minimum value $= -316$ at $x = 5$.

(ii) Let $\quad f(x) = 3x^4 - 2x^3 - 6x^2 + 6x + 1 \quad$ ...(1)

$\Rightarrow \quad f'(x) = 12x^3 - 6x^2 - 12x + 6 \quad$ ...(2)

and $\quad f''(x) = 36x^2 - 12x - 12 \quad$ ...(3)

$\therefore$ For a local maximum or minimum, we have

$$f'(x) = 0$$

$\Rightarrow \quad 12x^3 - 6x^2 - 12x + 6 = 0 \quad [\because$ By using (2)]

$\Rightarrow \quad 6(2x^3 - x^2 - 2x + 1) = 0$

$\Rightarrow \quad 2x^3 - x^2 - 2x + 1 = 0$

$\Rightarrow \quad x^2(2x - 1) - 1(2x - 1) = 0$

$\Rightarrow \quad (2x - 1)(x^2 - 1) = 0$

$\Rightarrow \quad (2x - 1)(x - 1)(x + 1) = 0 \quad \Rightarrow \quad x = \frac{1}{2}, 1, -1.$

$\therefore$ The critical values for $f(x)$ in the ascending order are $x = -1, \frac{1}{2}, 1$

Now, $f''(x)$ at $x = -1$, $f''(-1) = 36(-1)^2 - 12(-1) - 12 \quad [\because$ By using (3)]

$$= 36 + 12 - 12 = 36 > 0$$

$\therefore$ $x = -1$ is a point of local minimum and the corresponding local minimum value is given by

$$f(-1) = 3(-1)^4 - 2(-1)^3 - 6(-1)^2 + 6(-1) + 1 \quad [\because \text{ By using (1)}]$$
$$= 3 + 2 - 6 - 6 + 1 = -6$$

$f''(x)$ at $x = \frac{1}{2}$, $f''\left(\frac{1}{2}\right) = 36\left(\frac{1}{2}\right)^2 - 12\left(\frac{1}{2}\right) - 12 \quad [\because$ By using (3)]

$$= 36\left(\frac{1}{4}\right) - 6 - 12 = 9 - 18 = -9 < 0$$

$\therefore$ $x = \frac{1}{2}$ is a point of local maximum and the corresponding local maximum value is given by

$$f\left(\frac{1}{2}\right) = 3\left(\frac{1}{2}\right)^4 - 2\left(\frac{1}{2}\right)^3 - 6\left(\frac{1}{2}\right)^2 + 6\left(\frac{1}{2}\right) + 1 \qquad [\because \text{ By using (1)}]$$

$$= 3\left(\frac{1}{16}\right) - 2\left(\frac{1}{8}\right) - 6\left(\frac{1}{4}\right) + 3 + 1$$

$$= \frac{3}{16} - \frac{1}{4} - \frac{3}{2} + 4 = \frac{3 - 4 - 24 + 64}{16} = \frac{39}{16}$$

$f''(x)$ at $x = 1$, $f''(1) = 36(1)^2 - 12(1) - 12$ $\qquad [\because$ By using (3)]

$$= 36 - 12 - 12 = 12 > 0$$

$\therefore$ $x = 1$ is a point of local minimum and the corresponding local minimum value is given by

$$f(1) = 3(1)^4 - 2(1)^3 - 6(1)^2 + 6(1) + 1 \qquad [\because \text{ By using (1)}]$$

$$= 3 - 2 - 6 + 6 + 1 = 2$$

$\therefore$ Local minimum value $= -6$ at $x = -1$

Local maximum value $= \frac{39}{16}$ at $x = \frac{1}{2}$

Local minimum value $= 2$ at $x = 1$.

(*iii*) Let $\qquad f(x) = \frac{-3}{4}x^4 + 2x^3 + \frac{9}{2}x^2 + 100$ $\qquad$ ...(1)

$\Rightarrow$ $\qquad f'(x) = -3x^3 + 6x^2 + 9x$ $\qquad$ ...(2)

and $\qquad f''(x) = -9x^2 + 12x + 9$ $\qquad$ ...(3)

$\therefore$ For a local maximum or minimum, we have

$$f'(x) = 0$$

$\Rightarrow$ $\qquad -3x^3 + 6x^2 + 9x = 0$ $\qquad [\because$ By using (2)]

$\Rightarrow$ $\qquad -3x(x^2 - 2x - 3) = 0$

$\Rightarrow$ $\qquad x(x^2 - 2x - 3) = 0$

$\Rightarrow$ $\qquad x(x+1)(x-3) = 0$

$\Rightarrow$ $\qquad x = 0, -1, 3$

$\therefore$ The critical values for $f(x)$ in ascending order are $x = 0, -1, 3$.

Now, $f''(x)$ at $x = 0$.

$\Rightarrow$ $\qquad f''(0) = -9(0)^2 + 12(0) + 9 = 9 > 0$ $\qquad [\because$ By using (3)]

$\therefore$ $x = 0$ is a point of local minimum and the corresponding local minimum value is given by

$$f(0) = \frac{-3}{4}(0)^4 + 2(0)^3 + \frac{9}{2}(0)^2 + 100 = 100 \qquad [\because \text{ By using (1)}]$$

$f''(x)$ at $x = -1$, $f''(-1) = -9(-1)^2 + 12(-1) + 9$ [$\because$ By using (3)]

$$= -9 - 12 + 9 = -12 < 0$$

$\therefore$ $x = -1$ is a point of local maximum and the corresponding local maximum value is given by

$$f(-1) = \frac{-3}{4}(-1)^4 + 2(-1)^3 + \frac{9}{2}(-1)^2 + 100 \quad [\because \text{ By using (1)}]$$

$$= \frac{-3}{4} - 2 + \frac{9}{2} + 100 = \frac{-3}{4} + \frac{9}{2} + \frac{98}{1}$$

$$= \frac{-3 + 18 + 392}{4} = \frac{407}{4}$$

$f''(x)$ at $x = 3$, $f''(3) = -9(3)^2 + 12(3) + 9$ [$\because$ By using (3)]

$$= -81 + 36 + 9 = -36 < 0$$

$\therefore$ $x = 3$ is a point of local maximum and the corresponding local maximum value is given by

$$f(3) = \frac{-3}{4}(3)^4 + 2(3)^3 + \frac{9}{2}(3)^2 + 100 \quad [\because \text{ By using (1)}]$$

$$= \frac{-3}{4}(81) + 54 + \frac{81}{2} + 100$$

$$= \frac{-243}{4} + \frac{81}{2} + 154 = \frac{-243 + 162 + 616}{4} = \frac{535}{4}$$

$\therefore$ Local minimum value = 100 at $x = 0$

Local maximum value = $\frac{407}{4}$ at $x = -1$

Local maximum value = $\frac{535}{4}$ at $x = 3$.

(*iv*) Let $f(x) = \frac{-3}{4}x^4 - 8x^3 + \frac{45}{2}x^2 + 105$

Please try yourself.

[**Hint.** Proceed as in Part (*iii*).] [**Ans.** Local maximum value = 105 at $x = 0$

Local minimum value = $\frac{231}{4}$ at $x = -3$

Local maximum value = $\frac{295}{4}$ at $x = -5$]

(*v*) Let $f(x) = 6x^5 - 45x^4 + 80x^3 + 6$ ...(1)

$\Rightarrow$ $f'(x) = 30x^4 - 180x^3 + 240x^2$ ...(2)

and $f''(x) = 120x^3 - 540x^2 + 480x$ ...(3)

$\therefore$ For a local maximum or minimum, we have

$$f'(x) = 0$$

$\Rightarrow$ $30x^4 - 180x^3 + 240x^2 = 0$ [$\because$ By using (2)]

$\Rightarrow \quad 30x^2 (x^2 - 6x + 8) = 0$

$\Rightarrow \quad 30x^2 (x - 2)(x - 4) = 0$

$\Rightarrow \quad x = 0, 2, 4.$

$\therefore$ The critical values for $f(x)$ in ascending order are $x = 0, 2, 4$.

Now, $f''(x)$ at $x = 0$

$\Rightarrow \quad f''(0) = 120(0)^3 - 540(0)^2 + 480(0) = 0$ [$\because$ By using (3)]

$\therefore$ The second derivative test fails.

So, let us apply the first derivative test.

$\therefore$ When x is slightly less than 0, say (-0.1)

$\therefore \quad f'(-0.1) = 30(-0.1)^4 - 180(-0.1)^3 + 240(-0.1)^2$ [$\because$ By using (2)]

$= 0.003 + 0.18 + 2.4 = 2.583 > 0$ *i.e.*, +ve

When x is slightly greater than 0 ; say (0.1)

$\therefore \quad f'(0.1) = 30(0.1)^4 - 180(0.1)^3 + 240(0.1)^2$ [$\because$ By using (2)]

$= 0.003 - 0.18 + 2.4 = 2.223 > 0$ *i.e.*, +ve

$\therefore$ $f'(x)$ does not changes sign as increases through 0.

$\therefore$ $x = 0$ is a point of inflexion *i.e.*, $x = 0$ is neither a local maximum nor a local minimum.

$f''(x)$ at $x = 2$, $f''(2) = 120(2)^3 - 540(2)^2 + 480(2)$ [$\because$ By using (3)]

$= 120(8) - 540(4) + 960 = 960 - 2160 + 960 = -240 < 0$

$\therefore$ $x = 2$ is a point of local maximum and the corresponding local maximum value is given by

$f(2) = 6(2)^5 - 45(2)^4 + 80(2)^3 + 6$ [$\because$ By using (1)]

$= 6(32) - 45(16) + 80(8) + 6$

$= 192 - 720 + 640 + 6 = 118$

$f''(x)$ at $x = 4$, $f''(4) = 120(4)^3 - 540(4)^2 + 480(4)$ [$\because$ By using (3)]

$= 120(64) - 540(16) + 1920$

$= 7680 - 8640 + 1920 = 960 > 0$

$\therefore$ $x = 4$ is a point of local minimum and the corresponding local minimum value is given by

$f(4) = 6(4)^5 - 45(4)^4 + 80(4)^3 + 6$ [$\because$ By using (1)]

$= 6(1024) - 45(256) + 80(64) + 6$

$= 6144 - 11520 + 5120 + 6 = -250$

$\therefore$ Local maximum value $= 118$ at $x = 2$

Local minimum value $= -250$ at $x = 4$.

Example 4. *Find the local maximum and local minimum values, if any, of the following functions :*

(*i*) $(x + 3)^3 (x - 4)^4$ (*ii*) $(x - 1)^3 (x + 1)^2$

(*iii*) $(x - 2)^4 (x + 1)^3$ (*iv*) $(x - 3)^5 (x + 1)^2$

(*v*) $(x + 1)(x + 2)^{1/3}$; $x \geq -2$.

Solution. (*i*) Let $f(x) = (x + 3)^3 (x - 4)^4$ ...(1)

$\Rightarrow \quad f'(x) = (x + 3)^3 . 4(x - 4)^3 . 1 + (x - 4)^4 . 3(x + 3)^2 . 1$

$\Rightarrow \quad f'(x) = (x + 3)^2 . (x - 4)^3 . [4(x + 3) + 3(x - 4)]$

$\Rightarrow \quad f'(x) = (x + 3)^2 . (x - 4)^3 . [4x + 12 + 3x - 12]$

$\Rightarrow \quad f'(x) = (x + 3)^2 . (x - 4)^3 . 7x$ ...(2)

$\therefore$ For a local maximum or minimum, we have

$$f'(x) = 0$$

$\Rightarrow$ $7x(x + 3)^2 (x - 4)^3 = 0$

$\Rightarrow$ $x = 0, -3, 4.$

$\therefore$ The critical values for $f(x)$ in ascending order are $x = -3, 0, 4$.

For $x = -3$

When x is slightly less than -3, say (-3.1)

$\Rightarrow$ $f'(-3.1) = (-3.1 + 3)^2 (-3.1 - 4)^3 . 7(-3.1)$ [$\because$ By using (2)]

$= (+\text{ve}) (-\text{ve}) (+\text{ve}) (-\text{ve}) = +\text{ve}$

When x is slightly greater than -3, say (-2.9)

$\therefore$ $f'(-2.9) = (-2.9 + 3)^2 (-2.9 - 4)^3 .7(-2.9)$ [$\because$ By using (2)]

$= (+\text{ve}) (-\text{ve}) (+\text{ve}) (-\text{ve}) = +\text{ve}.$

$\therefore$ $f'(x)$ does not changes sign as x increases through -3.

$\therefore$ $x = -3$ is a point of inflexion *i.e.*, $x = -3$ is neither a local maximum nor a local minimum.

For $x = 0$

When x is slightly less than 0, say (-0.1)

$\Rightarrow$ $f'(-0.1) = (-0.1 + 3)^2 (-0.1 - 4)^3 . 7(-0.1)$ [$\because$ By using (2)]

$= (+\text{ve}) (-\text{ve}) (+\text{ve}) (-\text{ve}) = +\text{ve}$

When x is slightly greater than 0, say (0.1)

$\Rightarrow$ $f'(0.1) = (0.1 + 3)^2 (0.1 - 4)^3 . 7(0.1)$ [$\because$ By using (2)]

$= (+\text{ve}) (-\text{ve}) (+\text{ve}) (+\text{ve}) = -\text{ve}$

$\therefore$ $f'(x)$ changes sign from (+ve) to (–ve) as x increases through 0.

$\therefore$ $x = 0$ is a point of local maximum and the corresponding local maximum value is given by

$f(0) = (0 + 3)^3 (0 - 4)^4$ [$\because$ By using (1)]

$= (27) (256) = 6912$

For $x = 4$

When x is slightly less than 4, say (3.9)

$\therefore$ $f'(3.9) = (3.9 + 3)^2 (3.9 - 4)^3 . 7(3.9)$ [$\because$ By using (2)]

$= (+\text{ve}) (-\text{ve}) (+\text{ve}) (+\text{ve}) = -\text{ve}$

When x is slightly greater than 4, say (4.1)

$\therefore$ $f'(4.1) = (4.1 + 3)^2 (4.1 - 4)^3 . 7(4.1)$ [$\because$ By using (2)]

$= (+\text{ve}) (+\text{ve}) (+\text{ve}) (+\text{ve}) = +\text{ve}.$

$\Rightarrow$ $f'(x)$ changes sign from (–ve) to (+ve) as increases through 4.

$\therefore$ $x = 4$ is a point of local minimum and the corresponding local minimum value is given by

$f(4) = (4 + 3)^3 (4 - 4)^4 = (7)^3 (0) = 0$ [$\because$ By using (1)]

$\therefore$ Local maximum value = 6912 at $x = 0$

Local minimum value = 0 at $x = 4$.

(*ii*) Let $f(x) = (x - 1)^3 (x + 1)^2$ ...(1)

$$\Rightarrow \quad f'(x) = (x-1)^3 \,.\, 2(x+1) \,.\, 1 + (x+1)^2 \,.\, 3(x-1)^2 \,.\, 1$$
$$= (x+1)(x-1)^2 [2(x-1) + 3(x+1)]$$
$$= (x+1)(x-1)^2 [2x - 2 + 3x + 3]$$
$$f'(x) = (x+1)(x-1)^2 (5x+1) \quad ...(2)$$

$\therefore$ For a local maximum or minimum, we have

$$f'(x) = 0$$
$$\Rightarrow \quad (x+1)(x-1)^2(5x+1) = 0$$
$$\Rightarrow \quad x = -1, 1, \frac{-1}{5}$$

$\therefore$ The critical values for $f(x)$ in the ascending order are $x = -1, -0.2, 1$

$\therefore$ For $x = -1$

When x is slightly less than -1, say (-1.1)

$$\therefore \quad f'(-1.1) = (-1.1+1)(-1.1-1)^2 [5(-1.1)+1] \quad [\because \text{ By using (2)}]$$
$$= (-ve)(+ve)(-ve) = +ve$$

When x is slightly greater than -1, say (-0.9)

$$\therefore \quad f'(-0.9) = (-0.9+1)(-0.9-1)^2 [5(-0.9)+1] \quad [\because \text{ By using (2)}]$$
$$= (+ve)(+ve)(-ve) = -ve$$

$\therefore$ $f'(x)$ changes sign from (+ve) to (–ve) as increases through -1.

$\therefore$ $x = -1$ is a point of local maximum and the corresponding local maximum value is given by

$$f(-1) = (-1+1)(-1-1)^2 [5(-1)+1] = 0$$

For $x = -0.2$

When x is slightly less than -0.2 say (-0.3)

$$\therefore \quad f'(-0.3) = (-0.3+1)(-0.3-1)^2 [5(-0.3)+1]$$
$$= (+ve)(+ve)(-ve) = -ve$$

When x is slightly greater than (-0.2), say (-0.1)

$$\therefore \quad f'(-0.1) = (-0.1+1)(-0.1-1)^2 [5(-0.1)+1]$$
$$= (+ve)(+ve)(+ve) = +ve$$

$\therefore$ $f(x)$ changes sign from (–ve) to (+ve) as x increases through $x = -0.2$.

$\therefore$ $x = -0.2$ or $\frac{-1}{5}$ is a point of local minimum and the corresponding local minimum value is given by

$$f(-0.2) = (-0.2-1)^3(-0.2+1)^2 = (-1.2)^3(0.8)^2$$
$$= (-1.728)(0.64) = -1.106 \text{ or } \frac{-3456}{3125}$$

For $x = 1$

When x is slightly less than 1, say (0.9)

$$\therefore \quad f'(0.9) = (0.9+1)(0.9-1)^2 [5(0.9)+1]$$
$$= (+ve)(+ve)(+ve) = +ve$$

When x is slightly greater than 1, say (1.1)

$$\therefore \quad f'(1.1) = (1.1+1)(1.1-1)^2 [5(1.1)+1] = (+ve)(+ve)(+ve) = +ve$$

$\therefore$ $f(x)$ does not changes sign as x increases through $x = 1$.

$\therefore$ $x = 1$ is a point of inflexion *i.e.*, $x = 1$ is neither a local maximum nor a local minimum.

$\therefore$ Local minimum value $= -1.106$ or $\dfrac{-3456}{3125}$ at $x = -0.2$

Local maximum value $= 0$ at $x = -1$.

(*iii*) Let $f(x) = (x-2)^4 (x+1)^3$

Please try yourself.

[**Hint.** Proceed as in part (*ii*).] [**Ans.** Local maximum value $= 18.35$ at $x = \dfrac{2}{7}$ or 0.2

Local minimum value $= 0$ at $x = 2$

Point of inflexion at $x = -1$]

(*iv*) Let $f(x) = (x-3)^5 (x+1)^2$

Please try yourself.

[**Hint.** Proceed as in part (*ii*).] [**Ans.** Local maximum value $= 0$ at $x = -1$

Local minimum value $= \left(\dfrac{-20}{7}\right)^5 \left(\dfrac{8}{7}\right)^2$ at $x = \dfrac{1}{7}$

Point of inflexion at $x = 3$]

(*v*) Let $f(x) = (x+1)(x+2)^{1/3};\ x \ge -2$ ...(1)

$$\Rightarrow \quad f'(x) = (x+1) \cdot \frac{1}{3}(x+2)^{-2/3} \cdot 1 + (x+2)^{1/3} \cdot 1$$

$$= \frac{(x+1)}{3(x+2)^{2/3}} + (x+2)^{1/3}$$

$$= \frac{x+1+3(x+2)}{3(x+2)^{2/3}} = \frac{4x+7}{3(x+2)^{2/3}} \qquad ...(2)$$

$\therefore$ For a local maximum or minimum, we have

$$f'(x) = 0$$

$$\Rightarrow \quad \frac{4x+7}{3(x+2)^{2/3}} = 0 \Rightarrow 4x+7 = 0 \Rightarrow x = \frac{-7}{4}$$

$\therefore$ The critical value for $f(x)$ is $x = \dfrac{-7}{4}$.

$\therefore$ When x is slightly less than $\dfrac{-7}{4}$, say (-1.8)

$$\therefore \quad f'(-1.8) = \frac{4(-1.8)+7}{3(-1.8+2)^{2/3}} = \frac{(-\text{ve})}{(+\text{ve})(+\text{ve})} = -\text{ve}$$

When x is slightly greater than $\dfrac{-7}{4}$, say (-1.7)

$$\therefore \quad f'(-1.7) = \frac{4(-1.7)+7}{3(-1.7+2)^{2/3}} = \frac{(+\text{ve})}{(+\text{ve})(+\text{ve})} = (+\text{ve})$$

$\therefore$ $f'(x)$ changes sign from (–ve) to (+ve) as increases through $\frac{-7}{4}$.

$\therefore$ $x = \frac{-7}{4}$ is a point of local minimum and the corresponding local minimum value is given by

$$f\left(\frac{-7}{4}\right) = \left(\frac{-7}{4}+1\right)\left(\frac{-7}{4}+2\right)^{1/3}$$

$$= \left(\frac{-7+4}{4}\right)\left(\frac{-7+8}{4}\right)^{1/3} = \left(\frac{-3}{4}\right)\left(\frac{1}{4}\right)^{1/3} = \frac{-3}{(4)^{4/3}}.$$

Example 5. *Find the local maximum and local minimum values, if any, of the following functions :*

(*i*) $x\sqrt{1-x}$; $x \le 1$

(*ii*) $\frac{x}{(x-1)(x-4)}$; $1 < x < 4$

(*iii*) $\frac{x^2}{x^4+1}$

(*iv*) $\frac{(x-1)(x-6)}{x-10}$; $x \ne 10$

(*v*) $\frac{x^4}{x-1}$; $x \ne 1$

(*vi*) $\frac{x}{2}+\frac{2}{x}$; $x > 0$

Solution. (*i*) Let $f(x) = x\sqrt{1-x}; x \le 1$ ...(1)

$$\Rightarrow \quad f'(x) = x \cdot \frac{1}{2\sqrt{1-x}} \cdot (-1) + \sqrt{1-x} \cdot 1$$

$$= \frac{-x}{2\sqrt{1-x}} + \sqrt{1-x}$$

$$f'(x) = \frac{-x+2(1-x)}{2\sqrt{1-x}} = \frac{2-3x}{2\sqrt{1-x}} \quad \text{...(2)}$$

$\therefore$ For a local maximum or minimum, we have

$$f'(x) = 0$$

$$\Rightarrow \quad \frac{2-3x}{2\sqrt{1-x}} = 0 \Rightarrow 2-3x = 0 \Rightarrow x = \frac{2}{3} \quad [\because \text{ By using (2)}]$$

$\therefore$ The critical value for $f(x)$ is $x = \frac{2}{3}$.

When x is slightly less than $\frac{2}{3}$, say (0.6)

$$\Rightarrow \quad f'(0.6) = \frac{2-3(0.6)}{2\sqrt{1-0.6}} = \frac{(+\text{ve})}{(+\text{ve})\,(+\text{ve})} = +\text{ve}$$

When x is slightly greater than $\frac{2}{3}$, say (0.7)

$$\Rightarrow \qquad f'(0.7) = \frac{2-3(0.7)}{2\sqrt{1-0.7}} = \frac{(-\text{ve})}{(+\text{ve})\,(+\text{ve})} = -\text{ve}$$

$\therefore$ $f'(x)$ changes sign from (+ve) to (–ve) as x increases through $\frac{2}{3}$.

$\therefore$ $x = \frac{2}{3}$ is a point of local maximum and the corresponding local maximum value is given by

$$\Rightarrow \qquad f\left(\frac{2}{3}\right) = \frac{2}{3}\sqrt{1-\frac{2}{3}} = \frac{2}{3}\sqrt{\frac{1}{3}} = \frac{2}{3\sqrt{3}}$$

$\therefore$ Local maximum value $= \frac{2}{3\sqrt{3}}$ at $x = \frac{2}{3}$.

(*ii*) Let $\qquad f(x) = \frac{x}{(x-1)(x-4)};\ 1 < x < 4 \qquad$ ...(1)

$$\Rightarrow \qquad f(x) = \frac{x}{x^2-5x+4}$$

$$\Rightarrow \qquad f'(x) = \frac{(x^2-5x+4).1 - x(2x-5)}{(x^2-5x+4)^2}$$

$$= \frac{x^2-5x+4-2x^2+5x}{(x^2-5x+4)^2}$$

$$f'(x) = \frac{4-x^2}{(x^2-5x+4)^2} \qquad \text{...(2)}$$

$\therefore$ For a local maximum or minimum, we have

$$f'(x) = 0$$

$$\Rightarrow \qquad \frac{4-x^2}{(x^2-5x+4)^2} = 0 \qquad [\because \text{ By using (2)}]$$

$\Rightarrow \qquad (4-x^2) = 0 \ \Rightarrow\ (2-x)(2+x) = 0$

$\Rightarrow \qquad x = 2, -2$

$\therefore$ The critical value for $f(x)$ is $x = 2$ $\qquad [\because\ 1 < x < 4]$

For $x = 2$:

$\therefore$ When x is slightly less than 2, say (1.9)

$$\Rightarrow \qquad f'(x) = \frac{[4-(1.9)^2]}{[(1.9)^2-5(1.9)+4]^2} = \frac{(+\text{ve})}{(+\text{ve})} = (+\text{ve})$$

When x is slightly greater than 2, say (2.1)

$$\Rightarrow \qquad f'(x) = \frac{[4-(2.1)^2]}{[(2.1)^2-5(2.1)+4]^2} = \frac{(-\text{ve})}{(+\text{ve})} = (-\text{ve})$$

$\therefore$ $f'(x)$ changes sign from (+ve) to (–ve) as x increases through 2.

⇒ $x = 2$ is a point of local maximum and the corresponding local maximum value is given by

$$f(2) = \frac{2}{(2-1)(2-4)} = \frac{2}{(1)(-2)} = -1$$

∴ Local maximum value $= -1$ at $x = 2$.

(*iii*) Let $$f(x) = \frac{x^2}{x^4+1} \quad \text{...(1)}$$

$$\Rightarrow \quad f'(x) = \frac{(x^4+1).2x - x^2(4x^3)}{(x^4+1)^2} = \frac{2x^5+2x-4x^5}{(x^4+1)^2} = \frac{2x-2x^5}{(x^4+1)^2}$$

$$f'(x) = \frac{2x(1-x^4)}{(x^4+1)^2} \quad \text{...(2)}$$

∴ For a local maximum or minimum, we have

$$f'(x) = 0$$

$$\Rightarrow \quad \frac{2x(1-x^4)}{(x^4+1)^2} = 0 \quad [\because \text{ By using (2)}]$$

$$\Rightarrow \quad \frac{2x(1-x)(1+x)(1+x^2)}{(x^4+1)^2} = 0$$

$$\Rightarrow \quad 2x(1-x)(1+x)(1+x^2) = 0$$

⇒ $x = 0, 1, -1$ (only real values).

∴ The critical values for $f(x)$ in ascending order are $x = -1, 0, 1$.

For $x = -1$

When x is slightly less than -1, say (-1.1)

$$\Rightarrow \quad f'(-1.1) = \frac{2(-1.1)[1-(-1.1)^4]}{[(-1.1)^4+1]^2} = \frac{(+ve)(-ve)(-ve)}{(+ve)} = (+ve)$$

When x is slightly greater than -1, say (-0.9)

$$\Rightarrow \quad f'(-0.9) = \frac{2(-0.9)[1-(-0.9)^4]}{[(-0.9)^4+1]^2} = \frac{(+ve)(-ve)(+ve)}{(+ve)} = (-ve)$$

∴ $f'(x)$ changes sign from (+ve) to (–ve) as x increases through -1.

∴ $x = -1$ is a point of local maximum and the corresponding local maximum value is given by

$$f(-1) = \frac{(-1)^2}{(-1)^4+1} = \frac{1}{1+1} = \frac{1}{2}$$

For $x = 0$:

When x is slightly less than 0, say (-0.1)

$$\Rightarrow \quad f'(-0.1) = \frac{2(-0.1)[1-(-0.1)^4]}{[(-0.1)^4+1]^2} = \frac{(+ve)(-ve)(+ve)}{(+ve)} = -ve$$

When x is slightly greater than 0, say (0.1)

$$\therefore \quad f'(0.1) = \frac{2\,(0.1)\,[1-(0.1)^4]}{[(0.1)^4+1]^2} = \frac{(+\text{ve})\,(+\text{ve})\,(+\text{ve})}{(+\text{ve})} = (+\text{ve})$$

$\therefore$ $f'(x)$ changes sign from (–ve) to (+ve) as x increases through 0.

$\therefore$ $x = 0$ is a point of local minimum and the corresponding local minimum value is given by

$$f(0) = \frac{(0)^2}{(0)^4+1} = \frac{0}{1} = 0$$

For $x = 1$

When x is slightly less than 1, say (0.9)

$$\therefore \quad f'(0.9) = \frac{2(0.9)\,[1-(0.9)^4]}{[(0.9)^4+1]^2} = \frac{(+\text{ve})\,(+\text{ve})\,(+\text{ve})}{(+\text{ve})} = (+\text{ve})$$

When x is slightly greater than 1, say (1.1)

$$\therefore \quad f'(1.1) = \frac{2(1.1)\,[1-(1.1)^4]}{[(1.1)^4+1]^2} = \frac{(+\text{ ve})\,(+\text{ ve})\,(-\text{ ve})}{(+\text{ ve})} = (-\text{ve})$$

$\therefore$ $f'(x)$ changes sign from (+ve) to (–ve) as x increases through $x = 1$.

$\therefore$ $x = 1$ is a point of local maximum and the corresponding local maximum value is given by

$$f(1) = \frac{(1)^2}{(1)^4+1} = \frac{1}{1+1} = \frac{1}{2}$$

$\therefore$ Local minimum value = 0 at $x = 0$

Local maximum value $= \frac{1}{2}$ at $x = -1$

Local maximum value $= \frac{1}{2}$ at $x = 1$.

(*iv*) Let $\quad f(x) = \dfrac{(x-1)\,(x-6)}{x-10},\; x \neq 10 \quad$ ...(1)

$$\Rightarrow \quad f(x) = \frac{x^2-7x+6}{x-10}$$

$$\Rightarrow \quad f'(x) = \frac{(x-10)\,(2x-7)-(x^2-7x+6)\,.\,1}{(x-10)^2}$$

$$= \frac{(2x^2-20x-7x+70)-(x^2-7x+6)}{(x-10)^2} = \frac{x^2-20x+64}{(x-10)^2}$$

$$\Rightarrow \quad f'(x) = \frac{(x-4)\,(x-16)}{(x-10)^2} \quad \text{...(2)}$$

∴ For a local maximum or minimum, we have

$$f'(x) = 0$$

$$\Rightarrow \quad \frac{(x-4)(x-16)}{(x-10)^2} = 0$$

$$\Rightarrow \quad (x-4)(x-16) = 0$$

$$\Rightarrow \quad x = 4, 16$$

∴ The critical values for $f(x)$ in ascending order are $x = 4, 16$.

For $x = 4$

When x is slightly less than 4, say (3.9)

$$\Rightarrow \quad f'(3.9) = \frac{(3.9-4)(3.9-16)}{(3.9-10)^2} = \frac{(-\text{ve})(-\text{ve})}{(+\text{ve})} = +\text{ve}$$

When x is slightly greater than 4, say (4.1)

$$\Rightarrow \quad f'(4.1) = \frac{(4.1-4)(4.1-16)}{(4.1-10)^2} = \frac{(+\text{ve})(-\text{ve})}{(+\text{ve})} = -\text{ve}$$

∴ $f'(x)$ changes sign from +ve to –ve as increases through 4.

∴ $x = 4$ is a point of local maximum and the corresponding local maximum value is given by

$$f(4) = \frac{(4-1)(4-6)}{(4-10)} = \frac{(3)(-2)}{(-6)} = 1$$

For $x = 16$

When x is slightly less than 16, say (15.9)

$$\therefore \quad f'(15.9) = \frac{(15.9-4)(15.9-16)}{(15.9-10)^2} = \frac{(+\text{ve})(-\text{ve})}{(+\text{ve})} = -\text{ve}$$

When x is slightly greater than 16, say (16.1)

$$\therefore \quad f'(16.1) = \frac{(16.1-4)(16.1-16)}{(16.1-10)^2} = \frac{(+\text{ve})(+\text{ve})}{(+\text{ve})} = +\text{ve}$$

∴ $f(x)$ changes sign from (–ve) to (+ve) as x increases through 16.

∴ $x = 16$ is a point of local minimum and the corresponding local minimum value is given by

$$f(16) = \frac{(16-1)(16-6)}{(16-10)} = \frac{(15)(10)}{(6)} = \frac{150}{6} = 25$$

∴ Local maximum value = 1 at $x = 4$

Local minimum value = 25 at $x = 16$.

(*v*) Let $$f(x) = \frac{x^4}{x-1}, \; x \neq 1 \qquad \text{...(1)}$$

Please try yourself.

[**Ans.** Local maximum value = 0 at $x = 0$

Local minimum value $= \dfrac{256}{27}$ at $x = \dfrac{4}{3}$]

(*vi*) Let $f(x) = \frac{x}{2} + \frac{2}{x};\ x > 0$ ...(1)

$\Rightarrow \quad f'(x) = \frac{1}{2} - \frac{2}{x^2}$ ...(2)

$\Rightarrow \quad f''(x) = \frac{4}{x^3}$ ...(3)

$\therefore$ For a local maximum or minimum, we have

$f'(x) = 0$

$\Rightarrow \quad \frac{1}{2} - \frac{2}{x^2} = 0 \Rightarrow \frac{2}{x^2} = \frac{1}{2}$ [$\because$ By using (2)]

$\Rightarrow \quad x^2 = 4 \Rightarrow x = \pm 2$

$\therefore$ The critical value for $f(x)$ is $x = 2$ [$\because\ x > 0$]

Now, $f''(x)$ at $x = 2$

$\Rightarrow \quad f''(2) = \frac{4}{(2)^3} = \frac{4}{8} = \frac{1}{2} > 0$ [$\because$ By using (3)]

$\therefore$ $x = 2$ is a point of local minimum and the corresponding local minimum value is given by

$$f(2) = \frac{2}{2} + \frac{2}{2} = 1 + 1 = 2$$

$\therefore$ Local minimum value = 2 at $x = 2$.

Example 6. *Find the local maximum and local minimum values, if any, of the following functions :*

(*i*) $(1 - x)^2 e^x$ (*ii*) $x^2 e^x$

(*iii*) $\sin x$ (*iv*) $\cos x\ ;\ 0 < x < \pi$

Solution. (*i*) Let $f(x) = (1 - x)^2 e^x$ ...(1)

$\Rightarrow \quad f'(x) = (1 - x)^2 . e^x + e^x . 2(1 - x) . (-1)$

$= e^x . (1 - x)[1 - x - 2] = e^x . (1 - x)(-1 - x)$

$= e^x . (x - 1)(x + 1)$

$f'(x) = e^x (x^2 - 1)$ ...(2)

and $f''(x) = e^x . (2x) + (x^2 - 1) . e^x$

$f''(x) = e^x (2x + x^2 - 1)$ ...(3)

$\therefore$ For a local maximum or minimum, we have

$f'(x) = 0$

$\Rightarrow \quad e^x (x^2 - 1) = 0$

$\Rightarrow \quad x^2 - 1 = 0 \Rightarrow x = \pm 1$

$\therefore$ The critical values for $f(x)$ in the ascending order are $x = -1, 1$

Now, $f''(x)$ at $x = -1$

$\Rightarrow \quad f''(-1) = e^{-1} [2(-1) + (-1)^2 - 1]$

$= \frac{1}{e}(-2 + 1 - 1) = \frac{-2}{e} < 0$

∴ $x = -1$ is a point of local maximum and the corresponding local maximum value is given by

$$f(-1) = [1 - (-1)]^2 \cdot e^{-1} = \frac{(2)^2}{e} = \frac{4}{e}.$$

Also, $f''(x)$ at $x = 1$

∴ $f''(1) = e^1[2(1) + (1)^2 - 1] = e\,(2) = 2e > 0$

∴ $x = 1$ is a point of local minimum and the corresponding local minimum value is given by

$$f(1) = (1 - 1)^2 e^1 = 0$$

∴ Local maximum value $= \frac{4}{e}$ at $x = -1$

Local minimum value = 0 at $x = 1$.

(*ii*) Let $f(x) = x^2 e^x$ ...(1)

⇒ $f'(x) = x^2 \cdot e^x + e^x + e^x \cdot (2x)$

⇒ $f'(x) = xe^x (x + 2)$ ...(2)

and
$$f''(x) = \frac{d}{dx}[e^x (x^2 + 2x)]$$
$$= e^x (2x + 2) + (x^2 + 2x) \cdot e^x$$
$$= e^x [2x + 2 + x^2 + 2x]$$

⇒ $f''(x) = e^x (x^2 + 4x + 2)$ ...(3)

∴ For a local maximum or minimum, we have

$$f'(x) = 0$$

⇒ $xe^x (x + 2) = 0$

⇒ $x = 0, -2$ (only finite values)

∴ The critical values for $f(x)$ in the ascending order are $x = -2, 0$.

Now, $f''(x)$ at $x = -2$

⇒ $$f''(-2) = e^{-2}[(-2)^2 + 4(-2) + 2] = \frac{1}{e^2}(4 - 8 + 2) = \frac{-2}{e^2} < 0$$

∴ $x = -2$ is a point of local maximum and the corresponding local maximum value is given by

$$f(-2) = (-2)^2 e^{-2} = \frac{4}{e^2}$$

Also, $f''(x)$ at $x = 0$

⇒ $f''(0) = e^0 (0 + 0 + 2) = 2 > 0$

∴ $x = 0$ is a point of local minimum and the corresponding local minimum value is given by

$$f(0) = (0)^2 e^0 = 0$$

∴ Local maximum value $= \frac{4}{e^2}$ at $x = -2$

Local minimum value = 0 at $x = 0$

(*iii*) Let $f(x) = \sin x$ ...(1)

⇒ $f'(x) = \cos x$ ...(2)

and $f''(x) = -\sin x$ ...(3)

$\therefore$ For a local maximum or minimum, we have

$$f'(x) = 0$$

$$\Rightarrow \quad \cos x = 0$$

$$\Rightarrow \quad x = (2n + 1)\frac{\pi}{2};\ n \in Z$$

$\therefore$ The critical values for $f(x)$ are

$$x = (2n + 1)\frac{\pi}{2};\ n \in Z$$

Now, $f''(x)$ at $\quad x = (2n + 1)\dfrac{\pi}{2}$

$$f''\left((2n+1)\frac{\pi}{2}\right) = -\sin\left((2n+1)\frac{\pi}{2}\right) = -\sin\left(n\pi + \frac{\pi}{2}\right)$$

$$= -(-1)^n \sin\frac{\pi}{2} = (-1)^{n+1} \qquad [\because \sin(n\pi + \theta) = (-1)^n \sin\theta]$$

$$= (-1)^{n+1}$$

Now, two case arises

Case I. When n is an even integer

$\Rightarrow$ $(n + 1)$ is an odd number

$$\Rightarrow \quad (-1)^{n+1} = -1$$

$$\Rightarrow \quad f''\left((2n+1)\frac{\pi}{2}\right) = -1 < 0$$

$$\therefore \quad x = (2n + 1)\frac{\pi}{2}$$

is a point of local maximum and the corresponding local maximum value is given by

$$f\left((2n+1)\frac{\pi}{2}\right) = \sin\left((2n+1)\frac{\pi}{2}\right)$$

$$= \sin\left(n\pi + \frac{\pi}{2}\right) = (-1)^n . 1 = 1 \qquad [\because n \text{ is even}]$$

Case II. When n is an odd integer

$\Rightarrow$ $(n + 1)$ is an even number

$$\Rightarrow \quad (-1)^{n+1} = +1$$

$$\Rightarrow \quad f''\left((2n+1)\frac{\pi}{2}\right) = 1 > 0$$

$$\therefore \quad x = (2n + 1)\frac{\pi}{2}$$

is a point of local minimum and the corresponding local minimum value is given by

$$f\left((2n+1)\frac{\pi}{2}\right) = \sin\left((2n+1)\frac{\pi}{2}\right) = \sin\left(n\pi + \frac{\pi}{2}\right) = (-1)^n . 1 = -1 \quad [\because n \text{ is odd}]$$

$\therefore$ Local maximum value = 1 at $x = (2n + 1)\,\dfrac{\pi}{2}$; n is even

Local minimum value = – 1 at $x = (2n + 1)\,\dfrac{\pi}{2}$; n is odd.

(*iv*) Let $\quad f(x) = \cos x \,;\, 0 < x < \pi \qquad$ ...(1)

$\Rightarrow \quad f'(x) = -\sin x \qquad$ ...(2)

$\therefore$ For a local maximum or minimum, we have

$$f'(x) = 0$$

i.e., $\quad -\sin x = 0 \;\Rightarrow\; \sin x = 0$

Which is not possible for $0 < x < \pi$.

$\therefore$ $f(x)$ has neither a local maximum nor a local minimum for $0 < x < \pi$.

Example 7. *Find the local maximum and local minimum values of the following functions :*

(*i*) $\sin x - \cos x \,;\, 0 < x < 2\pi$ $\qquad$ (*ii*) $\sin x + \cos x \,;\, 0 < x < \dfrac{\pi}{2}$

(*iii*) $2 \sin x + \cos 2x \,;\, 0 < x < \pi$ $\qquad$ (*iv*) $\sin^4 x + \cos^4 x \,;\, 0 < x < \dfrac{\pi}{2}$

(*v*) $2 \sin x + \sin 2x \,;\, 0 \le x \le 2\pi$ $\qquad$ (*vi*) $\sin 2x - x \,;\, 0 \le x \le 2\pi$.

Solution. (*i*) Let $\quad f(x) = \sin x - \cos x \,;\, 0 < x < 2\pi \qquad$ ...(1)

$\Rightarrow \quad f'(x) = \cos x + \sin x \qquad$ ...(2)

and $\quad f''(x) = -\sin x + \cos x \qquad$ ...(3)

$\therefore$ For a local maximum or minimum, we have

$$f'(x) = 0$$

$\Rightarrow \quad \cos x + \sin x = 0 \qquad$ [$\because$ By using (2)]

$\Rightarrow \quad \cos x = -\sin x$

$\Rightarrow \quad -\tan x = 1 \;\Rightarrow\; \tan x = -1$

$$\Rightarrow \quad x = \left(\pi - \frac{\pi}{4}\right), \left(2\pi - \frac{\pi}{4}\right)$$

$$\Rightarrow \quad x = \frac{3\pi}{4}, \frac{7\pi}{4}.$$

$\therefore$ The critical values for $f(x)$ in the ascending order are $x = \dfrac{3\pi}{4}, \dfrac{7\pi}{4}$

Now, $f''(x)$ at $x = \dfrac{3\pi}{4}$

$$\Rightarrow \quad f''\left(\frac{3\pi}{4}\right) = -\sin\frac{3\pi}{4} + \cos\frac{3\pi}{4}$$

$$= -\sin\left(\pi - \frac{\pi}{4}\right) + \cos\left(\pi - \frac{\pi}{4}\right)$$

$$= -\sin\frac{\pi}{4} - \cos\frac{\pi}{4} = -\frac{1}{\sqrt{2}} - \frac{1}{\sqrt{2}} = \frac{-2}{\sqrt{2}} < 0$$

$\therefore \quad x = \dfrac{3\pi}{4}$ is a point of local maximum and the corresponding local maximum value is given by

$$f\left(\frac{3\pi}{4}\right) = \sin\frac{3\pi}{4} - \cos\frac{3\pi}{4}$$

$$= \frac{1}{\sqrt{2}} - \left(-\frac{1}{\sqrt{2}}\right) = \frac{1}{\sqrt{2}} + \frac{1}{\sqrt{2}} = \frac{2}{\sqrt{2}} = \sqrt{2}$$

Also, $f''(x)$ at $x = \dfrac{7\pi}{4}$.

$$\Rightarrow \quad f''\left(\frac{7\pi}{4}\right) = -\sin\frac{7\pi}{4} + \cos\frac{7\pi}{4}$$

$$= -\sin\left(2\pi - \frac{\pi}{4}\right) + \cos\left(2\pi - \frac{\pi}{4}\right)$$

$$= +\sin\frac{\pi}{4} + \cos\frac{\pi}{4} = \frac{1}{\sqrt{2}} + \frac{1}{\sqrt{2}} = \frac{2}{\sqrt{2}} > 0$$

$\therefore \quad x = \dfrac{7\pi}{4}$ is a point of local minimum and the corresponding local minimum value is given by

$$f\left(\frac{7\pi}{4}\right) = \sin\frac{7\pi}{4} - \cos\frac{7\pi}{4} = \frac{-1}{\sqrt{2}} - \frac{1}{\sqrt{2}} = \frac{-2}{\sqrt{2}} = -\sqrt{2}$$

$\therefore \quad$ Local maximum value $= \sqrt{2}$ at $x = \dfrac{3\pi}{4}$

Local minimum value $= -\sqrt{2}$ at $x = \dfrac{7\pi}{4}$.

(*ii*) Let $\quad f(x) = \sin x + \cos x \,;\, 0 < x < \dfrac{\pi}{2}$ ...(1)

$f'(x) = \cos x - \sin x$ ...(2)

and $\quad f''(x) = -\sin x - \cos x$ ...(3)

$\therefore \quad$ For a local maximum or minimum, we have

$$f'(x) = 0$$

$$\Rightarrow \quad \cos x - \sin x = 0$$

$$\Rightarrow \quad \sin x = \cos x \quad \Rightarrow \quad \tan x = 1$$

$$\Rightarrow \quad x = \frac{\pi}{4} \qquad \left(\because \quad 0 < x < \frac{\pi}{4}\right)$$

$\therefore \quad$ The critical value for $f(x)$ is $x = \dfrac{\pi}{4}$.

Now, $f''(x)$ at $x = \dfrac{\pi}{4}$

$$\Rightarrow \quad f''\left(\frac{\pi}{4}\right) = -\sin\frac{\pi}{4} - \cos\frac{\pi}{4} = \frac{-1}{\sqrt{2}} - \frac{1}{\sqrt{2}} = \frac{-2}{\sqrt{2}} = -\sqrt{2} < 0$$

$\therefore \quad x = \dfrac{\pi}{4}$ is a point of local maximum and the corresponding local maximum value is given by

$$f\left(\frac{\pi}{4}\right) = \sin\frac{\pi}{4} + \cos\frac{\pi}{4} = \frac{1}{\sqrt{2}} + \frac{1}{\sqrt{2}} = \frac{2}{\sqrt{2}} = \sqrt{2}$$

$\therefore \quad$ Local maximum value $= \sqrt{2}$ at $x = \dfrac{\pi}{4}$.

(iii) Let $\quad f(x) = 2\sin x + \cos 2x\,;\ 0 < x < \pi \quad$ ...(1)

$\Rightarrow \quad f'(x) = 2\cos x - 2\sin 2x \quad$ ...(2)

and $\quad f''(x) = -2\sin x - 4\cos 2x \quad$ ...(3)

$\therefore \quad$ For a local maximum or minimum, we have

$$f'(x) = 0$$

$\Rightarrow \quad 2\cos x - 2\sin 2x = 0 \quad$ [$\because$ By using (2)]

$\Rightarrow \quad \cos x - \sin 2x = 0$

$\Rightarrow \quad \cos x - 2\sin x\cos x = 0 \quad$ [$\because$ $\sin 2A = 2\sin A\cos A$]

$\Rightarrow \quad \cos x\,(1 - 2\sin x) = 0$

$\Rightarrow$ either $\quad \cos x = 0 \quad$ or $\quad 1 - 2\sin x = 0$

$\Rightarrow \quad x = \dfrac{\pi}{2} \quad$ or $\quad 2\sin x = 1 \Rightarrow \sin x = \dfrac{1}{2}$

$\Rightarrow \quad x = \dfrac{\pi}{6} \quad$ or $\quad \dfrac{5\pi}{6}$.

$\therefore \quad$ The critical values for $f(x)$ in the ascending order are

$$x = \frac{\pi}{6}, \frac{\pi}{2}, \frac{5\pi}{6}$$

Now, $f''(x)$ at $x = \dfrac{\pi}{6}$

$$\Rightarrow \quad f''\left(\frac{\pi}{6}\right) = -2\,[\sin x + 2\cos 2x]$$

$$= -2\left[\sin\frac{\pi}{6} + 2\cos\frac{\pi}{3}\right] = -2\left(\frac{1}{2} + 2\,.\,\frac{1}{2}\right)$$

$$= -2\left(\frac{1}{2} + 1\right) = -3 < 0$$

$\therefore \quad x = \dfrac{\pi}{6}$ is a point of local maximum and the corresponding local maximum value is given by

$$f\left(\frac{\pi}{6}\right) = 2\sin\frac{\pi}{6} + \cos\frac{2\pi}{6} = 2\left(\frac{1}{2}\right) + \frac{1}{2} = \frac{3}{2}$$

Also, $f''(x)$ at $x = \frac{\pi}{2}$

$$\Rightarrow \quad f''\left(\frac{\pi}{2}\right) = -2\,[\sin x + 2\cos 2x] = -2\left[\sin\frac{\pi}{2} + 2\cos\frac{2\pi}{2}\right]$$

$$= -2\,[1 + 2(-1)] = -2(1-2) = 2 > 0$$

$\therefore \quad x = \frac{\pi}{2}$ is a point of local minimum and the corresponding local minimum value is given by

$$f\left(\frac{\pi}{2}\right) = 2\sin\frac{\pi}{2} + \cos\frac{2\pi}{2} = 2 - 1 = 1$$

$f''(x)$ at $x = \frac{5\pi}{6}$.

$$\Rightarrow \quad f''\left(\frac{5\pi}{6}\right) = -2(\sin x + 2\cos 2x)$$

$$= -2\left[\sin\frac{5\pi}{6} + 2\cos\frac{5\pi}{3}\right] = -2\left[\frac{1}{2} + 2\left(\frac{1}{2}\right)\right] = -3 < 0$$

$\therefore \quad x = \frac{5\pi}{6}$ is a point of local maximum and the corresponding local maximum value is given by

$$f\left(\frac{5\pi}{6}\right) = 2\sin\frac{5\pi}{6} + \cos\frac{5\pi}{3} = 2\left(\frac{1}{2}\right) + \frac{1}{2} = \frac{3}{2}$$

$\therefore \quad$ Local maximum value $= \frac{3}{2}$ at $x = \frac{\pi}{6}$

Local minimum value $= 1$ at $x = \frac{\pi}{2}$

Local maximum value $= \frac{3}{2}$ at $x = \frac{5\pi}{6}$.

(*iv*) Let $\quad f(x) = \sin^4 x + \cos^4 x\,;\ 0 < x < \frac{\pi}{2}$ ...(1)

$$\Rightarrow \quad f'(x) = 4\sin^3 x\cos x - 4\cos^3 x\sin x$$

$$= -4\cos x\sin x\,(\cos^2 x - \sin^2 x)$$

$$= -2\,(2\sin x\cos x).\cos 2x \qquad \left[\because \begin{array}{r}\cos^2 A - \sin^2 A = \cos 2A\\ 2\sin A\cos A = \sin 2A\end{array}\right]$$

$$= -2\sin 2x\cos 2x$$

$\Rightarrow \qquad f'(x) = -\sin 4x$ ...(2)

and $\qquad f''(x) = -4\cos 4x$ ...(3)

$\therefore$ For a local maximum or minimum, we have

$$f'(x) = 0$$

$$\Rightarrow \quad -\sin 4x = 0 \quad \Rightarrow \quad \sin 4x = 0$$

$$\Rightarrow \quad 4x = \pi \qquad \left[\because \ 0 < x < \frac{\pi}{2}\right]$$

$$\Rightarrow \quad x = \frac{\pi}{4}$$

$\therefore$ The critical value for $f(x)$ is $x = \frac{\pi}{4}$.

Now, $f''(x)$ at $x = \frac{\pi}{4}$

$$\Rightarrow \quad f''\left(\frac{\pi}{4}\right) = -4\cos 4\left(\frac{\pi}{4}\right) = -4\cos\pi = (-4)(-1) = 4 > 0$$

$\therefore$ $x = \frac{\pi}{4}$ is a point of local minimum and the corresponding local minimum value is given by

$$f\left(\frac{\pi}{4}\right) = \sin^4\frac{\pi}{4} + \cos^4\frac{\pi}{4}$$

$$= \left(\frac{1}{\sqrt{2}}\right)^4 + \left(\frac{1}{\sqrt{2}}\right)^4 = \frac{1}{4} + \frac{1}{4} = \frac{1+1}{4} = \frac{2}{4} = \frac{1}{2}$$

$\therefore$ Local minimum value $= \frac{1}{2}$ at $x = \frac{\pi}{4}$.

(*v*) Let $\qquad f(x) = 2\sin x + \sin 2x \; ; 0 \le x \le 2\pi$ ...(1)

Please try yourself.

[**Ans.** Local maximum value $= \frac{3\sqrt{3}}{2}$ at $x = \frac{\pi}{3}$

Local minimum value $= \frac{-3\sqrt{3}}{2}$ at $x = \frac{5\pi}{3}$

Point of inflexion at $x = \pi$]

(*vi*) Let $\qquad f(x) = \sin 2x - x \; ; 0 \le x \le 2\pi$ ...(1)

$\Rightarrow \qquad f'(x) = 2\cos 2x - 1$ ...(2)

and $\qquad f''(x) = -4\sin 2x$ ...(3)

$\therefore$ For a local maximum or minimum, we have

$$f'(x) = 0 \qquad [\because \text{ By using (2)}]$$

$$\Rightarrow \quad 2\cos 2x - 1 = 0 \quad \Rightarrow \quad 2\cos 2x = 1$$

$$\Rightarrow \quad \cos 2x = \frac{1}{2} \quad \Rightarrow \quad 2x = \frac{\pi}{3}, \frac{5\pi}{3}$$

$$\Rightarrow \qquad x = \frac{\pi}{6}, \frac{5\pi}{6} \qquad [\because \ 0 \le x \le 2\pi]$$

$\therefore$ The critical values for $f(x)$ are $x = \frac{\pi}{6}, \frac{5\pi}{6}$.

Now, $f''(x)$ at $x = \frac{\pi}{6}$

$$\Rightarrow \qquad f''\left(\frac{\pi}{6}\right) = -4\sin\frac{2\pi}{6} = -4\sin\frac{\pi}{3} = -4\left(\frac{\sqrt{3}}{2}\right) = -2\sqrt{3} < 0$$

$\therefore$ $x = \frac{\pi}{6}$ is point of local maximum and the corresponding local maximum value is given by

$$f\left(\frac{\pi}{6}\right) = \sin\frac{\pi}{3} - \frac{\pi}{6} = \left(\frac{\sqrt{3}}{2} - \frac{\pi}{6}\right)$$

Also, $f''(x)$ at $x = \frac{5\pi}{6}$

$$\Rightarrow \qquad f''\left(\frac{5\pi}{6}\right) = -4\sin\frac{5\pi}{3}$$

$$= -4\sin\left(2\pi - \frac{\pi}{3}\right) = 4\sin\frac{\pi}{3} = 4\left(\frac{\sqrt{3}}{2}\right) = 2\sqrt{3} > 0$$

$\therefore$ $x = \frac{5\pi}{6}$ is a point of local minimum and the corresponding local minimum value is given by

$$f\left(\frac{5\pi}{6}\right) = \sin\frac{5\pi}{3} - \frac{5\pi}{6} = \left(\frac{-\sqrt{3}}{2} - \frac{5\pi}{6}\right)$$

$\therefore$ Local maximum value $= \left(\frac{\sqrt{3}}{2} - \frac{\pi}{6}\right)$ at $x = \frac{\pi}{6}$

Local minimum value $= \left(\frac{-\sqrt{3}}{2} - \frac{5\pi}{6}\right)$ at $x = \frac{5\pi}{6}$.

Example 8. *Find the local maximum and local minimum values of the following functions :*

(i) $3\sin^2 x + 4\cos^2 x$ (ii) $2\cos x + x \,;\, 0 \le x \le \pi$

(iii) $\sin x\,(1 + \cos x) \,;\, 0 \le x \le \frac{\pi}{2}$ (iv) $\sin x + \frac{1}{2}\cos 2x \,;\, 0 < x < \frac{\pi}{2}$

(v) $2\sin x - x \,;\, \frac{-\pi}{2} \le x \le \frac{\pi}{2}$ (vi) $e^x \sin x \,;\, 0 < x < 2\pi$.

Solution. (*i*) Let $f(x) = 3\sin^2 x + 4\cos^2 x$ ...(1)

$\Rightarrow$ $f'(x) = 6\sin x\cos x - 8\cos x\sin x$

$= -2\sin x\cos x$

$\Rightarrow$ $f'(x) = -\sin 2x$ ...(2) [$\because$ $\sin 2A = 2\sin A\cos A$]

and $f''(x) = -2\cos 2x$ ...(3)

$\therefore$ For a local maximum or minimum, we have

$$f'(x) = 0$$

$\Rightarrow$ $-\sin 2x = 0$

$\Rightarrow$ $2x = 0$ or π

$\Rightarrow$ $x = 0$ or $\dfrac{\pi}{2}$

$\therefore$ The critical values for $f(x)$ in ascending order are $x = 0, \dfrac{\pi}{2}$.

Now, $f''(x)$ at $x = 0$

$\Rightarrow$ $f''(0) = -2\cos 0 = -2 < 0$

$\therefore$ $x = 0$ is a point of local maximum and the corresponding local maximum value is given by

$$f(0) = 3\sin^2 0 + 4\cos^2 0 = 0 + 4(1) = 4$$

Also, $f''(x)$ at $x = \dfrac{\pi}{2}$.

$\Rightarrow$ $f''\left(\dfrac{\pi}{2}\right) = -2\cos 2\left(\dfrac{\pi}{2}\right) = -2\cos\pi = -2(-1) = 2 > 0$

$\therefore$ $x = \dfrac{\pi}{2}$ is a point of local minimum and the corresponding local minimum value is given by

$$f\left(\frac{\pi}{2}\right) = 3\sin^2\frac{\pi}{2} + 4\cos^2\frac{\pi}{2}$$

$$= 3(1)^2 + 4(0) = 3 + 0 = 3$$

$\therefore$ Local maximum value $= 4$ at $x = 0$

Local minimum value $= 3$ at $x = \dfrac{\pi}{2}$.

(*ii*) Let $f(x) = 2\cos x + x\,;\ 0 \le x \le \pi$ ...(1)

$\Rightarrow$ $f'(x) = -2\sin x + 1$ ...(2)

and $f''(x) = -2\cos x$ ...(3)

$\therefore$ For a local maximum or minimum, we have

$$f'(x) = 0$$

$\Rightarrow$ $-2\sin x + 1 = 0 \Rightarrow -2\sin x = -1$

$\Rightarrow$ $\sin x = \dfrac{1}{2}$

$\Rightarrow$ $x = \dfrac{\pi}{6}$ or $\dfrac{5\pi}{6}$ [$\because$ $x \in [0, \pi]$]

$\therefore$ The critical values for $f(x)$ in the ascending order are $x = \frac{\pi}{6}, \frac{5\pi}{6}$.

Now, $f''(x)$ at $x = \frac{\pi}{6}$

$$\Rightarrow \quad f''\left(\frac{\pi}{6}\right) = -2\cos\frac{\pi}{6} = -2\left(\frac{\sqrt{3}}{2}\right) = -\sqrt{3} < 0$$

$\therefore$ $x = \frac{\pi}{6}$ is a point of local maximum and the corresponding local maximum value is given by

$$f\left(\frac{\pi}{6}\right) = 2\cos\frac{\pi}{6} + \frac{\pi}{6} = 2.\frac{\sqrt{3}}{2} + \frac{\pi}{6} = \left(\sqrt{3} + \frac{\pi}{6}\right).$$

Also, $f''(x)$ at $x = \frac{5\pi}{6}$

$$\Rightarrow \quad f''\left(\frac{5\pi}{6}\right) = -2\cos\left(\frac{5\pi}{6}\right) = -2\left(\frac{-\sqrt{3}}{2}\right) = \sqrt{3} > 0$$

$\therefore$ $x = \frac{5\pi}{6}$ is a point of local minimum and the corresponding local minimum value is given by

$$f\left(\frac{5\pi}{6}\right) = 2\cos\frac{5\pi}{6} + \frac{5\pi}{6}$$

$$= 2\left(\frac{-\sqrt{3}}{2}\right) + \frac{5\pi}{6} = \left(-\sqrt{3} + \frac{5\pi}{6}\right)$$

$\therefore$ Local maximum value $= \left(\frac{\pi}{6} + \sqrt{3}\right)$ at $x = \frac{\pi}{6}$

Local minimum value $= \left(\frac{5\pi}{6} - \sqrt{3}\right)$ at $x = \frac{5\pi}{6}$.

(iii) Let $\quad f(x) = \sin x\,(1 + \cos x)\,;\ 0 \le x \le \frac{\pi}{2}$ ...(1)

$$\Rightarrow \quad f'(x) = \sin x\,(-\sin x) + (1 + \cos x)\,.\,\cos x$$
$$= -\sin^2 x + \cos x + \cos^2 x$$
$$= -(1 - \cos^2 x) + \cos x + \cos^2 x$$
$$= -1 + \cos^2 x + \cos x + \cos^2 x$$
$$= 2\cos^2 x + \cos x - 1$$
$$= 2\cos^2 x + 2\cos x - \cos x - 1$$
$$= 2\cos x\,(\cos x + 1) - 1\,(\cos x + 1)$$
$$f'(x) = (\cos x + 1)\,(2\cos x - 1) \quad \text{...(2)}$$

$\Rightarrow \quad f''(x) = (\cos x + 1) . (-2 \sin x) + (2 \cos x - 1) . (-\sin x)$

$= -2 \sin x \cos x - 2 \sin x - 2 \sin x \cos x + \sin x$

$\Rightarrow \quad f''(x) = -4 \sin x \cos x - \sin x$ ...(3)

$\therefore$ For a local maximum or minimum, we have

$$f'(x) = 0$$

$\Rightarrow$ $(\cos x + 1)(2 \cos x - 1) = 0$ [$\because$ By using (2)]

$\Rightarrow \quad \cos x = -1 \quad \text{or} \quad \cos x = \frac{1}{2}$

$\Rightarrow \quad x = \frac{\pi}{3}$ $\left[\because \ 0 \le x \le \frac{\pi}{2}\right]$

$\therefore$ The critical value for $f(x)$ is $x = \frac{\pi}{3}$.

Now, $f''(x)$ at $x = \frac{\pi}{3}$

$$\Rightarrow \quad f''\left(\frac{\pi}{3}\right) = -4 \sin\frac{\pi}{3} \cos\frac{\pi}{3} - \sin\frac{\pi}{3}$$

$$= -4\left(\frac{\sqrt{3}}{2}\right)\left(\frac{1}{2}\right) - \frac{\sqrt{3}}{2} = -\sqrt{3} - \frac{\sqrt{3}}{2} = \frac{-3\sqrt{3}}{2} < 0$$

$\therefore \quad x = \frac{\pi}{3}$ is a point of local maximum and the corresponding local maximum value is given by

$$f\left(\frac{\pi}{3}\right) = \sin\frac{\pi}{3}\left(1 + \cos\frac{\pi}{3}\right) = \frac{\sqrt{3}}{2}\left(1 + \frac{1}{2}\right) = \left(\frac{\sqrt{3}}{2}\right)\left(\frac{3}{2}\right) = \frac{3\sqrt{3}}{4}$$

$\therefore$ Local maximum value $= \frac{3\sqrt{3}}{4}$ at $x = \frac{\pi}{3}$.

(*iv*) Let $\quad f(x) = \sin x + \frac{1}{2} \cos 2x \, ; \, 0 \le x \le \frac{\pi}{2}$ ...(1)

$\Rightarrow \quad f'(x) = \cos x - \sin 2x$ ...(2)

and $\quad f''(x) = -\sin x - 2 \cos 2x$ ...(3)

$\therefore$ For a local maximum or minimum, we have

$$f'(x) = 0$$

$\Rightarrow \quad \cos x - \sin 2x = 0$

$\Rightarrow \quad \cos x - 2 \sin x \cos x = 0$ [$\because$ $\sin 2A = 2 \sin A \cos A$]

$\Rightarrow \quad \cos x (1 - 2 \sin x) = 0$

$\Rightarrow \quad \cos x = 0 \quad \text{or} \quad 1 - 2 \sin x = 0$

$\Rightarrow \quad 2 \sin x = 1 \ \Rightarrow \ \sin x = \frac{1}{2}$

$\Rightarrow \quad x = \frac{\pi}{2} \quad \text{or} \quad \frac{\pi}{6}$ $\left[\because \ 0 \le x \le \frac{\pi}{2}\right]$

$\therefore$ The critical values for $f(x)$ in the ascending order are $x = \frac{\pi}{6}, \frac{\pi}{2}$.

Now, $f''(x)$ at $x = \frac{\pi}{6}$

$$\Rightarrow \quad f''\left(\frac{\pi}{6}\right) = -\sin\frac{\pi}{6} - 2\cos\frac{2\pi}{6}$$

$$= \frac{-1}{2} - 2\cos\frac{\pi}{3} = \frac{-1}{2} - 2.\left(\frac{1}{2}\right) = \frac{-1}{2} - 1 = \frac{-3}{2} < 0$$

$\therefore$ $x = \frac{\pi}{6}$ is a point of local maximum and the corresponding local maximum value is given by

$$f\left(\frac{\pi}{6}\right) = \sin\frac{\pi}{6} + \frac{1}{2}\cos\frac{2\pi}{6}$$

$$= \frac{1}{2} + \frac{1}{2}\left(\frac{1}{2}\right) = \frac{1}{2} + \frac{1}{4} = \frac{2+1}{4} = \frac{3}{4}.$$

Also, $f''(x)$ at $x = \frac{\pi}{2}$

$$\Rightarrow \quad f''\left(\frac{\pi}{2}\right) = -\sin\frac{\pi}{2} - 2\cos\pi$$

$$= -1 - 2(-1) = -1 + 2 = 1 > 0$$

$\therefore$ $x = \frac{\pi}{2}$ is a point of local minimum and the corresponding local minimum value is given by

$$f\left(\frac{\pi}{2}\right) = \sin\frac{\pi}{2} + \frac{1}{2}\cos\pi = 1 + \frac{1}{2}(-1) = 1 - \frac{1}{2} = \frac{1}{2}$$

$\therefore$ Local maximum value $= \frac{3}{4}$ at $x = \frac{\pi}{6}$

Local minimum value $= \frac{1}{2}$ at $x = \frac{\pi}{2}$.

(*v*) Let $\quad f(x) = 2\sin x - x\ ;\ \frac{-\pi}{2} \le x \le \frac{\pi}{2}$

Please try yourself.

[**Ans.** Local maximum value $= \left(\sqrt{3} - \frac{\pi}{3}\right)$ at $x = \frac{\pi}{3}$

Local minimum value $= \left(\frac{\pi}{3} - \sqrt{3}\right)$ at $x = \frac{-\pi}{3}$]

(*vi*) Let $\quad f(x) = e^x \sin x\ ;\ 0 < x < 2\pi \quad$ ...(1)

$$\Rightarrow \quad f'(x) = e^x . \cos x + \sin x . e^x$$

$$\Rightarrow \quad f'(x) = e^x (\cos x + \sin x) \quad ...(2)$$

and

$$f''(x) = e^x (-\sin x + \cos x) + (\cos x + \sin x) e^x$$

$$= e^x (-\sin x + \cos x + \cos x + \sin x)$$

$$\Rightarrow \quad f''(x) = 2e^x \cos x \quad ...(3)$$

$\therefore$ For a local maximum or minimum, we have

$$f'(x) = 0$$

$$\Rightarrow \quad e^x (\cos x + \sin x) = 0$$

$$\Rightarrow \quad e^x = 0$$

or

$$\cos x + \sin x = 0 \quad \text{(Not possible)}$$

$$\Rightarrow \quad \cos x = -\sin x$$

$$\Rightarrow \quad -\tan x = 1$$

$$\Rightarrow \quad \tan x = -1$$

$$\Rightarrow \quad x = \frac{3\pi}{4}, \frac{7\pi}{4} \qquad (\because \ 0 < x < 2\pi)$$

$\therefore$ The critical values for $f(x)$ in the ascending order are $x = \frac{3\pi}{4}, \frac{7\pi}{4}$.

Now, $f''(x)$ at $x = \frac{3\pi}{4}$

$$\Rightarrow \quad f''\left(\frac{3\pi}{4}\right) = 2\,e^{3\pi/4} \cos\left(\frac{3\pi}{4}\right) = 2\,e^{3\pi/4}\left(\frac{-1}{\sqrt{2}}\right) < 0$$

$\therefore$ $x = \frac{3\pi}{4}$ is a point of local maximum and the corresponding local maximum value is given by

$$f\left(\frac{3\pi}{4}\right) = e^{3\pi/4} . \sin\frac{3\pi}{4} = e^{3\pi/4}\left(\frac{+1}{\sqrt{2}}\right) = \frac{+1}{\sqrt{2}} e^{3\pi/4}.$$

Also, $f''(x)$ at $x = \frac{7\pi}{4}$

$$\Rightarrow \quad f''\left(\frac{7\pi}{4}\right) = 2\,e^{7\pi/4} \cos\left(\frac{7\pi}{4}\right) = 2\,e^{7\pi/4}\left(\frac{1}{\sqrt{2}}\right) > 0$$

$\therefore$ $x = \frac{7\pi}{4}$ is a point of local minimum and the corresponding local minimum value is given by

$$f\left(\frac{7\pi}{4}\right) = e^{7\pi/4} . \sin\frac{7\pi}{4} = e^{7\pi/4}\left(\frac{-1}{\sqrt{2}}\right) = \frac{-1}{\sqrt{2}} e^{7\pi/4}$$

$\therefore$ Local maximum value $= \frac{1}{\sqrt{2}} e^{3\pi/4}$ at $x = \frac{3\pi}{4}$

Local minimum value $= \frac{-1}{\sqrt{2}} e^{7\pi/4}$ at $x = \frac{7\pi}{4}$.

Example 9. *If* $y = \dfrac{ax - b}{(x-1)(x-4)}$ *has a turning point P (2, – 1), find the values of a and b. Also, show that y is maximum at point P (2, – 1).*

Solution. We have, $y = \dfrac{(ax-b)}{(x-1)(x-4)} = \dfrac{ax-b}{x^2 - 5x + 4}$ ...(1)

$$\Rightarrow \quad \frac{dy}{dx} = \frac{(x^2 - 5x + 4).(a) - (ax - b)(2x - 5)}{(x^2 - 5x + 4)^2}$$

$$\Rightarrow \quad \frac{dy}{dx} = \frac{a(x^2 - 5x + 4) - (2ax^2 - 2bx - 5ax - 5b)}{(x^2 - 5x + 4)^2}$$

$$\Rightarrow \quad \frac{dy}{dx} = \frac{a(x^2 - 5x + 4 - 2x^2 + 5x) + b(2x - 5)}{(x^2 - 5x + 4)^2}$$

$$\Rightarrow \quad \frac{dy}{dx} = \frac{a(4 - x^2) + b(2x - 5)}{(x^2 - 5x + 4)^2} \quad ...(2)$$

It is given that point P(2, – 1) is a turning point.

∴ For a local maximum or minimum, we have

$$\left.\frac{dy}{dx}\right]_{\text{at }(2,\,-1)} = 0$$

$$\Rightarrow \quad \frac{a[4 - (2)^2] + b[2(2) - 5]}{[(2)^2 - 5(2) + 4]^2} = 0$$

$$\Rightarrow \quad a(4 - 4) + b(4 - 5) = 0$$

$$\Rightarrow \quad -b = 0$$

$$\Rightarrow \quad b = 0 \quad ...(3)$$

At P(2, – 1), from equation (1), we have

$$\Rightarrow \quad -1 = \frac{a(2) - b}{[(2)^2 - 5(2) + 4]}$$

$$\Rightarrow \quad -1 = \frac{2a - b}{4 - 10 + 4}$$

$$\Rightarrow \quad -1 = \frac{2a - b}{-2} \Rightarrow 2a - b = 2 \quad ...(4)$$

Put $b = 0$ in equation (4), we have

$$\Rightarrow \quad 2a - 0 = 2 \Rightarrow 2a = 2 \Rightarrow a = 1$$

Substituting these values of a and b in equation (1), we have

$$y = \frac{x - 0}{(x^2 - 5x + 4)} = \frac{x}{(x^2 - 5x + 4)}$$

$$\Rightarrow \quad \frac{dy}{dx} = \frac{(x^2 - 5x + 4) \cdot 1 - x(2x - 5)}{(x^2 - 5x + 4)^2}$$

$$= \frac{x^2 - 5x + 4 - 2x^2 + 5x}{(x^2 - 5x + 4)^2} = \frac{4 - x^2}{(x^2 - 5x + 4)^2} \quad \text{...(5)}$$

For a local maximum or minimum, we have

$$\frac{dy}{dx} = 0$$

$$\Rightarrow \quad \frac{4 - x^2}{(x^2 - 5x + 4)^2} = 0 \Rightarrow 4 - x^2 = 0$$

$$\Rightarrow \quad (2 - x)(2 + x) = 0$$

$$\Rightarrow \quad x = 2, -2$$

Since, $x = 2$ is given to be a critical value for y.

$\therefore$ For $x = 2$

When x is slightly less than 2, say (1.9)

$$\Rightarrow \quad \frac{dy}{dx} = \frac{4 - (1.9)^2}{[(1.9)^2 - 5(1.9) + 4]^2} = \frac{(+\text{ve})}{(+\text{ve})} = +\text{ve}$$

When x is slightly greater than 2, say (2.1)

$$\Rightarrow \quad \frac{dy}{dx} = \frac{4 - (2.1)^2}{[(2.1)^2 - 5(2.1) + 4]^2} = \frac{(-\text{ve})}{(+\text{ve})} = -\text{ve}$$

$\therefore$ $\frac{dy}{dx}$ changes sign from +ve to –ve as x increases through 2.

$\therefore$ $x = 2$ is a point of local maximum, and y is maximum.

Example 10. *Find the maximum values of the following functions :*

(*i*) $f(x) = \frac{\log x}{x}; x > 0$ (*ii*) $f(x) = x^{1/x}$

(*iii*) $f(x) = \left(\frac{1}{x}\right)^x$.

Solution. (*i*) We have

$$f(x) = \frac{\log x}{x}; x > 0 \quad \text{...(1)}$$

$$\Rightarrow \quad f'(x) = \frac{x \cdot \frac{1}{x} - \log x \cdot 1}{x^2}$$

$$\Rightarrow \qquad f'(x) = \frac{1 - \log x}{x^2} \qquad \ldots(2)$$

and
$$f''(x) = \frac{x^2 \cdot \left(\frac{-1}{x}\right) - (1 - \log x) \cdot (2x)}{x^4}$$

$$= \frac{-x - 2x + 2x \log x}{x^4} = \frac{-3x + 2x \log x}{x^4}$$

$$\Rightarrow \qquad f''(x) = \frac{2 \log x - 3}{x^3} \qquad \ldots(3)$$

$\therefore$ For a local maximum or minimum, we have

$$f'(x) = 0$$

$$\Rightarrow \qquad \frac{1 - \log x}{x^2} = 0 \qquad [\because \text{ By using (2)}]$$

$$\Rightarrow \qquad 1 - \log x = 0 \;\Rightarrow\; \log x = 1 \;\Rightarrow\; x = e^1 \;\Rightarrow\; x = e.$$

$\therefore$ The critical value for $f(x)$ is $x = e$.

Now, $f''(x)$ at $x = e$

$$\Rightarrow \qquad f''(e) = \frac{2 \log e - 3}{e^3} = \frac{2 - 3}{e^3} \qquad [\because \log e = 1]$$

$$= \frac{-1}{e^3} < 0.$$

$\therefore$ $x = e$ is a point of local maximum and the corresponding local maximum value is given by

$$f(e) = \frac{\log e}{e} = \frac{1}{e} \qquad [\because \log e = 1]$$

$\therefore$ Local maximum value $= \dfrac{1}{e}$ at $x = e$.

(*ii*) We have $\qquad f(x) = x^{1/x} \qquad \ldots(1)$

Taking logarithms on both sides, we get

$$\Rightarrow \qquad \log f(x) = \log x^{1/x}$$

$$\Rightarrow \qquad \log f(x) = \frac{1}{x} \log x \qquad \ldots(2) \qquad [\because \log m^n = n \log m]$$

Differentiating both sides w.r.t. x, we have

$$\frac{1}{f(x)} \cdot f'(x) = \frac{1}{x} \cdot \frac{1}{x} + \log x \left(\frac{-1}{x^2}\right)$$

$$\Rightarrow \qquad f'(x) = f(x)\left[\frac{1}{x^2} - \frac{1}{x^2} \log x\right]$$

$$\Rightarrow \qquad f'(x) = \frac{x^{1/x}(1 - \log x)}{x^2} \qquad \ldots(3) \qquad [\because f(x) = x^{1/x}]$$

$\therefore$ For a local maximum or minimum, we have

$$f'(x) = 0 \qquad [\because \text{ By using (3)}]$$

$$\Rightarrow \quad \frac{x^{1/x}(1 - \log x)}{x^2} = 0$$

$$\Rightarrow \quad x^{1/x}(1 - \log x) = 0 \Rightarrow \log x = 1 \Rightarrow x = e^1$$

$$\Rightarrow \quad x = e$$

$\therefore$ The critical value for $f(x)$ is $x = e$.

When x is slightly less than e, *i.e.*, $x < e$, then, $\log x < 1$.

$$\Rightarrow \quad f'(x) = \frac{(+\text{ve})\,(+\text{ve})}{(+\text{ve})} = +\text{ ve}$$

When x is slightly greater than e, *i.e.*, $x > e$, then, $\log x > 1$.

$$\Rightarrow \quad f'(x) = \frac{(+\text{ve})\,(-\text{ve})}{(+\text{ve})} = -\text{ ve}$$

$\therefore$ $f'(x)$ changes sign from (+ve) to (–ve) as x increases through e.

$\therefore$ $x = e$ is a point of local maximum and the corresponding local maximum value is given by

$$f(e) = e^{1/e}$$

$\therefore$ Local maximum value $= e^{1/e}$ at $x = e$.

(*iii*) We have $\quad f(x) = \left(\frac{1}{x}\right)^x = (x^{-1})^x$

$$\Rightarrow \quad f(x) = x^{-x} \qquad \text{...(1)}$$

Taking logarithms on both sides, we get

$$\Rightarrow \quad \log f(x) = \log (x^{-x})$$

$$\Rightarrow \quad \log f(x) = -x \log x \qquad [\because \log m^n = n \log m]$$

Differentiating both sides w.r.t. x, we get

$$\frac{1}{f(x)} \cdot f'(x) = -x \cdot \frac{1}{x} + \log x\,(-1)$$

$$\Rightarrow \quad f'(x) = f(x)\,[-1 - \log x] \qquad \text{...(2)}$$

And $\quad f''(x) = f(x)\left[\frac{-1}{x}\right] + [-1 - \log x] \cdot f'(x)$

$$= -\left(\frac{1}{x}\right)\left(\frac{1}{x}\right)^x - (1 + \log x) \cdot f'(x)$$

$$f''(x) = -\left(\frac{1}{x}\right)^{x+1} - (1 + \log x) \cdot f'(x) \qquad \text{...(3)}$$

$\therefore$ For a local maximum or minimum, we have

$$f'(x) = 0$$

$\Rightarrow$ $f(x)\,[-1 - \log x] = 0$

$\Rightarrow$ $x^{-x}(-1 - \log x) = 0$

$\Rightarrow$ $-1 - \log x = 0$

$\Rightarrow$ $1 + \log x = 0$

$\Rightarrow$ $\log x = -1 = -\log e$

$\Rightarrow$ $\log x = \log\left(\dfrac{1}{e}\right)$

$\Rightarrow$ $x = \dfrac{1}{e}$

$\therefore$ The critical value for $f(x)$ is $x = \dfrac{1}{e}$.

Now, $f''(x)$ at $x = \dfrac{1}{e}$

$\Rightarrow$ $-\left(\dfrac{1}{1/e}\right)^{\frac{1}{e}+1} - \left(1 + \log\dfrac{1}{e}\right).f'\left(\dfrac{1}{e}\right)$

$\Rightarrow$ $-(e)^{\frac{1}{e}+1} - \left(1 + \log e^{-1}\right).f'\left(\dfrac{1}{e}\right)$

$\Rightarrow$ $-(e)^{\frac{1}{e}+1} - (1 - \log e).f'\left(\dfrac{1}{e}\right)$

$\Rightarrow$ $-e^{\frac{1}{e}+1} - (1 - 1).f'\left(\dfrac{1}{e}\right)$

$\Rightarrow$ $-e^{\frac{1}{e}+1} - 0 = -e^{\frac{1}{e}+1} < 0.$

$\therefore$ $x = \dfrac{1}{e}$ is a point of local maximum and the corresponding local maximum value is given by

$$f\left(\frac{1}{e}\right) = \left(\frac{1}{1/e}\right)^{1/e} = e^{1/e}$$

$\therefore$ Local maximum value $= e^{1/e}$ at $x = \dfrac{1}{e}$.

Example 11. *Show that $\sin^p\theta \,.\, \cos^q\theta$ attains a maximum, when $\theta = \tan^{-1}\sqrt{\left(\dfrac{p}{q}\right)}$, $p \geq 0$, $q \geq 0$.*

Solution. Let $y = \sin^p \theta \,.\, \cos^q \theta$ ...(1)

$$\Rightarrow \quad \frac{dy}{d\theta} = \sin^p \theta \,.\, q \cos^{q-1} \theta\,(-\sin \theta) + \cos^q \theta \,.\, p \sin^{p-1} \theta\,(\cos \theta)$$

$$= -q \cos^{q-1} \theta \sin^{p+1} \theta + p \sin^{p-1} \theta \cos^{q+1} \theta$$

$$= \sin^{p-1} \theta \cos^{q-1} \theta\,(p \cos^2 \theta - q \sin^2 \theta)$$

$$= \frac{\sin^p \theta \cos^q \theta}{\sin \theta \cos \theta}(p \cos^2 \theta - q \sin^2 \theta)$$

$$= \sin^p \theta \cos^q \theta \left(\frac{p \cos^2 \theta}{\sin \theta \cos \theta} - \frac{q \sin^2 \theta}{\sin \theta \cos \theta}\right)$$

$$\frac{dy}{d\theta} = \sin^p \theta \cos^q \theta\,(p \cot \theta - q \tan \theta) \quad ...(2)$$

$$\Rightarrow \quad \frac{dy}{d\theta} = y.\,(p \cot \theta - q \tan \theta) \quad [\because \text{ By using (1)}]$$

$$\Rightarrow \quad \frac{d^2y}{d\theta^2} = y\,.\,(-p \operatorname{cosec}^2 \theta - q \sec^2 \theta) + (p \cot \theta - q \tan \theta)\,.\,\frac{dy}{d\theta} \quad ...(3)$$

$\therefore$ For a local maximum or minimum, we have

$$\frac{dy}{d\theta} = 0$$

$$\Rightarrow \quad \sin^p \theta \cos^q \theta\,(p \cot \theta - q \tan \theta) = 0 \quad [\because \text{ By using (2)}]$$

$$\Rightarrow \quad \sin^p \theta = 0 \quad \text{or} \quad \cos^q \theta = 0 \quad \text{or} \quad p \cot \theta - q \tan \theta = 0$$

$$\Rightarrow \quad \theta = 0 \quad \text{or} \quad \theta = \frac{\pi}{2} \quad \text{or} \quad \frac{p}{\tan \theta} = q \tan \theta$$

$$\Rightarrow \quad \frac{p}{q} = \tan^2 \theta$$

$$\Rightarrow \quad \tan \theta = \sqrt{\frac{p}{q}}$$

$$\Rightarrow \quad \theta = \tan^{-1} \sqrt{\frac{p}{q}}$$

Now, $\dfrac{d^2y}{dx^2}$ at $\theta = \tan^{-1} \sqrt{\dfrac{p}{q}}$

$$\Rightarrow \quad \frac{d^2y}{d\theta^2} = y(-p \operatorname{cosec}^2 \theta - q \sec^2 \theta) + 0 \quad \left[\because \frac{dy}{d\theta} = 0 \text{ at } \theta = \tan^{-1} \sqrt{\frac{p}{q}}\right]$$

$$= -y(p \operatorname{cosec}^2 \theta + q \sec^2 \theta) < 0$$

$\therefore$ $\theta = \tan^{-1} \sqrt{\dfrac{p}{q}}$ is a point of local maximum.

Example 12. *Find the absolute maximum and absolute minimum values of the following functions on the given intervals.*

(*i*) $x^2 - 3x$ *on* $[0, 2]$ (*ii*) $-2x^2 + 4x - 7$ *on* $[0, 5]$

(*iii*) $x^3 - 48x + 17$ *on* $[0, 5]$ (*iv*) $\left(\frac{1}{2} - x\right)^2 + x^3$ *on* $\left[-2, \frac{5}{2}\right]$

(*v*) $(x-1)^{1/3}(x-2)$ *on* $[1, 9]$.

Solution. (*i*) Let $f(x) = x^2 - 3x$ on $[0, 2]$ ...(1)

$\Rightarrow$ $f'(x) = 2x - 3$ ...(2)

$\therefore$ For a maximum or minimum, we have

$$f'(x) = 0$$

$$\Rightarrow \quad 2x - 3 = 0 \Rightarrow x = \frac{3}{2}.$$

$\therefore$ The critical value for $f(x)$ on $[0, 2]$ is $x = \frac{3}{2}$

Now, we find the values of $f(x)$ at $x = 0, \frac{3}{2}, 2$.

$$\therefore \quad f(0) = (0)^2 - 3(0) = 0$$

$$f\left(\frac{3}{2}\right) = \left(\frac{3}{2}\right)^2 - 3\left(\frac{3}{2}\right) = \frac{9}{4} - \frac{9}{2} = \frac{-9}{4}$$

$$f(2) = (2)^2 - 3(2) = 4 - 6 = -2$$

$\therefore$ Absolute maximum value $= 0$ at $x = 0$

Absolute minimum value $= \frac{-9}{4}$ at $x = \frac{3}{2}$.

(*ii*) Let $f(x) = -2x^2 + 4x - 7$ on $[0, 5]$ ...(1)

$\Rightarrow$ $f'(x) = -4x + 4$ ...(2)

$\therefore$ For a maximum or minimum, we have

$$f'(x) = 0$$

$$\Rightarrow \quad -4x + 4 = 0 \Rightarrow -4x = -4 \Rightarrow x = 1$$

$\therefore$ The critical value for $f(x)$ on $[0, 5]$ is $x = 1$.

Now, we find the values of $f(x)$ at $x = 0, 1, 5$.

$$\therefore \quad f(0) = -2(0)^2 + 4(0) - 7 = -7$$

$$f(1) = -2(1)^2 + 4(1) - 7 = -5$$

$$f(5) = -2(5)^2 + 4(5) - 7$$

$$= -2(25) + 20 - 7 = -50 + 13 = -37$$

$\therefore$ Absolute maximum value $= -5$ at $x = 1$

Absolute minimum value $= -37$ at $x = 5$.

(*iii*) Let $f(x) = x^3 - 48x + 17$ on $[0, 5]$ ...(1)

$\Rightarrow$ $f'(x) = 3x^2 - 48$ ...(2)

$\therefore$ For a maximum or minimum, we have

$$f'(x) = 0$$

$\Rightarrow \quad 3x^2 - 48 = 0 \Rightarrow 3x^2 = 48 \Rightarrow x^2 = 16 \Rightarrow x = \pm 4 (\because \; -4 \notin [0, 5])$

$\therefore$ The critical value for $f(x)$ on [0, 5] is $x = 4$.

Now, we find the values of $f(x)$ at $x = 0, 4, 5$.

$$\therefore \quad f(0) = (0)^3 - 48(0) + 17 = 17$$

$$f(4) = (4)^3 - 48(4) + 17 = 64 - 192 + 17 = -111$$

$$f(5) = (5)^3 - 48(5) + 17 = 125 - 240 + 17 = -98$$

$\therefore$ Absolute maximum value = 17 at $x = 0$

Absolute minimum value = – 111 at $x = 4$.

(*iv*) Let $$f(x) = \left(\frac{1}{2} - x\right)^2 + x^3 \text{ on } [-2, 2.5] \qquad ...(1)$$

$$\Rightarrow \quad f'(x) = 2\left(\frac{1}{2} - x\right)(-1) + 3x^2$$

$$f'(x) = -1 + 2x + 3x^2 \qquad ...(2)$$

$\therefore$ For a maximum or minimum, we have

$$f'(x) = 0$$

$$\Rightarrow \quad 3x^2 + 2x - 1 = 0$$

$$\Rightarrow \quad (3x - 1)(x + 1) = 0$$

$$\Rightarrow \quad x = \frac{1}{3}, -1$$

$\therefore$ The critical values for $f(x)$ on [– 2, 2.5] are $x = -1, \frac{1}{3}$.

Now, we find the values of $f(x)$ at $x = -2, -1, \frac{1}{3}, 2.5$

$$f(-2) = \left[\frac{1}{2} - (-2)\right]^2 + (-2)^3 = \left(\frac{1}{2} + 2\right)^2 + (-8)$$

$$= \left(\frac{5}{2}\right)^2 - 8 = \frac{25}{4} - 8 = \frac{25 - 32}{4} = \frac{-7}{4} = -1.75$$

$$f(-1) = \left[\frac{1}{2} - (-1)\right]^2 + (-1)^3 = \left[\frac{1}{2} + 1\right]^2 + (-1)$$

$$= \left(\frac{3}{2}\right)^2 - 1 = \frac{9}{4} - 1 = \frac{9 - 4}{4} = \frac{5}{4} = 1.25$$

$$f\left(\frac{1}{3}\right) = \left(\frac{1}{2} - \frac{1}{3}\right)^2 + \left(\frac{1}{3}\right)^3 = \left(\frac{1}{6}\right)^2 + \frac{1}{27}$$

$$= \frac{1}{36} + \frac{1}{27} = \frac{3 + 4}{108} = \frac{7}{108} = 0.065$$

$$f(2.5) = \left(\frac{1}{2} - 2.5\right)^2 + (2.5)^3 = \left(\frac{1-5}{2}\right)^2 + 15.625$$

$$= 4 + 15.625 = 19.625$$

∴ Absolute maximum value = 19.625 at $x = 2.5$

Absolute minimum value = – 1.75 at $x = -2$.

(v) Let $f(x) = (x-1)^{1/3}(x-2)$ on [1, 9] ...(1)

$$\Rightarrow \quad f'(x) = (x-1)^{1/3} \,.\, 1 + (x-2) \,.\, \frac{1}{3}(x-1)^{-2/3} \,.\, 1$$

$$= (x-1)^{1/3} + \frac{(x-2)}{3(x-1)^{2/3}}$$

$$= \frac{3(x-1)+x-2}{3(x-1)^{2/3}} = \frac{3x-3+x-2}{3(x-1)^{2/3}} = \frac{4x-5}{3(x-1)^{2/3}} \quad ...(2)$$

∴ For a maximum or minimum, we have

$$f'(x) = 0$$

$$\Rightarrow \quad \frac{4x-5}{3(x-1)^{2/3}} = 0 \Rightarrow 4x - 5 = 0 \Rightarrow x = \frac{5}{4}$$

∴ The critical value for $f(x)$ on [1, 9] is $x = \frac{5}{4}$.

Now, we find the values of $f(x)$ at $x = 1, \frac{5}{4}, 9$

$$\therefore \quad f(1) = (1-1)^{1/3} \,.\, (1-2) = 0$$

$$f\left(\frac{5}{4}\right) = \left(\frac{5}{4} - 1\right)^{1/3}\left(\frac{5}{4} - 2\right) = \left(\frac{1}{4}\right)^{1/3}\left(\frac{-3}{4}\right) = \frac{-3}{4^{1/3} \,.\, 4} = \frac{-3}{4^{4/3}}$$

$$f(9) = (9-1)^{1/3}(9-2) = 8^{1/3} \times 7 = 2 \times 7 = 14$$

∴ Absolute maximum value = 14 at $x = 9$

Absolute minimum value = $\frac{-3}{4^{4/3}}$ at $x = \frac{5}{4}$.

Example 13. *Find the absolute maximum and absolute minimum values of the following functions on the given intervals :*

(i) $x^{2/3} + (x-2)^{2/3}$ on [0, 2] (ii) $x^3 - 12x^2 + 36x + 18$ on [1, 10]
(iii) $3x^4 - 8x^3 + 12x^2 - 48x + 25$ on [0, 3] (iv) $3x^4 - 8x^3 + 12x^2 - 48x + 1$ on [1, 4]
(v) $3x^4 - 2x^3 - 6x^2 + 6x + 1$ on [0, 2].

Solution. (i) Let $f(x) = x^{2/3} + (x-2)^{2/3}$ on [0, 2] ...(1)

$$f'(x) = \frac{2}{3}x^{-1/3} + \frac{2}{3}(x-2)^{-1/3} = \frac{2}{3}\left[\frac{1}{x^{1/3}} + \frac{1}{(x-2)^{1/3}}\right]$$

$$\Rightarrow \quad f'(x) = \frac{2}{3}\left[\frac{(x-2)^{1/3} + x^{1/3}}{x^{1/3}(x-2)^{1/3}}\right] \quad ...(2)$$

$\therefore$ For a maximum or minimum, we have

$$f'(x) = 0$$

$$\Rightarrow \quad \frac{2}{3}\left[\frac{(x-2)^{1/3} + x^{1/3}}{x^{1/3}\,.\,(x-2)^{1/3}}\right] = 0$$

$$\Rightarrow \quad (x-2)^{1/3} + x^{1/3} = 0$$

$$\Rightarrow \quad (x-2)^{1/3} = -x^{1/3}$$

$$\Rightarrow \quad x - 2 = -x$$

$$\Rightarrow \quad 2x = 2 \Rightarrow x = 1$$

$\therefore$ The critical value for $f(x)$ on $[0, 2]$ is $x = 1$.

Now, we find the values of $f(x)$ at $x = 0, 1, 2$.

$$\therefore \quad f(0) = (0)^{2/3} + (0-2)^{2/3} = [(-2)^2]^{1/3} = 4^{1/3}$$

$$f(1) = (1)^{2/3} + (1-2)^{2/3} = 1 + [(-1)^2]^{1/3} = 1 + 1^{1/3} = 2$$

$$f(2) = (2)^{2/3} + (2-2)^{2/3} = 2^{2/3} + 0 = (2^2)^{1/3} = 4^{1/3}$$

$\therefore$ Absolute maximum value = 2 at $x = 1$

Absolute minimum value = $4^{1/3}$ at $x = 0, 2$.

(ii) Let $f(x) = x^3 - 12x^2 + 36x + 18$ on $[1, 10]$

$$\Rightarrow \quad f'(x) = 3x^2 - 24x + 36$$

$\therefore$ For a maximum and minimum, we have

$$f'(x) = 0$$

$$\Rightarrow \quad 3x^2 - 24x + 36 = 0$$

$$\Rightarrow \quad 3(x^2 - 8x + 12) = 0$$

$$\Rightarrow \quad 3(x-6)(x-2) = 0$$

$$\Rightarrow \quad x = 2, 6$$

$\therefore$ The critical values for $f(x)$ on $[1, 10]$ are $x = 2, 6$

Now, we find the values of $f(x)$ at $x = 1, 2, 6, 10$.

$$\therefore \quad f(1) = (1)^3 - 12(1)^2 + 36(1) + 18 = 1 - 12 + 36 + 18 = 43$$

$$f(2) = (2)^3 - 12(2)^2 + 36(2) + 18 = 8 - 48 + 72 + 18 = 50$$

$$f(6) = (6)^3 - 12(6)^2 + 36(6) + 18 = 216 - 432 + 216 + 18 = 18$$

$$f(10) = (10)^3 - 12(10)^2 + 36(10) + 18 = 1000 - 1200 + 360 + 18 = 178$$

$\therefore$ Absolute maximum value = 178 at $x = 10$

Absolute minimum value = 18 at $x = 6$.

(iii) Let $f(x) = 3x^4 - 8x^3 + 12x^2 - 48x + 25$ on $[0, 3]$ ...(1)

$$\Rightarrow \quad f'(x) = 12x^3 - 24x^2 + 24x - 48 \qquad ...(2)$$

$\therefore$ For a maximum or minimum, we have

$$f'(x) = 0$$

$$\Rightarrow \quad 12x^3 - 24x^2 + 24x - 48 = 0$$

$$\Rightarrow \quad 12(x^3 - 2x^2 + 2x - 4) = 0$$

$$\Rightarrow \quad 12[x^2(x-2) + 2(x-2)] = 0$$

$$\Rightarrow \quad (x-2)(x^2+2) = 0$$

$\Rightarrow$ $x = 2$, $x^2 = -2$ [Rejecting, as it gives only imaginary values of x]

$\therefore$ The critical value for $f(x)$ on $[0, 3]$ is $x = 2$.

Now, we find the values of $f(x)$ at $x = 0, 2, 3$.

$$\therefore \quad f(0) = 3(0)^4 - 8(0)^3 + 12(0)^2 - 48(0) + 25 = 25$$

$$f(2) = 3(2)^4 - 8(2)^3 + 12(2)^2 - 48(2) + 25$$
$$= 3(16) - 8(8) + 12(4) - 96 + 25$$
$$= 48 - 64 + 48 - 96 + 25 = -39$$

$$f(3) = 3(3)^4 - 8(3)^3 + 12(3)^2 - 48(3) + 25$$
$$= 243 - 216 + 108 - 144 + 25 = 16$$

$\therefore$ Absolute maximum value = 25 at $x = 0$

Absolute minimum value = – 39 at $x = 2$.

(*iv*) Let $\quad f(x) = 3x^4 - 8x^3 + 12x^2 - 48x + 1$ on $[1, 4]$ ...(1)

Please try yourself.

[**Hint.** Proceed as in part (*iii*)]

[**Ans.** Absolute maximum value = 257 at $x = 4$
Absolute minimum value = – 59 at $x = 2$]

(*v*) Let $\quad f(x) = 3x^4 - 2x^3 - 6x^2 + 6x + 1$ on $[0, 2]$

Please try yourself.

[**Hint.** Proceed as in part (*iii*)]

[**Ans.** Absolute maximum value = 21 at $x = 2$
Absolute minimum value = 1 at $x = 0$]

Example 14. *Find the absolute maximum and absolute minimum values of the following functions on the given intervals :*

(*i*) $x^{50} - x^{20}$ *on* $[0, 1]$ (*ii*) $x^{30} - x^{10}$ *on* $[0, 1]$.

Solution. (*i*) Let $\quad f(x) = x^{50} - x^{20}$ on $[0, 1]$ ...(1)

$$\Rightarrow \quad f'(x) = 50x^{49} - 20x^{19}$$

$$= 20x^{19}\left[\frac{5}{2}x^{30} - 1\right] \quad ...(2)$$

$\therefore$ For a maximum or minimum, we have

$$f'(x) = 0$$

$$\Rightarrow \quad 20x^{19}\left[\frac{5}{2}x^{30} - 1\right] = 0$$

$$\Rightarrow \quad x = 0, \ \frac{5}{2}x^{30} = 1$$

$$\Rightarrow \quad x^{30} = \frac{2}{5} \Rightarrow x = \left(\frac{2}{5}\right)^{1/30}$$

$\therefore$ The critical values for $f(x)$ on $[0, 1]$ are $x = 0, \left(\frac{2}{5}\right)^{1/30}$.

Now, we find the values of $f(x)$ at $x = 0, \left(\frac{2}{5}\right)^{1/30}, 1$.

$$\Rightarrow \quad f(0) = (0)^{50} - (0)^{20} = 0 - 0 = 0$$

$$f\left(\frac{2}{5}\right)^{1/30} = \left[\left(\frac{2}{5}\right)^{1/30}\right]^{50} - \left[\left(\frac{2}{5}\right)^{1/30}\right]^{20}$$

$$= \left(\frac{2}{5}\right)^{5/3} - \left(\frac{2}{5}\right)^{2/3} = \left(\frac{2}{5}\right)^{2/3}\left[\frac{2}{5} - 1\right]$$

$$= \left(\frac{2}{5}\right)^{2/3}\left(\frac{2-5}{5}\right) = \frac{-3}{5}\left(\frac{2}{5}\right)^{2/3}$$

$$f(1) = (1)^{50} - (1)^{20} = 1 - 1 = 0$$

∴ **Absolute maximum value = 0 at $x = 0, 1$**

Absolute minimum value = $\frac{-3}{5}\left(\frac{2}{5}\right)^{2/3}$ **at** $x = \left(\frac{2}{5}\right)^{1/30}$.

(ii) Let $f(x) = x^{30} - x^{10}$ on $[0, 1]$ ...(1)

⇒ $f'(x) = 30x^{29} - 10x^9 = 10x^9[3x^{20} - 1]$ ...(2)

∴ For a maximum or minimum, we have

$$f'(x) = 0$$

⇒ $10x^9[3x^{20} - 1] = 0$

⇒ $x = 0,\ 3x^{20} = 1 \Rightarrow x^{20} = \frac{1}{3}$

⇒ $x = \left(\frac{1}{3}\right)^{1/20}$

∴ The critical values for $f(x)$ on $[0, 1]$ are $x = 0, \left(\frac{1}{3}\right)^{1/20}$

Now, we find the values of $f(x)$ at $x = 0, \left(\frac{1}{3}\right)^{1/20}, 1.$

⇒ $f(0) = (0)^{30} - (0)^{10} = 0 - 0 = 0.$

⇒ $f\left[\left(\frac{1}{3}\right)^{1/20}\right] = \left[\left(\frac{1}{3}\right)^{1/20}\right]^{30} - \left[\left(\frac{1}{3}\right)^{1/20}\right]^{10}$

$$= \left(\frac{1}{3}\right)^{3/2} - \left(\frac{1}{3}\right)^{1/2} = \left(\frac{1}{3}\right)^{1/2}\left[\frac{1}{3} - 1\right]$$

$$= \frac{1}{\sqrt{3}}\left(\frac{1-3}{3}\right) = \frac{-2}{3\sqrt{3}}$$

$$f(1) = (1)^{30} - (1)^{10} = 1 - 1 = 0$$

∴ **Absolute maximum value = 0 at $x = 0, 1$**

Absolute minimum value = $\frac{-2}{3\sqrt{3}}$ **at** $x = \left(\frac{1}{3}\right)^{1/20}$.

Example 15. *Find the absolute maximum and absolute minimum values of the following functions on the given intervals :*

(*i*) $\sin x$; $[\pi, 2\pi]$ (*ii*) $\sin 2x$; $[0, 2\pi]$

(*iii*) $\sin x + \cos 2x$; $\left[0, \frac{\pi}{2}\right]$ (*iv*) $\sin x + \frac{1}{2}\cos 2x$; $\left[0, \frac{\pi}{2}\right]$

(*v*) $x + \sin 2x$; $[0, 2\pi]$ (*vi*) $2\cos 2x - \cos 4x$; $[0, \pi]$.

Solution. (*i*) Let $f(x) = \sin x$; $[0, \pi]$ ...(1)

$\Rightarrow$ $f'(x) = \cos x$ ...(2)

$\therefore$ For a maximum or minimum, we have

$$f'(x) = 0$$

$\Rightarrow$ $\cos x = 0$

$\Rightarrow$ $x = \frac{3\pi}{2}$ $[\because \; x \in [\pi, 2\pi]]$

$\therefore$ The critical value for $f(x)$ on $[\pi, 2\pi]$ is $x = \frac{3\pi}{2}$

Now, we find the values of $f(x)$ at $x = \pi, \frac{3\pi}{2}, 2\pi$

$\Rightarrow$ $f(\pi) = \sin \pi = 0$

$$f\left(\frac{3\pi}{2}\right) = \sin\left(\frac{3\pi}{2}\right) = \sin\left(\pi + \frac{\pi}{2}\right) = -\sin\frac{\pi}{2} = -1$$

$$f(2\pi) = \sin(2\pi) = 0$$

$\therefore$ Absolute maximum value $= 0$ at $x = \pi, 2\pi$

Absolute minimum value $= -1$ at $x = \frac{3\pi}{2}$.

(*ii*) Let $f(x) = \sin 2x$ on $\left[0, \frac{\pi}{2}\right]$ ...(1)

$\Rightarrow$ $f'(x) = 2\cos 2x$ ...(2)

$\therefore$ For a maximum or minimum, we have

$$f'(x) = 0$$

$\Rightarrow$ $2\cos 2x = 0 \Rightarrow \cos 2x = 0$ $\left(\because \; x \in \left[0, \frac{\pi}{2}\right]\right)$

$\Rightarrow$ $2x = \frac{\pi}{2} \Rightarrow x = \frac{\pi}{4}$

$\therefore$ The critical value for $f(x)$ on $\left[0, \frac{\pi}{2}\right]$ is $x = \frac{\pi}{4}$

Now, we find the values of $f(x)$ at $x = 0, \frac{\pi}{4}, \frac{\pi}{2}$.

$$f(0) = \sin 0 = 0$$

$$f\left(\frac{\pi}{4}\right) = \sin 2\left(\frac{\pi}{4}\right) = \sin\frac{\pi}{2} = 1$$

$$f\left(\frac{\pi}{2}\right) = \sin 2\left(\frac{\pi}{2}\right) = \sin\pi = 0$$

$\therefore$ Absolute maximum value = 1 at $x = \frac{\pi}{4}$

Absolute minimum value = 0 at $x = 0, \frac{\pi}{2}$.

(*iii*) Let $\quad f(x) = \sin x + \cos 2x \quad$ on $\left[0, \frac{\pi}{2}\right]$ ...(1)

$\Rightarrow \quad f'(x) = \cos x - 2\sin 2x$

$\therefore$ For a maximum or minimum, we have

$$f'(x) = 0$$

$\Rightarrow \quad \cos x - 2\sin 2x = 0$

$\Rightarrow \quad \cos x - 2\sin 2x = 0 \quad$ [$\because$ $\sin 2A = 2\sin A\cos A$]

$\Rightarrow \quad \cos x - 4\sin x\cos x = 0$

$\Rightarrow \quad \cos x(1 - 4\sin x) = 0$

$\Rightarrow \quad \cos x = 0, \quad 1 - 4\sin x = 0$

$\Rightarrow \quad x = \frac{\pi}{2}, \quad 4\sin x = 1 \quad \Rightarrow \quad \sin x = \frac{1}{4}$

$\Rightarrow \quad x = \sin^{-1}\left(\frac{1}{4}\right)$

$\therefore$ The critical values for $f(x)$ on $\left[0, \frac{\pi}{2}\right]$ are $x = \frac{\pi}{2}, \sin^{-1}\left(\frac{1}{4}\right)$

Now, we find the values of $f(x)$ at $x = 0, \sin^{-1}\left(\frac{1}{4}\right), \frac{\pi}{2}$.

$\therefore \quad f(0) = \sin 0 + \cos 0 = 0 + 1 = 1$

$$f\left(\sin^{-1}\left(\frac{1}{4}\right)\right) = \sin\left(\sin^{-1}\left(\frac{1}{4}\right)\right) + \cos 2.\left(\sin^{-1}\frac{1}{4}\right)$$

$$= \frac{1}{4} + \frac{7}{8}$$

$$\left[\begin{aligned} \because \cos 2x &= 1 - 2\sin^2 x \\ &= 1 - 2(\sin x)^2 \\ &= 1 - 2\left(\sin\left(\sin^{-1}\frac{1}{4}\right)\right)^2 \\ &= \left[1 - 2\left(\frac{1}{4}\right)^2\right] \\ &= 1 - \frac{2}{16} = 1 - \frac{1}{8} = \frac{7}{8} \end{aligned}\right]$$

$$\frac{2+7}{8} = \frac{9}{8}$$

$$f\left(\frac{\pi}{2}\right) = \sin\frac{\pi}{2} + \cos\pi = 1 - 1 = 0$$

$\therefore$ Absolute maximum value $= \frac{9}{8}$ at $x = \sin^{-1}\left(\frac{1}{4}\right)$

Absolute minimum value $= 0$ at $x = \frac{\pi}{2}$.

(*iv*) Let $$f(x) = \sin x + \frac{1}{2}\cos 2x \quad \text{on } \left[0, \frac{\pi}{2}\right] \qquad ...(1)$$

Please try yourself.

$\left[\right.$ **Ans.** Absolute maximum value $= \frac{3}{4}$ at $x = \frac{\pi}{6}$

Absolute minimum value $= \frac{1}{2}$ at $x = 0, \frac{\pi}{2}$ $\left.\right]$

(*v*) Let $$f(x) = x + \sin 2x \quad \text{on } [0, 2\pi] \qquad ...(1)$$

$$\Rightarrow \quad f'(x) = 1 + 2\cos 2x \qquad ...(2)$$

$\therefore$ For a maximum or minimum, we have

$$f'(x) = 0$$

$$\Rightarrow \quad 1 + 2\cos 2x = 0$$

$$\Rightarrow \quad \cos 2x = -\frac{1}{2} \qquad [\because \ x \in [0, 2\pi]]$$

$$\Rightarrow \quad 2x = \frac{2\pi}{3} \quad \text{or} \quad 2x = \frac{4\pi}{3}$$

$$\Rightarrow \quad x = \frac{\pi}{3} \quad \text{or} \quad x = \frac{2\pi}{3}$$

$\therefore$ The critical values for $f(x)$ on $[0, 2\pi]$ are $x = \frac{\pi}{3}, \frac{2\pi}{3}$.

Now, we find the values of $f(x)$ at $x = 0, \frac{\pi}{3}, \frac{2\pi}{3}, 2\pi$.

$$\therefore \quad f(0) = 0 + \sin 0 = 0 + 0 = 0$$

$$f\left(\frac{\pi}{3}\right) = \frac{\pi}{3} + \sin\frac{2\pi}{3} = \frac{\pi}{3} + \frac{\sqrt{3}}{2}$$

$$f\left(\frac{2\pi}{3}\right) = \frac{2\pi}{3} + \sin\frac{4\pi}{3} = \frac{2\pi}{3} - \frac{\sqrt{3}}{2}$$

$$f(2\pi) = 2\pi + \sin 4\pi = 2\pi + 0 = 2\pi$$

$\therefore$ Absolute maximum value $= 2\pi$ at $x = 2\pi$

Absolute minimum value $= 0$ at $x = 0$.

(*vi*) Let $$f(x) = 2\cos 2x - \cos 4x \text{ on } [0, \pi] \qquad ...(1)$$

$$\Rightarrow \quad f'(x) = -4\sin 2x + 4\sin 4x \qquad ...(2)$$

$\therefore$ For a maximum or minimum, we have

$$f'(x) = 0$$

$$\Rightarrow \quad -4\sin 2x + 4\sin 4x = 0$$

$$\Rightarrow \quad -4\sin 2x + 4(2\sin 2x\cos 2x) = 0 \qquad \left[\begin{array}{l} \because \sin 2A = 2\sin A\cos A \\ \Rightarrow \sin 4A = 2\sin 2A\cos 2A \end{array}\right]$$

$$\Rightarrow \quad 4\sin 2x\,[2\cos 2x - 1] = 0$$

$$\Rightarrow \quad \sin 2x = 0 \quad \text{or} \quad 2\cos 2x - 1 = 0$$

$$\Rightarrow \quad 2x = 0, \pi, 2\pi \quad \text{or} \quad 2\cos 2x = 1$$

$$\Rightarrow \quad x = 0, \frac{\pi}{2}, \pi \quad \text{or} \quad \cos 2x = \frac{1}{2} \qquad (\because x \in [0, \pi])$$

$$\Rightarrow \quad 2x = \frac{\pi}{3}, \frac{5\pi}{3} \quad \Rightarrow \quad x = \frac{\pi}{6}, \frac{5\pi}{6}$$

$\therefore$ The critical values for $f(x)$ on $[0, \pi]$ are

$$x = 0, \frac{\pi}{6}, \frac{\pi}{2}, \frac{5\pi}{6}, \pi.$$

Now, we find the values of $f(x)$ at $x = 0, \frac{\pi}{6}, \frac{\pi}{2}, \frac{5\pi}{6}, \pi$.

$$\Rightarrow \quad f(0) = 2\cos 0 - \cos 0 = 2(1) - 1 = 1.$$

$$\Rightarrow \quad f\left(\frac{\pi}{6}\right) = 2\cos\frac{\pi}{3} - \cos\frac{2\pi}{3} = 2\left(\frac{1}{2}\right) - \left(-\frac{1}{2}\right) = 1 + \frac{1}{2} = \frac{3}{2}$$

$$\Rightarrow \quad f\left(\frac{\pi}{2}\right) = 2\cos\pi - \cos 2\pi = 2(-1) - (1) = -2 - 1 = -3$$

$$\Rightarrow \quad f\left(\frac{5\pi}{6}\right) = 2\cos\frac{5\pi}{3} - \cos\left(\frac{10\pi}{3}\right) = 2\cos\left(2\pi - \frac{\pi}{3}\right) - \cos\left(2\pi + \frac{4\pi}{3}\right)$$

$$= 2\cos\frac{\pi}{3} - \cos\frac{4\pi}{3}$$

$$= 2\cos\frac{\pi}{3} - \cos\left(\pi + \frac{\pi}{3}\right)$$

$$= 2\cos\frac{\pi}{3} + \cos\frac{\pi}{3}$$

$$= 2\left(\frac{1}{2}\right) + \frac{1}{2} = 1 + \frac{1}{2} = \frac{3}{2}$$

$$\Rightarrow \quad f(\pi) = 2\cos 2\pi - \cos 4\pi = 2(1) - 1 = 1.$$

$\therefore$ Absolute maximum value $= \frac{3}{2}$ at $x = \frac{\pi}{6}, \frac{5\pi}{6}$.

Absolute minimum value $= -3$ at $x = \frac{\pi}{2}$.

Example 16. *It is given that at $x = 1$, the function $f(x) = x^4 - 62x^2 + ax + 9$ attains its maximum value on the interval $[0, 2]$. Find the value of a.*

Solution. We have

$$f(x) = x^4 - 62x^2 + ax + 9 \qquad \ldots(1)$$

$$\Rightarrow \quad f'(x) = 4x^3 - 124x + a \qquad \ldots(2)$$

Since, $f(x)$ attains its maximum value at $x = 1$.

$\therefore \quad f'(1) = 0$

$\Rightarrow \quad f'(1) = 4(1)^3 - 124(1) + a = 0$

$\Rightarrow \quad 4 - 124 + a = 0$

$\Rightarrow \quad a = 120.$

6.13 APPLICATIONS OF MAXIMA AND MINIMA TO PRACTICAL PROBLEMS

The knowledge of finding maximum or minimum values of functions is found very useful in solving various kinds of problems. In problems of geometry and algebra, some times we have, the data in two variables which are connected by some given relation. In such problems, it is necessary to eliminate one variable with the help of the given relation.

In solving applied problems on maxima and minima the following results will be very useful.

1. Area of square = $(\text{side})^2 = x^2$
2. Perimeter of square = $4 \times$ side = $4x$
3. Area of rectangle of sides x and $y = xy$
4. Perimeter of rectangle of sides x and $y = 2(x + y)$.
5. Area of trapezium = $\frac{1}{2}$ (sum of parallel sides) × (distance between them).
6. Area of circle of radius $r = \pi r^2$
7. Perimeter (circumference) of circle of radius $r = 2\pi r$
8. Volume of sphere of radius $r = \frac{4}{3}\pi r^3$.
9. Surface area of sphere of radius $r = 4\pi r^2$.
10. For a right circular cylinder of base radius r and height h :
    (*i*) Volume = $\pi r^2 h$
    (*ii*) Curved surface area = $2\pi rh$
    (*iii*) Total surface area = $2\pi rh + 2\pi r^2$.
11. For a right circular cone of base radius r, height h and slant height l : where $l = \sqrt{r^2 + h^2}$
    (*i*) Volume = $\frac{1}{3}\pi r^2 h$
    (*ii*) Curved surface area = πrl
    (*iii*) Total surface area = $\pi r^2 + \pi rl$.
12. For a cuboid of edges of lengths x, y and z :
    (*i*) Volume = xyz
    (*ii*) Surface area = $2(xy + yz + zx)$.
13. For a cube of edge of length x :
    (*i*) Volume = x^3
    (*ii*) Surface Area = $6x^2$.
14. Area of an equilateral triangle = $\frac{\sqrt{3}}{4}(\text{side})^2$.
15. Area of triangle = $\sqrt{s(s-a)(s-b)(s-c)}$. [$\because$ By Hero's Formula]

    where $s = \frac{a+b+c}{2}$ (semi-perimeter).

Example 17. (*i*) *Determine two positive numbers whose sum is 15 and the sum of whose squares is minimum.*

(*ii*) *Divide 30 into two parts such that their product is maximum.*

(*iii*) *Find two numbers whose sum is 24 and whose product is as large as possible.*

Solution. (*i*) Let the two numbers be x and y.

$$\Rightarrow \quad x + y = 15 \quad \text{...(1)}$$

Let S be the sum of squares of numbers :

$$\therefore \quad S = x^2 + y^2 \quad [\because\ x + y = 15 \Rightarrow y = (15 - x)]$$

$$S = x^2 + (15 - x)^2$$

$$\Rightarrow \quad \frac{dS}{dx} = 2x + 2(15 - x)(-1) = 2x - 30 + 2x$$

$$\frac{dS}{dx} = 4x - 30 \quad \text{...(2)}$$

$$\text{and} \quad \frac{d^2S}{dx^2} = 4 \quad \text{...(3)}$$

$\therefore$ For maximum or minimum, we must have

$$\frac{dS}{dx} = 0$$

$$\Rightarrow \quad 4x - 30 = 0 \quad \Rightarrow \quad 4x = 30 \quad \Rightarrow \quad x = \frac{15}{2}.$$

$$\text{Now,} \quad \left.\frac{d^2S}{dx^2}\right]_{x=\frac{15}{2}} = 4 > 0$$

$\therefore$ S is minimum ; when $x = \frac{15}{2}$.

$$\therefore \text{ From (1),} \quad y = 15 - x$$

$$\Rightarrow \quad y = 15 - \frac{15}{2} = \frac{15}{2}.$$

Hence, the sum of squares is minimum when the numbers are $\frac{15}{2}, \frac{15}{2}$.

(*ii*) Let the two numbers be x and y.

$$\therefore \quad x + y = 30 \quad \text{...(1)}$$

Let P be the product of two numbers,

$$\Rightarrow \quad P = xy \quad [\because\ x + y = 30 \Rightarrow y = 30 - x]$$

$$\Rightarrow \quad P = x(30 - x)$$

$$\Rightarrow \quad P = 30x - x^2$$

$$\Rightarrow \quad \frac{dP}{dx} = 30 - 2x \quad \text{...(2)}$$

$$\text{and} \quad \frac{d^2P}{dx^2} = -2 \quad \text{...(3)}$$

$\therefore$ For a maximum or minimum, we must have

$$\frac{dP}{dx} = 0$$

$\Rightarrow$ $30 - 2x = 0$ $\Rightarrow$ $2x = 30$ $\Rightarrow$ $x = 15.$

Now, $\left.\frac{d^2P}{dx^2}\right]_{x=15} = -2 < 0$

$\therefore$ P is maximum, when $x = 15$.

$\therefore$ From (1), $y = 30 - x$

$\Rightarrow$ $y = 30 - 15 = 15.$

Hence, the product is maximum when the numbers are 15, 15.

(*iii*) Let the two numbers be x and y.

$\therefore$ $x + y = 24$ ...(1)

Let P be the product of two numbers.

$\Rightarrow$ $P = xy$ $[\because\ x + y = 24 \Rightarrow y = 24 - x]$

$\Rightarrow$ $P = x(24 - x)$

$\Rightarrow$ $P = 24x - x^2$

$\Rightarrow$ $\frac{dP}{dx} = 24 - 2x$ ...(2)

and $\frac{d^2P}{dx^2} = -2$ ...(3)

$\therefore$ For maximum or minimum, we must have

$\frac{dP}{dx} = 0$ $\Rightarrow$ $24 - 2x = 0$

$\Rightarrow$ $2x = 24$ $\Rightarrow$ $x = 12.$

Now, $\left.\frac{d^2P}{dx^2}\right]_{x=12} = -2 < 0$

$\therefore$ P is maximum, when $x = 12$.

$\therefore$ From (1), $y = 24 - x = 24 - 12 = 12$

Hence, the product is maximum when the numbers are 12 and 12.

Example 18. (*i*) *Amongest all pairs of positive numbers with product 64, find those whose sum is the least.*

(*ii*) *Amongest all pairs of positive numbers with product 256, find those whose sum is the least.*

(*iii*) *Find two positive numbers whose sum is 14 and the sum of whose squares is minimum.*

Solution. (*i*) Let the two numbers be x and y.

$\Rightarrow$ $xy = 64$ ...(1)

Let S be the sum of two numbers

$\Rightarrow$ $S = x + y$ $\left[\because\ xy = 64 \Rightarrow y = \frac{64}{x}\right]$

$$\Rightarrow \qquad S = x + \frac{64}{x}$$

$$\Rightarrow \qquad \frac{dS}{dx} = 1 - \frac{64}{x^2} \qquad ...(2)$$

and
$$\frac{d^2S}{dx^2} = + \frac{128}{x^3} \qquad ...(3)$$

$\therefore$ For a maximum or minimum, we must have

$$\frac{dS}{dx} = 0$$

$$\Rightarrow \qquad 1 - \frac{64}{x^2} = 0 \quad \Rightarrow \quad \frac{x^2 - 64}{x^2} = 0$$

$$\Rightarrow \qquad x^2 - 64 = 0 \quad \Rightarrow \quad x^2 = 64 \quad \Rightarrow \quad x = 8 \qquad [\because \; x > 0]$$

Now, $$\left.\frac{d^2S}{dx^2}\right]_{x=8} = \frac{128}{(8)^3} = \frac{128}{512} = \frac{1}{4} > 0$$

$\therefore$ S is minimum, when $x = 8$.

$\therefore$ From (1), $$y = \frac{64}{x}$$

$$\Rightarrow \qquad y = \frac{64}{8} = 8$$

Hence, sum of two numbers is least when the numbers are 8, 8.

(*ii*) Let the two numbers be x and y.

$$\therefore \qquad xy = 256 \qquad ...(1)$$

Let S be the sum of two numbers.

$$\Rightarrow \qquad S = x + y \qquad \left[\because \; xy = 256 \;\Rightarrow\; y = \frac{256}{x}\right]$$

$$\Rightarrow \qquad S = x + \frac{256}{x}$$

$$\Rightarrow \qquad \frac{dS}{dx} = 1 - \frac{256}{x^2} \qquad ...(2)$$

and
$$\frac{d^2S}{dx^2} = \frac{512}{x^3} \qquad ...(3)$$

$\therefore$ For maximum or minimum, we must have

$$\frac{dS}{dx} = 0$$

$$\Rightarrow \qquad 1 - \frac{256}{x^2} = 0 \quad \Rightarrow \quad \frac{x^2 - 256}{x^2} = 0 \quad \Rightarrow \quad x^2 - 256 = 0$$

$$\Rightarrow \qquad x^2 = 256 \quad \Rightarrow \quad x = 16 \qquad [\because \; x > 0]$$

Now, $\left.\dfrac{d^2S}{dx^2}\right]_{x=16} = \dfrac{512}{(16)^3} = \dfrac{512}{4096} = \dfrac{1}{8} > 0$

$\therefore$ S is minimum, when $x = 16$.

$\therefore$ From (1), $y = \dfrac{256}{x} = \dfrac{256}{16} = 16$

Hence, sum of two numbers is minimum, when the numbers are 16, 16.

(*iii*) Let the two numbers be x and y.

$\therefore \quad x + y = 14 \quad \ldots(1)$

Let S be the sum of squares of two numbers.

$\Rightarrow \quad S = x^2 + y^2 \qquad [\because \; x + y = 14 \;\Rightarrow\; y = 14 - x]$

$\Rightarrow \quad S = x^2 + (14 - x)^2$

$\Rightarrow \quad \dfrac{dS}{dx} = 2x + 2(14 - x)(-1) = 2x - 28 + 2x$

$\Rightarrow \quad \dfrac{dS}{dx} = 4x - 28 \quad \ldots(2)$

and $\quad \dfrac{d^2S}{dx^2} = 4 \quad \ldots(3)$

$\therefore$ For maximum or minimum, we must have

$$\frac{dS}{dx} = 0 \;\Rightarrow\; 4x - 28 = 0$$
$$\Rightarrow\; 4x = 28$$
$$\Rightarrow\; x = 7$$

Now, $\left.\dfrac{d^2S}{dx^2}\right]_{x=7} = 4 < 0$

$\therefore$ S is minimum, when $x = 7$.

$\therefore$ From (1), $y = 14 - x$

$\Rightarrow \quad y = 14 - 7 = 7$

Hence, the sum of squares of two numbers is minimum, when the numbers are 7, 7.

Example 19. (*i*) *Find two positive numbers x and y such that $x + y = 60$ and xy^3 is maximum.*

(ii) Divide 10 into two parts such that the product of square of one with cube of the other may be greatest.

(iii) Find two positive numbers x and y such that $x + y = 35$ and x^2y^5 is maximum.

(iv) Divide 4 into two positive numbers such that the sum of the squares of one and the cube of the other is minimum.

Solution. (*i*) Let the two numbers be x and y.

$\Rightarrow \quad x + y = 60 \quad \ldots(1)$

Let $\quad P = xy^3 \qquad [\because \; x + y = 60 \;\Rightarrow\; x = 60 - y]$

$\Rightarrow \quad P = (60 - y)\, y^3 = 60y^3 - y^4$

$$\Rightarrow \qquad \frac{dP}{dy} = 180y^2 - 4y^3 \qquad ...(2)$$

and
$$\frac{d^2P}{dy^2} = 360y - 12y^2 \qquad ...(3)$$

$\therefore$ For a maximum or minimum, we must have

$$\frac{dP}{dy} = 0$$

$\Rightarrow \quad 180y^2 - 4y^3 = 0$

$\Rightarrow \quad 4y^2(45 - y) = 0$

$\Rightarrow \quad y = 0, \quad y = 45.$

Since, $y = 0$ is not possible.

$\therefore \quad y = 45.$

Now,
$$\left.\frac{d^2P}{dy^2}\right]_{y=45} = 360(45) - 12(45)^2$$

$$= 16200 - 24300 = -8100 < 0$$

$\therefore$ P is minimum, when $y = 45$.

$\therefore$ From (1), $\quad x = 60 - y.$

$\Rightarrow \quad x = 60 - 45 = 15$

Hence, the product P is minimum when the numbers are 15, 45.

(*ii*) Let the two numbers be x and y.

$\therefore \quad x + y = 10 \qquad ...(1)$

Let $\quad P = x^2y^3$

$\Rightarrow \quad P = (10 - y)^2 . y^3 \qquad [\because \; x + y = 10 \;\Rightarrow\; x = (10 - y)]$

$\Rightarrow \quad P = (100 + y^2 - 20y)\, y^3 = y^5 - 20y^4 + 100y^3$

$$\Rightarrow \qquad \frac{dP}{dy} = 5y^4 - 80y^3 + 300y^2 \qquad ...(2)$$

and
$$\frac{d^2P}{dy^2} = 20y^3 - 240y^2 + 600y \qquad ...(3)$$

$\therefore$ For a maximum or minimum, we must have

$$\frac{dP}{dy} = 0$$

$\Rightarrow \quad 5y^4 - 80y^3 + 300y^2 = 0$

$\Rightarrow \quad 5y^2(y^2 - 16y - 60) = 0$

$\Rightarrow \quad 5y^2(y - 6)(y - 10) = 0$

$\Rightarrow \quad y = 0, 6, 10.$

Since, $y = 0$ and $y = 10$ are not possible.

$\therefore \quad y = 6$

Now,
$$\left.\frac{d^2P}{dy^2}\right]_{y=6} = 20(6)^3 - 240(6)^2 + 600(6)$$

$$= 20\,(216) - 240\,(36) + 3600$$
$$= 4320 - 8640 + 3600 = -720 < 0$$

$\therefore$ P is maximum, when $y = 6$

$\therefore$ From (1), $x = 10 - y$

$\Rightarrow$ $x = 10 - 6 = 4$

Hence, the product P is maximum, when the numbers are 4, 6.

(*iii*) Let the two numbers be x and y.

$\therefore$ $x + y = 35$ ...(1)

Let $P = x^2y^5$

$\Rightarrow$ $P = (35 - y)^2y^5$ $[\because\ x + y = 35 \Rightarrow x = (35 - y)]$

$$\Rightarrow \frac{dP}{dy} = (35 - y)^2.\,5y^4 + y^5.\,2\,(35 - y)\,(-1)$$

$$\Rightarrow \frac{dP}{dy} = y^4\,(35 - y)\,[5\,(35 - y) - 2y] = y^4\,(35 - y)\,[175 - 5y - 2y]$$

$$\Rightarrow \frac{dP}{dy} = y^4\,(35 - y)\,(175 - 7y) \quad ...(2)$$

and
$$\frac{d^2P}{dy^2} = y^4\,(35 - y)\,(-7) + (35 - y)\,(175 - 7y).\,4y^3 + y^4(175 - 7y)\,(-1)$$

$$\Rightarrow \frac{d^2P}{dy^2} = -7y^4(35 - y) + 28y^3\,(35 - y)\,(25 - y) - 7y^4\,(25 - y) \quad ...(3)$$

$\therefore$ For a maximum or minimum, we must have

$$\frac{dP}{dy} = 0$$

$\Rightarrow$ $y^4(35 - y)\,(175 - 7y) = 0$

$\Rightarrow$ $y = 0,\ y = 35,\ 175 - 7y = 0$

$\Rightarrow$ $7y = 175 \Rightarrow y = 25.$

Since, $y = 0$ and $y = 35$ are not possible.

$\therefore$ $y = 25$

Now,
$$\left.\frac{d^2P}{dy^2}\right]_{y=25} = -7\,(25)^4\,(35 - 25) + 0 - 0$$
$$= -7(390625)\,(10) = -27343750 < 0$$

$\therefore$ P is maximum, when $y = 25$.

$\therefore$ From (1), $x = 35 - y$

$\Rightarrow$ $x = 35 - 25 = 10$

Hence, the product P is maximum, when the numbers are 10, 25.

(*iv*) Let the two numbers are x and y.

$\therefore$ $x + y = 4$ ...(1)

Let $S = x^2 + y^3$

$\Rightarrow$ $S = (4 - y)^2 + y^3$

$$\Rightarrow \quad \frac{dS}{dy} = 2(4-y)(-1) + 3y^2$$

$$\Rightarrow \quad \frac{dS}{dy} = -8 + 2y + 3y^2 \qquad ...(2)$$

and $$\frac{d^2S}{dy^2} = 2 + 6y$$

$\therefore$ For maximum or minimum, we must have

$$\frac{dS}{dy} = 0$$

$$\Rightarrow \quad 3y^2 + 2y - 8 = 0$$

$$\Rightarrow \quad (3y-4)(y+2) = 0 \quad \Rightarrow \quad y = \frac{4}{3}, -2$$

Since, $y = -2$ is not possible.

$$\therefore \quad y = \frac{4}{3}.$$

Now, $$\left.\frac{d^2S}{dy^2}\right]_{y=\frac{4}{3}} = 2 + 6\left(\frac{4}{3}\right) = 2 + 8 = 10 > 0$$

$\therefore$ The sum S is minimum, when $y = \frac{4}{3}$.

$\therefore$ From (1), $x = 4 - y = 4 - \frac{4}{3} = \frac{12-4}{3} = \frac{8}{3}$

Hence, the sum S is minimum, when the numbers are $\frac{8}{3}, \frac{4}{3}$.

Example 20. (*i*) *Show that among the rectangles of given perimeter, the square has the greatest area.*

(*ii*) *Show that among rectangle of given area, the square has the smaller perimeter.*

Solution. (*i*) Let length and breadth of rectangle be x and y respectively.

$\therefore$ Perimeter of rectangle = P = $2(x + y)$ ...(1)

and Area of rectangle = A = xy

$$\left[\begin{array}{l} \because \quad 2(x+y) = P \\ \Rightarrow \quad 2x + 2y = P \\ \Rightarrow \quad 2y = P - 2x \\ \Rightarrow \quad y = \frac{(P-2x)}{2} \\ \Rightarrow \quad y = \left(\frac{P}{2} - x\right) \end{array}\right]$$

$$\Rightarrow \quad A = x\left(\frac{P}{2} - x\right)$$

$$\Rightarrow \quad A = \frac{xP}{2} - x^2$$

$$\Rightarrow \quad \frac{dA}{dx} = \frac{P}{2} - 2x \qquad ...(2)$$

and $$\frac{d^2A}{dx^2} = -2 \quad ...(3)$$

$\therefore$ For maximum or minimum, we must have

$$\frac{dA}{dx} = 0 \Rightarrow \frac{P}{2} - 2x = 0 \Rightarrow 2x = \frac{P}{2}$$

$$\Rightarrow \quad x = \frac{P}{4}.$$

Now, $$\left.\frac{d^2A}{dx^2}\right]_{x=\frac{P}{4}} = -2 < 0$$

$\therefore$ Area is maximum when $x = \frac{P}{4}$

$\therefore$ From (1), $$y = \frac{P}{2} - x$$

$$\Rightarrow \quad y = \frac{P}{2} - \frac{P}{4} = \frac{2P - P}{4} = \frac{P}{4}$$

$$\Rightarrow \quad x = y$$

Hence, Area of the rectangle is maximum, when $x = y$ *i.e.*, the rectangle is a square.

(*ii*) Let length and breadth of rectangle be x and y respectively.

$\therefore$ Area of rectangle = A = xy ...(1)

and Perimeter of rectangle = P = $2(x + y)$

$$\Rightarrow \quad P = 2x + 2y \qquad \left[\because xy = A \Rightarrow y = \frac{A}{x}\right]$$

$$\Rightarrow \quad P = 2x + \frac{2A}{x}$$

$$\Rightarrow \quad \frac{dP}{dx} = 2 - \frac{2}{x^2}A \quad ...(2)$$

and $$\frac{d^2P}{dx^2} = \frac{4A}{x^3} \quad ...(3)$$

$\therefore$ For a maximum or minimum, we must have

$$\frac{dP}{dx} = 0$$

$$\Rightarrow \quad 2 - \frac{2}{x^2}A = 0 \Rightarrow \frac{2}{x^2}A = 2$$

$$\Rightarrow \quad x^2 = A \Rightarrow x = \sqrt{A} \qquad [\because x > 0]$$

Now, $$\left.\frac{d^2P}{dx^2}\right]_{x=\sqrt{A}} = \frac{4A}{(\sqrt{A})^3} = \frac{4A}{A\sqrt{A}} = \frac{4}{\sqrt{A}} > 0$$

$\therefore$ Perimeter is minimum, when $x = \sqrt{A}$.

$\therefore$ From (1), $\quad y = \dfrac{A}{x} \Rightarrow y = \dfrac{A}{\sqrt{A}} = \sqrt{A}.$

$\Rightarrow \quad x = y = \sqrt{A}$

Hence, Perimeter of the rectangle is minimum, when $x = y$, *i.e.*, rectangle is a square.

Example 21. (*i*) *The perimeter of a triangle is 8 cm. One of the sides is 3 cm. Find the other two sides such that the triangle has maximum area.*

(*ii*) *Prove that the perimeter of a right angled triangle of given hypotenuse is maximum when triangle is isosceles.*

Solution. (*i*) Let *a, b, c* be the three sides of the triangle.

Since, Perimeter of triangle is 8.

$$\therefore \quad a + b + c = 8 \quad \text{...(1)}$$

$$\Rightarrow \text{(semi-perimeter) } (s) = \frac{a+b+c}{2} = \frac{8}{2} = 4.$$

$$\text{Let} \quad a = 3 \quad \text{...(2)}$$

$$\Rightarrow \quad 3 + b + c = 8$$

$$\Rightarrow \quad b + c = 5$$

$$\Rightarrow \quad c = 5 - b \quad \text{...(3)}$$

Let the area of the triangle is given by

$$A = \sqrt{s(s-a)(s-b)(s-c)} \quad [\because \text{ By Hero's Formula}]$$

$$= \sqrt{4(4-3)(4-b)[4-(5-b)]}$$

$$= \sqrt{4(1)(4-b)(-1+b)} = 2\sqrt{(4-b)(b-1)}$$

$$= 2\sqrt{-b^2+5b-4} = 2(-b^2+5b-4)^{1/2}$$

$$\Rightarrow \quad \frac{dA}{db} = 2.\frac{1}{2}(-b^2+5b-4)^{-\frac{1}{2}}.(-2b+5)$$

$$\Rightarrow \quad \frac{dA}{db} = \frac{(5-2b)}{\sqrt{-b^2+5b-4}} \quad \text{...(4)}$$

$\therefore$ For a maximum or minimum, we must have

$$\frac{dA}{db} = 0$$

$$\Rightarrow \quad \frac{5-2b}{\sqrt{-b^2+5b-4}} = 0$$

$$\Rightarrow \quad 5 - 2b = 0 \Rightarrow 2b = 5 \Rightarrow b = \frac{5}{2} = 2.5.$$

Now, when $\dfrac{dA}{db}$ is slightly less than 2.5, say (2.4)

$$\Rightarrow \quad \frac{dA}{db} = \frac{5-2(2.4)}{\sqrt{-(2.4)^2+5(2.4)-4}} = \frac{+ve}{+ve} = +ve.$$

When $\frac{dA}{db}$ is slightly greater than 2.5, say (2.6)

$$\Rightarrow \quad \frac{dA}{db} = \frac{5 - 2\,(2.6)}{\sqrt{-(2.6)^2 + 5\,(2.6) - 4}} = \frac{-ve}{+ve} = -ve$$

$\therefore$ $\frac{dA}{db}$ changes sign from (+ve) to (–ve), as b increases through $\frac{5}{2}$.

$\therefore$ Area is maximum, when $b = \frac{5}{2}$.

$$\therefore \quad a = 3$$

$$b = \frac{5}{2}.$$

From (3), $$c = 5 - b = 5 - \frac{5}{2} = \frac{5}{2}$$

$$\Rightarrow \quad b = c = \frac{5}{2}.$$

Hence, the area of the triangle is maximum, when $a = 3$, $b = \frac{5}{2}$, $c = \frac{5}{2}$.

(*ii*) Let ABC be a right-angled triangle at B.

Let BC = x, AB = y and AC = h.

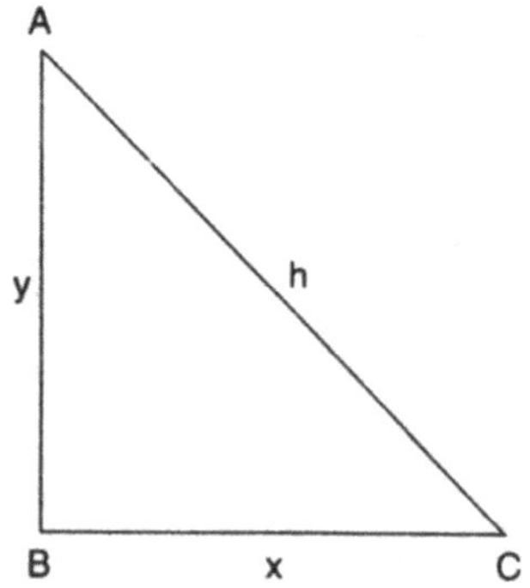

$\therefore$ By Pythagoras theorem, we have

$$(AC)^2 = (BC)^2 + (AB)^2$$

$$\Rightarrow \quad h^2 = x^2 + y^2 \quad ...(1)$$

$$\Rightarrow \quad y^2 = h^2 - x^2$$

$$\Rightarrow \quad y = \sqrt{h^2 - x^2} \quad ...(2)$$

Now, Let the Perimeter of triangle ABC is $P = x + y + h$ ...(3)

$$\Rightarrow \quad P = x + \sqrt{h^2 - x^2} + h \quad [\because \text{ By using (2)}]$$

$$\Rightarrow \quad \frac{dP}{dx} = 1 + \frac{1}{2\sqrt{h^2 - x^2}}\,(-2x) + 0$$

$$\Rightarrow \quad \frac{dP}{dx} = 1 - \frac{x}{\sqrt{h^2 - x^2}} \quad ...(4)$$

and $$\frac{d^2P}{dx^2} = 0 - \left[\frac{\sqrt{h^2 - x^2}\,(1) - x \cdot \frac{1}{2\sqrt{h^2 - x^2}}\,(-2x)}{\left(\sqrt{h^2 - x^2}\right)^2}\right]$$

$$= -\left[\frac{(h^2 - x^2) + x^2}{\sqrt{h^2 - x^2}\,(h^2 - x^2)}\right]$$

$$\Rightarrow \qquad \frac{d^2P}{dx^2} = -\frac{h^2}{(h^2 - x^2)^{3/2}} \qquad \text{...(5)}$$

$\therefore$ For a maximum or minimum, we must have

$$\frac{dP}{dx} = 0$$

$$\Rightarrow \qquad 1 - \frac{x}{\sqrt{h^2 - x^2}} = 0$$

$$\Rightarrow \qquad \frac{\sqrt{h^2 - x^2} - x}{\sqrt{h^2 - x^2}} = 0$$

$$\Rightarrow \qquad \sqrt{h^2 - x^2} - x = 0 \quad \Rightarrow \quad \sqrt{h^2 - x^2} = x$$

$$\Rightarrow \quad h^2 - x^2 = x^2$$

$$\Rightarrow \quad 2x^2 = h^2$$

$$\Rightarrow \quad x^2 = \frac{h^2}{2}$$

$$\Rightarrow \quad x = \frac{h}{\sqrt{2}} \qquad [\because \ x > 0]$$

Now,
$$\left.\frac{d^2P}{dx^2}\right]_{x=\frac{h}{\sqrt{2}}} = \frac{-h^2}{\left[h^2 - \frac{h^2}{2}\right]^{3/2}} = \frac{-h^2}{\left[\frac{2h^2 - h^2}{2}\right]^{3/2}}$$

$$= \frac{-h^2}{\left[\frac{h^2}{2}\right]^{3/2}} < 0$$

$\therefore$ Perimeter P of triangle is maximum, when $x = \frac{h}{\sqrt{2}}$.

$\therefore$ From (1), $y^2 = h^2 - x^2$

$$\Rightarrow \qquad y^2 = h^2 - \frac{h^2}{2} = \frac{h^2}{2}$$

$$\Rightarrow \qquad y = \frac{h}{\sqrt{2}}$$

$$\Rightarrow \qquad x = y = \frac{h}{\sqrt{2}}.$$

$\therefore$ The Perimeter P of the triangle ABC is maximum, when $x = y$ *i.e.*, the triangle is isosceles.

Example 22. (*i*) *Prove that the area of right angled triangle of given hypotenuse is maximum when the triangle is isosceles.*

(*ii*) *Show that the isosceles triangle of maximum area that can be inscribed in a given circle is an equilateral triangle.*

Solution. (*i*) Let ABC be a right-angled triangle at B.

Let BC = x, AB = y and AC = h.

$\therefore$ By Pythagoras theorem, we have

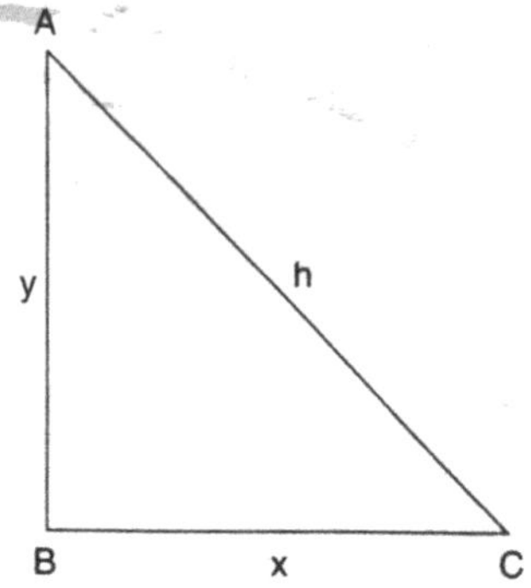

$$(AC)^2 = (BC)^2 + (AB)^2$$

$$\Rightarrow \quad h^2 = x^2 + y^2 \quad ...(1)$$

$$\Rightarrow \quad y^2 = h^2 - x^2$$

$$\Rightarrow \quad y = \sqrt{h^2 - x^2} \quad ...(2)$$

Now, Area of the Δ ABC is $\quad A = \frac{1}{2} \times \text{base} \times \text{altitude}$

$$A = \frac{1}{2} \times x \times y \quad [\because \text{ By using (2)}]$$

$$A = \frac{1}{2} x\sqrt{h^2 - x^2}$$

$$\Rightarrow \quad \frac{dA}{dx} = \frac{1}{2}\left[x \cdot \frac{1}{2\sqrt{h^2 - x^2}}(-2x) + \sqrt{h^2 - x^2} \cdot 1\right]$$

$$= \frac{1}{2}\left[\frac{-x^2}{\sqrt{h^2 - x^2}} + \frac{\sqrt{h^2 - x^2}}{1}\right] = \frac{1}{2}\left[\frac{-x^2 + h^2 - x^2}{\sqrt{h^2 - x^2}}\right]$$

$$\frac{dA}{dx} = \frac{1}{2}\left[\frac{h^2 - 2x^2}{\sqrt{h^2 - x^2}}\right] \quad ...(3)$$

and

$$\frac{d^2A}{dx^2} = \frac{1}{2}\left[\frac{\sqrt{h^2 - x^2}\,(-4x) - (h^2 - 2x^2) \cdot \frac{1}{2\sqrt{h^2 - x^2}}(-2x)}{\left(\sqrt{h^2 - x^2}\right)^2}\right]$$

$$= \frac{1}{2}\left[\frac{-4x\sqrt{h^2 - x^2} + x(h^2 - 2x^2)/\sqrt{h^2 - x^2}}{(h^2 - x^2)}\right]$$

$$= \frac{1}{2}\left[\frac{-4x(h^2 - x^2) + x(h^2 - 2x^2)}{(h^2 - x^2)^{3/2}}\right]$$

$$= \frac{1}{2}\left[\frac{-4xh^2 + 4x^3 + xh^2 - 2x^3}{(h^2 - x^2)^{3/2}}\right]$$

$$\Rightarrow \quad \frac{d^2A}{dx^2} = \frac{1}{2}\left[\frac{2x^3 - 3xh^2}{(h^2 - x^2)^{3/2}}\right] \quad ...(4)$$

$\therefore$ For a maximum or minimum, we must have

$$\frac{dA}{dx} = 0$$

$$\Rightarrow \quad \frac{1}{2}\left[\frac{h^2 - 2x^2}{\sqrt{h^2 - x^2}}\right] = 0$$

$$\Rightarrow \quad h^2 - 2x^2 = 0$$

$$\Rightarrow \quad 2x^2 = h^2 \Rightarrow x^2 = \frac{h^2}{2} \Rightarrow x = \frac{h}{\sqrt{2}} \qquad (\because \; x > 0)$$

Now,
$$\left.\frac{d^2A}{dx^2}\right]_{x=\frac{h}{\sqrt{2}}} = \frac{1}{2}\left[\frac{2\left(\frac{h}{\sqrt{2}}\right)^3 - 3\left(\frac{h}{\sqrt{2}}\right)h^2}{\left(h^2 - \frac{h^2}{2}\right)^{3/2}}\right]$$

$$= \frac{1}{2}\left[\frac{\frac{2h^3}{2\sqrt{2}} - \frac{3h^3}{\sqrt{2}}}{\left(\frac{h^2}{2}\right)^{3/2}}\right] = \frac{1}{2}\left[\frac{\frac{h^3}{\sqrt{2}} - \frac{3h^3}{\sqrt{2}}}{\left(\frac{h^2}{2}\right)^{3/2}}\right]$$

$$= \frac{1}{2}\left[\frac{-2h^3}{\sqrt{2}} \times \left(\frac{2}{h^2}\right)^{3/2}\right] = \frac{1}{2}\left(\frac{-2h^3}{\sqrt{2}}\right)\left(\frac{2\sqrt{2}}{h^3}\right) = -2 < 0$$

$\therefore$ The area of triangle ABC is maximum, when $x = \frac{h}{\sqrt{2}}$.

$\therefore$ From (1), $\quad y^2 = h^2 - x^2 = h^2 - \frac{h^2}{2} = \frac{h^2}{2}$

$$\Rightarrow \quad y = \frac{h}{\sqrt{2}}$$

$$\Rightarrow \quad x = y = \frac{h}{\sqrt{2}}$$

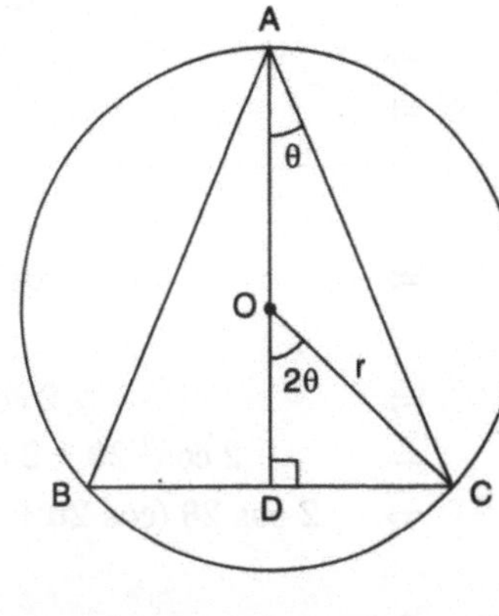

Hence, the area of a right angled triangle is maximum, when $x = y$ *i.e.*, the triangle is isosceles.

(*ii*) Let ABC be an isosceles triangle inscribed in a circle of radius r.

Let AB = AC.

$\therefore$ A lies on the diameter perpendicular to BC and passing through the mid-point of BC.

Let $\angle OAC = \theta$

$\Rightarrow \quad \angle DOC = 2 \angle OAC = 2\theta$

[$\because$ Angle at the centre of a circle is twice the angle made at the remaining part of the circle]

In right angled triangle DOC, we have

$$\sin 2\theta = \frac{DC}{OC} \quad \text{and} \quad \cos 2\theta = \frac{OD}{OC}$$

$$\Rightarrow \quad \sin 2\theta = \frac{DC}{r} \quad \text{and} \quad \cos 2\theta = \frac{OD}{r}.$$

$$\Rightarrow \quad DC = r \sin 2\theta \quad \text{and} \quad OD = r \cos 2\theta.$$

$$\therefore \quad BC = 2\,CD$$

$$BC = 2r \sin 2\theta \qquad ...(1)$$

and $$AD = AO + OD = r + r \cos 2\theta \qquad ...(2)$$

$\therefore$ Area of a triangle ABC $= A = \frac{1}{2} \times$ base $\times$ altitude

$$A = \frac{1}{2} \times BC \times AD$$

$$= \frac{1}{2} \times (2r \sin 2\theta)(r + r \cos 2\theta)$$

$$= r^2 \sin 2\theta (1 + \cos 2\theta) = r^2(\sin 2\theta + \sin 2\theta \cos 2\theta)$$

$$\Rightarrow \quad A = r^2 \left(\sin 2\theta + \frac{1}{2} \sin 4\theta\right)$$

$\left[\begin{array}{ll} \because & \sin 2A = 2 \sin A \cos A \\ \Rightarrow & \sin 4A = 2 \sin 2A \cos 2A \\ \Rightarrow & \frac{1}{2}\sin 4A = \sin 2A \cos 2A \end{array}\right]$

$$\Rightarrow \quad \frac{dA}{d\theta} = r^2 (2 \cos 2\theta + 2 \cos 4\theta) \qquad ...(3)$$

and $$\frac{d^2A}{d\theta^2} = r^2 (-4 \sin 2\theta - 8 \sin 4\theta) \qquad ...(4)$$

$\therefore$ For maximum or minimum, we must have

$$\frac{dA}{d\theta} = 0$$

$$\Rightarrow \quad r^2 (2 \cos 2\theta + 2 \cos 4\theta) = 0$$

$$\Rightarrow \quad 2 \cos 2\theta + 2 \cos 4\theta = 0$$

$$\Rightarrow \quad \cos 2\theta + \cos 4\theta = 0$$

$$\Rightarrow \quad \cos 2\theta + 2 \cos^2 2\theta - 1 = 0$$

$\left[\begin{array}{ll} \because & 1 + \cos 2A = 2 \cos^2 A \\ \Rightarrow & 1 + \cos 4A = 2 \cos^2 2A \\ \Rightarrow & \cos 4A = 2 \cos^2 2A - 1 \end{array}\right]$

$$\Rightarrow \quad 2 \cos^2 2\theta + \cos 2\theta - 1 = 0$$

$$\Rightarrow \quad 2 \cos^2 2\theta + 2 \cos 2\theta - \cos 2\theta - 1 = 0$$

$$\Rightarrow \quad 2 \cos 2\theta (\cos 2\theta + 1) - 1 (\cos 2\theta + 1) = 0$$

$$\Rightarrow \quad (2\cos 2\theta - 1)(\cos 2\theta + 1) = 0$$

$$\Rightarrow \quad \cos 2\theta = \frac{1}{2} \quad \text{or} \quad \cos 2\theta = -1$$

$$\Rightarrow \quad 2\theta = \frac{\pi}{3} \quad \text{or} \quad 2\theta = \pi$$

$$\Rightarrow \quad \theta = \frac{\pi}{6} \quad \text{or} \quad \theta = \frac{\pi}{2}. \qquad \left[\text{Since, } \theta = \frac{\pi}{2} \text{ is not possible}\right]$$

Now, $$\left.\frac{d^2A}{d\theta^2}\right]_{\theta=\frac{\pi}{6}} = -4r^2\left(\sin\frac{\pi}{3} + 2\sin\frac{2\pi}{3}\right)$$

$$= -4r^2\left[\frac{\sqrt{3}}{2} + 2\left(\frac{\sqrt{3}}{2}\right)\right] = -4r^2\left(\frac{\sqrt{3}}{2} + \sqrt{3}\right) = -6\sqrt{3}r^2 < 0$$

$\therefore$ Area is maximum, when $\theta = \frac{\pi}{6}$.

$\therefore$ From the figure, $\angle BAC = 2\theta = 2\left(\frac{\pi}{6}\right) = \frac{\pi}{3}$.

Also, $$\angle ABC = \angle ACB = \frac{1}{2}\left[\pi - \frac{\pi}{3}\right] = \frac{1}{2}\left(\frac{2\pi}{3}\right) = \frac{\pi}{3}.$$

$\therefore$ In Δ ABC, $\angle ABC = \angle ACB = \angle BAC = \frac{\pi}{3} = 60$.

$\therefore$ Δ ABC is an equilateral triangle.

Example 23. (*i*) *Show that the rectangle of maximum area that can be inscribed in a circle of radius r is a square of side* $\sqrt{2}r$.

(*ii*) *Show that the rectangle of maximum perimeter that can be inscribed in a circle of radius r is a square of side* $\sqrt{2}r$.

Solution. (*i*) Let a rectangle ABCD is inscribed in a circle of radius r.

Let length of rectangle $= 2x$

and breadth of rectangle $= 2y$

$\therefore$ Area of rectangle = length × breadth

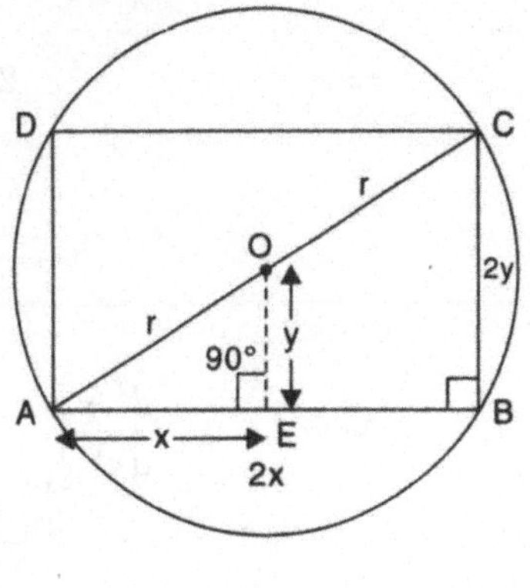

$$A = (2x)(2y)$$

$$\Rightarrow \quad A = 4xy \qquad ...(1)$$

Now, In Δ AOE, By Pythagoras theorem, we have

$$(OA)^2 = (AE)^2 + (OE)^2$$

$$\Rightarrow \quad r^2 = x^2 + y^2$$

$$\Rightarrow \quad y^2 = r^2 - x^2$$

$$\Rightarrow \quad y = \sqrt{r^2 - x^2} \qquad ...(2)$$

Putting this value of y in (1), we get

$$A = 4x\sqrt{r^2 - x^2}$$

$$\Rightarrow \quad \frac{dA}{dx} = 4\left[x.\frac{1}{2\sqrt{r^2-x^2}}.(-2x)+\sqrt{r^2-x^2}\,.\,1\right]$$

$$= 4\left[\frac{-x^2}{\sqrt{r^2-x^2}}+\sqrt{r^2-x^2}\right]$$

$$\Rightarrow \quad \frac{dA}{dx} = 4\left[\frac{-x^2+r^2-x^2}{\sqrt{r^2-x^2}}\right] = 4\left[\frac{r^2-2x^2}{\sqrt{r^2-x^2}}\right] \qquad ...(3)$$

and

$$\frac{d^2A}{dx^2} = 4\left[\frac{\sqrt{r^2-x^2}\,.(-4x)-(r^2-2x^2)\left(\frac{1}{2\sqrt{r^2-x^2}}(-2x)\right)}{(r^2-x^2)}\right]$$

$$= 4\left[\frac{-4x\sqrt{r^2-x^2}+\frac{x(r^2-2x^2)}{\sqrt{r^2-x^2}}}{(r^2-x^2)}\right]$$

$$= 4\left[\frac{-4x(r^2-x^2)+xr^2-2x^3}{(r^2-x^2)^{3/2}}\right]$$

$$= 4\left[\frac{-4xr^2+4x^3+xr^2-2x^3}{(r^2-x^2)^{3/2}}\right] = 4\left[\frac{2x^3-3xr^2}{(r^2-x^2)^{3/2}}\right] \qquad ...(4)$$

$\therefore$ For a maximum or minimum, we must have

$$\frac{dA}{dx} = 0$$

$$\Rightarrow \quad 4\left[\frac{r^2-2x^2}{\sqrt{r^2-x^2}}\right] = 0 \qquad \Rightarrow \qquad r^2-2x^2 = 0$$

$$\Rightarrow \quad 2x^2 = r^2 \qquad \Rightarrow \qquad x^2 = \frac{r^2}{2}$$

$$\Rightarrow \quad x = \frac{r}{\sqrt{2}}$$

Now,

$$\left.\frac{d^2A}{dx^2}\right]_{x=\frac{r}{\sqrt{2}}} = 4\left[\frac{2\left(\frac{r}{\sqrt{2}}\right)^3-3\left(\frac{r}{\sqrt{2}}\right)r^2}{\left(r^2-\frac{r^2}{2}\right)^{3/2}}\right] = 4\left[\frac{\frac{2r^3}{2\sqrt{2}}-\frac{3r^3}{\sqrt{2}}}{\left(\frac{r^2}{2}\right)^{3/2}}\right]$$

$$= 4\left[\frac{r^3}{\sqrt{2}}-\frac{3r^3}{\sqrt{2}}\right]\left[\left(\frac{2}{r^2}\right)^{3/2}\right] = 4\left[\frac{-2r^3}{\sqrt{2}}\right]\left[\frac{2\sqrt{2}}{r^3}\right] = -16 < 0$$

$\therefore$ Area of rectangle is maximum, when $x = \frac{r}{\sqrt{2}}$

$\therefore$ From (2), we have $y = \sqrt{r^2 - \frac{r^2}{2}} = \sqrt{\frac{r^2}{2}} = \frac{r}{\sqrt{2}}$

$$\Rightarrow \quad x = y = \frac{r}{\sqrt{2}}$$

$$\Rightarrow \quad 2x = 2y$$

Hence, the area of rectangle is maximum, when $2x = 2y$ *i.e.*, the rectangle is a square.

(*ii*) Let a rectangle ABCD is inscribed in a circle of radius r.

Let length of rectangle $= 2x$

and breadth of rectangle $= 2y$.

$\therefore$ Perimeter of rectangle $= 2\,(l + b)$

$$P = 2\,(2x + 2y)$$

$$P = 4\,(x + y) \quad \text{...(1)}$$

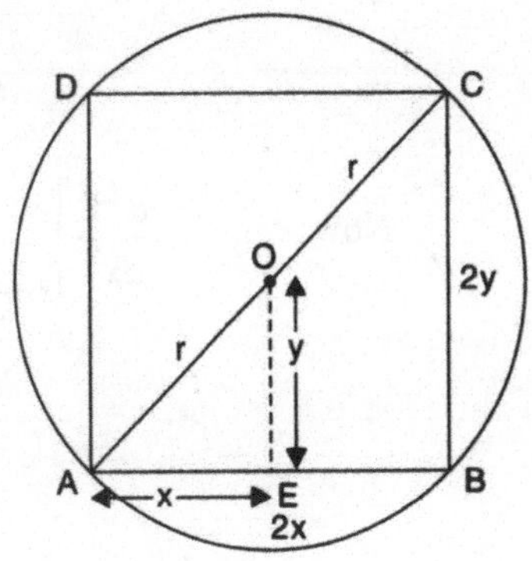

Now, In Δ AOE, By Pythagoras theorem, we have

$$(OA)^2 = (AE)^2 + (OE)^2$$

$$\Rightarrow \quad r^2 = x^2 + y^2$$

$$\Rightarrow \quad y^2 = r^2 - x^2 \quad \Rightarrow \quad y = \sqrt{r^2 - x^2} \quad \text{...(2)}$$

Putting this value of y in (1), we get

$$P = 4\,(x + \sqrt{r^2 - x^2})$$

$$\Rightarrow \quad \frac{dP}{dx} = 4\left[1 + \frac{1}{2\sqrt{r^2 - x^2}}(-2x)\right]$$

$$\Rightarrow \quad \frac{dP}{dx} = 4\left[1 - \frac{x}{\sqrt{r^2 - x^2}}\right] \quad \text{...(3)}$$

and

$$\frac{d^2P}{dx^2} = 4\left[0 - \frac{\sqrt{r^2 - x^2}\,.\,1 - \frac{x}{2\sqrt{r^2 - x^2}}(-2x)}{r^2 - x^2}\right]$$

$$= 4\left[-\frac{\sqrt{(r^2 - x^2)} + \frac{x^2}{\sqrt{r^2 - x^2}}}{(r^2 - x^2)}\right]$$

$$= -4\left[\frac{r^2 - x^2 + x^2}{(r^2 - x^2)^{3/2}}\right] = \frac{-4r^2}{(r^2 - x^2)^{3/2}} \quad \text{...(4)}$$

$\therefore$ For a maximum or minimum, we must have

$$\frac{dP}{dx} = 0$$

$$\Rightarrow \quad 4\left[1 - \frac{x}{\sqrt{r^2 - x^2}}\right] = 0$$

$$\Rightarrow \quad \frac{\sqrt{r^2 - x^2} - x}{\sqrt{r^2 - x^2}} = 0$$

$$\Rightarrow \quad x = \sqrt{r^2 - x^2}$$

$$\Rightarrow \quad x^2 = r^2 - x^2 \;\Rightarrow\; 2x^2 = r^2 \;\Rightarrow\; x = \frac{r}{\sqrt{2}}$$

Now,
$$\left.\frac{d^2P}{dx^2}\right]_{x=\frac{r}{\sqrt{2}}} = \frac{-4r^2}{\left(r^2 - \frac{r^2}{2}\right)^{3/2}} = \frac{-4r^2}{\left(\frac{r^2}{2}\right)^{3/2}}$$

$$= \frac{-4r^2}{1} \times \left(\frac{2}{r^2}\right)^{3/2} = -4r^2 \cdot \frac{2\sqrt{2}}{r^3} = \frac{-8\sqrt{2}}{r} < 0$$

$\therefore$ The perimeter P is maximum, when $x = \dfrac{r}{\sqrt{2}}$

$\therefore$ From (2), we have $y = \sqrt{r^2 - \dfrac{r^2}{2}} = \sqrt{\dfrac{r^2}{2}} = \dfrac{r}{\sqrt{2}}$

$$\Rightarrow \quad x = y = \frac{r}{\sqrt{2}} \;\Rightarrow\; 2x = 2y.$$

Hence, the perimeter of rectangle is maximum, when $2x = 2y$, *i.e.*, the rectangle is a square.

Example 24. (*i*) *Two sides of a triangle have lengths x and y and the angle between them is θ. What value of θ will maximize the area of the triangle. Also, find the maximum area of the triangle.*

(*ii*) *If the sum of the lengths of the hypotenuse and a side of a right angled triangle is given, show that the area of the triangle is maximum when the angle between them is* $\frac{\pi}{3}$.

Solution. (*i*) It is given that, the angle between the two sides of a triangle having lengths x and y is θ.

$\therefore$ Area of triangle $= A = \dfrac{1}{2}\, ab \sin\theta$ ...(1)

$$\Rightarrow \quad \frac{dA}{d\theta} = \frac{1}{2}\, ab \cos\theta \qquad ...(2)$$

and
$$\frac{d^2A}{d\theta^2} = -\frac{1}{2}\, ab \sin\theta \qquad ...(3)$$

$\therefore$ For a maximum or minimum, we must have

$$\frac{dA}{d\theta} = 0$$

$$\Rightarrow \quad \frac{1}{2} ab \cos\theta = 0 \quad \Rightarrow \quad \cos\theta = 0 \quad \Rightarrow \quad \theta = \frac{\pi}{2}.$$

Now, $\left.\frac{d^2A}{d\theta^2}\right]_{\theta=\frac{\pi}{2}} = -\frac{1}{2} ab \sin\frac{\pi}{2} = -\frac{1}{2} ab < 0$

$\therefore$ Area of the triangle is maximum, when $\theta = \frac{\pi}{2}$.

$\therefore$ From (1), we have

$$A = \frac{1}{2} ab \sin\frac{\pi}{2} = \frac{1}{2} ab.$$

(*ii*) Let ABC be a right angled triangle.

$\therefore$ It is given that

$$x + y = p \qquad \text{...(1)}$$

where $p \to$ constant.

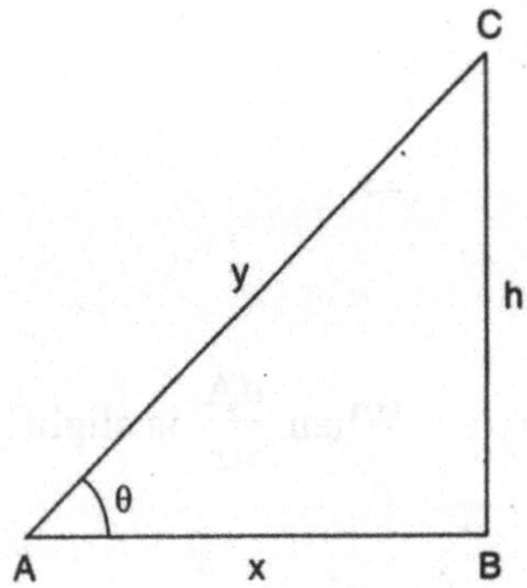

Let θ be the angle between the sides AB and AC.

$\therefore$ Area of triangle = $A = \frac{1}{2} \times AB \times BC$

$$A = \frac{1}{2} \times x \times h \qquad \text{...(2)}$$

Now, In Δ ABC, By Pythagoras theorem, we have

$$(AC)^2 = (AB)^2 + (BC)^2$$

$$\Rightarrow \quad y^2 = x^2 + h^2$$

$$\Rightarrow \quad h^2 = y^2 - x^2 \quad \Rightarrow \quad h = \sqrt{y^2 - x^2}$$

Putting this value of h in equation (2), we get

$$A = \frac{1}{2} x \sqrt{y^2 - x^2} \qquad [\because \ x + y = p \ \Rightarrow \ y = p - x]$$

$$A = \frac{1}{2} x \sqrt{(p-x)^2 - x^2} = \frac{1}{2} x \sqrt{p^2 + x^2 - 2px - x^2}$$

$$= \frac{1}{2} x \sqrt{p^2 - 2px}$$

$$\Rightarrow \quad A = \frac{1}{2} \sqrt{p^2x^2 - 2px^3} \qquad \text{...(3)}$$

$$\Rightarrow \quad \frac{dA}{dx} = \frac{1}{2} \cdot \frac{1}{2\sqrt{p^2x^2 - 2px^3}} \cdot (2p^2x - 6px^2)$$

$$= \frac{1}{4\sqrt{p^2x^2 - 2px^3}} \cdot 2p(px - 3x^2)$$

$$\Rightarrow \quad \frac{dA}{dx} = \frac{(px - 3x^2)}{2\sqrt{p^2x^2 - 2px^3}} \quad ...(4)$$

$\therefore$ For a maximum or minimum, we must have

$$\frac{dA}{dx} = 0$$

$$\Rightarrow \quad \frac{px - 3x^2}{2\sqrt{p^2x^2 - 2px^3}} = 0 \quad \Rightarrow \quad px - 3x^2 = 0$$

$$\Rightarrow \quad 3x^2 = px$$

$$\Rightarrow \quad 3x = p$$

$$\Rightarrow \quad x = \frac{p}{3}.$$

$\therefore$ When $\frac{dA}{dx}$ is slightly less than $x = \frac{p}{3}$, say $\left(\frac{p}{4}\right)$.

$$\Rightarrow \quad \frac{dA}{dx} = \frac{p\left(\frac{p}{4}\right) - 3\left(\frac{p}{4}\right)^2}{2\sqrt{p^2\left(\frac{p}{4}\right)^2 - 2p\left(\frac{p}{4}\right)^3}} = \frac{(+\text{ve})}{(+\text{ve})} = (+\text{ve})$$

When $\frac{dA}{dx}$ is slightly greater than $x = \frac{p}{3}$, say $\left(\frac{p}{2}\right)$:

$$\Rightarrow \quad \frac{dA}{dx} = \frac{p\left(\frac{p}{2}\right) - 3\left(\frac{p}{2}\right)^2}{2\sqrt{p^2\left(\frac{p}{2}\right)^2 - 2p\left(\frac{p}{2}\right)^3}} = \frac{(-\text{ve})}{(+\text{ve})} = (-\text{ve})$$

$\therefore$ $\frac{dA}{dx}$ changes sign from (+ve) to (–ve) as x increases through $\frac{p}{3}$.

$\therefore$ Area of triangle is maximum, when $x = \frac{p}{3}$.

$\therefore$ From (1), we have $y = p - x$

$$\Rightarrow \quad y = p - \frac{p}{3} = \frac{2p}{3}.$$

$\therefore$ In Δ ABC, $\cos\theta = \frac{x}{y} = \frac{p/3}{2p/3} = \frac{1}{2}$

$$\Rightarrow \quad \theta = \frac{\pi}{3}.$$

Example 25. (*i*) *If 40 square feet of sheet of metal is to be used in the construction of an open tank with a square base, find the dimensions for maximum volume.*

(*ii*) *A jet of an enemy is flying along the curve* $y = x^2 + 2$. *A soldier is placed at the point* (3, 2). *What is nearest distance between the soldier and the jet.*

Solution. (*i*) Let x be the length of edge of the square base of the tank and y be the height of the tank.

Then, the surface area of open tank is given by

$$x^2 + 4xy = 40 \quad \text{...(1)}$$

Let the volume of the open tank = $V = x^2y$ ...(2)

$$\Rightarrow \quad V = x^2\left(\frac{40 - x^2}{4x}\right) \qquad \left[\begin{array}{lr} \because & x^2 + 4xy = 40 \\ \Rightarrow & 4xy = 40 - x^2 \\ \Rightarrow & y = \dfrac{40 - x^2}{4x} \end{array}\right]$$

$$\Rightarrow \quad V = \frac{x}{4}(40 - x^2)$$

$$\Rightarrow \quad V = \frac{1}{4}(40x - x^3)$$

$$\Rightarrow \quad \frac{dV}{dx} = \frac{1}{4}(40 - 3x^2) \quad \text{...(3)}$$

and

$$\frac{d^2V}{dx^2} = \frac{1}{4}(0 - 6x) = -\frac{3}{2}x \quad \text{...(4)}$$

$\therefore$ For a maximum or minimum, we must have

$$\frac{dV}{dx} = 0$$

$$\Rightarrow \quad \frac{1}{4}(40 - 3x^2) = 0 \quad \Rightarrow \quad 40 - 3x^2 = 0$$

$$\Rightarrow \quad 3x^2 = 40 \qquad [\because \ x > 0]$$

$$\Rightarrow \quad x^2 = \frac{40}{3} \quad \Rightarrow \quad x = \sqrt{\frac{40}{3}}$$

Now,

$$\left.\frac{d^2V}{dx^2}\right]_{x=\sqrt{\frac{40}{3}}} = -\frac{3}{2}.\sqrt{\frac{40}{3}} < 0$$

$\therefore$ The volume of tank is maximum, when $x = \sqrt{\dfrac{40}{3}}$.

$\therefore$ From (1), we have

$$y = \frac{40 - x^2}{4x} = \frac{40 - \frac{40}{3}}{4\sqrt{\frac{40}{3}}} = \frac{\frac{120 - 40}{3}}{8\sqrt{\frac{10}{3}}}$$

$$= \frac{80}{3} \times \frac{1}{8}\sqrt{\frac{3}{10}} = \frac{10}{3}\sqrt{\frac{3}{10}} = \sqrt{\frac{10}{3}}.$$

Hence, the volume of the tank is maximum, when

$$x = \sqrt{\frac{40}{3}} \quad \text{and} \quad y = \sqrt{\frac{10}{3}}.$$

(*ii*) Let the equation of the curve is given by

$$y = x^2 + 2 \qquad \text{...(1)}$$

Let P(x, y) be the position of the jet on the curve.

Let Q (3, 2) be the position of the soldier.

$\therefore$ The distance PQ between the soldier and the jet is given by

$$h = PQ = \sqrt{(x-3)^2 + (y-2)^2} \qquad \left[\because \text{By Distance Formula : } PQ = \sqrt{(x_2 - x_1)^2 + (y_2 - y_1)^2}\right]$$

$$\Rightarrow \quad h = \sqrt{x^2 + 9 - 6x + y^2 + 4 - 4y}$$

$$\Rightarrow \quad h = \sqrt{x^2 + y^2 - 6x - 4y + 13} \qquad [\because y = (x^2 + 2)]$$

$$\Rightarrow \quad h = \sqrt{x^2 + (x^2+2)^2 - 6x - 4(x^2+2) + 13}$$

$$\Rightarrow \quad h = \sqrt{x^2 + x^4 + 4 + 4x^2 - 6x - 4x^2 - 8 + 13}$$

$$\Rightarrow \quad h = \sqrt{x^4 + x^2 - 6x + 9}$$

$$\Rightarrow \quad h^2 = x^4 + x^2 - 6x + 9 \qquad \text{...(2)}$$

$$\Rightarrow \quad \frac{d(h^2)}{dx} = 4x^3 + 2x - 6 \qquad \text{...(3)}$$

and

$$\frac{d^2(h^2)}{dx^2} = 12x^2 + 2 \qquad \text{...(4)}$$

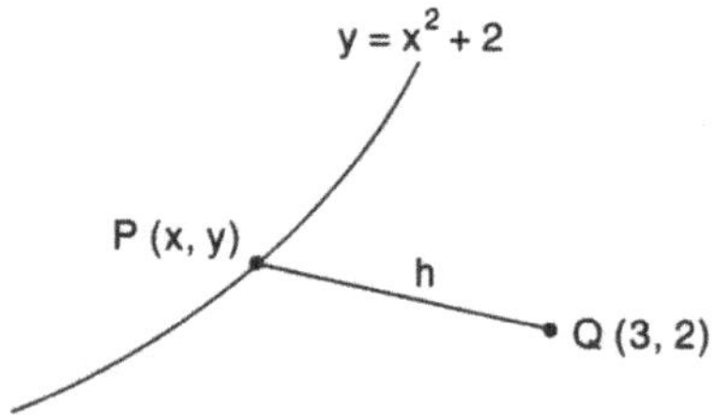

$\therefore$ For a maximum or minimum, we must have

$$\frac{d(h^2)}{dx} = 0 \quad \Rightarrow \quad 4x^3 + 2x - 6 = 0$$

$$\Rightarrow \quad (x-1)(4x^2 + 4x + 6) = 0$$

$$\Rightarrow \quad x = 1 \quad \text{or} \quad 4x^2 + 4x + 6 = 0$$

[As it gives, only imaginary values of x, hence rejected]

Now, $$\left.\frac{d^2(h^2)}{dx^2}\right]_{x=1} = 12\,(1)^2 + 2 = 14 > 0.$$

$\therefore$ h^2 and hence h is minimum, when $x = 1$.

$\therefore$ From (1), we have $\quad y = x^2 + 2$

$$\Rightarrow \quad y = (1)^2 + 2 = 3$$

$\therefore$ The position of jet is (1, 3)

$\Rightarrow$ The minimum distance between the soldier and the jet is equal to the distance between the points (3, 2) and (1, 3).

i.e.,

$$PQ = \sqrt{(3-1)^2 + (2-3)^2} = \sqrt{4+1} = \sqrt{5} \text{ units.}$$

Example 26. (*i*) *Show that the surface area of a closed cuboid with square base and given volume is minimum, when it is a cube.*

(*ii*) *A square piece of side 18 cm is to be made into a box without top by cutting a square from each corner and folding up the flaps to form a box. What should be the side of the square to be cut off, so that the volume of the box is maximum. Also, find the maximum volume.*

Solution. (*i*) Let the length of the edge of square base of cuboid is x and y be the height of the cuboid.

$\therefore$ The volume of the cuboid $= V = x^2y$ ...(1)

and the surface area of the cuboid

$$= S = 2(x^2 + xy + xy) = 2(x^2 + 2xy)$$

$$S = 2x^2 + 4xy$$

$$\Rightarrow \quad S = 2x^2 + 4x\left(\frac{V}{x^2}\right) \qquad \left[\because V = x^2y \Rightarrow y = \frac{V}{x^2}\right]$$

$$\Rightarrow \quad S = 2x^2 + \frac{4V}{x} \qquad ...(2)$$

$$\Rightarrow \quad \frac{dS}{dx} = 4x - \frac{4V}{x^2} \qquad ...(3)$$

and
$$\frac{d^2S}{dx^2} = 4 + \frac{8V}{x^3} \qquad ...(4)$$

$\therefore$ For a maximum or minimum, we have

$$\frac{dS}{dx} = 0$$

$$\Rightarrow \quad 4x - \frac{4V}{x^2} = 0$$

$$\Rightarrow \quad \frac{4V}{x^2} = 4x \qquad [\because V = x^2y]$$

$$\Rightarrow \quad V = x^3 \quad \Rightarrow \quad x^2y = x^3$$

$$\Rightarrow \quad y = x$$

Now,
$$\left.\frac{d^2S}{dx^2}\right]_{y=x} = 4 + \frac{8V}{x^3} \qquad [\because V = x^2y]$$

$$= 4 + \frac{8x^2y}{x^3} \qquad [\because y = x]$$

$$= 4 + \frac{8x^3}{x^3} = 4 + 8 = 12 > 0$$

$\therefore$ The surface area of cuboid is minimum, when $y = x$.

i.e., when it is a cube. *i.e.*, when, length, breadth and height are equal.

(*ii*) Let each side of the square to be cut off $= x$ cm.

Side of square sheet $= 18$ cm

$\therefore$ length of box $= 18 - 2x$ cm

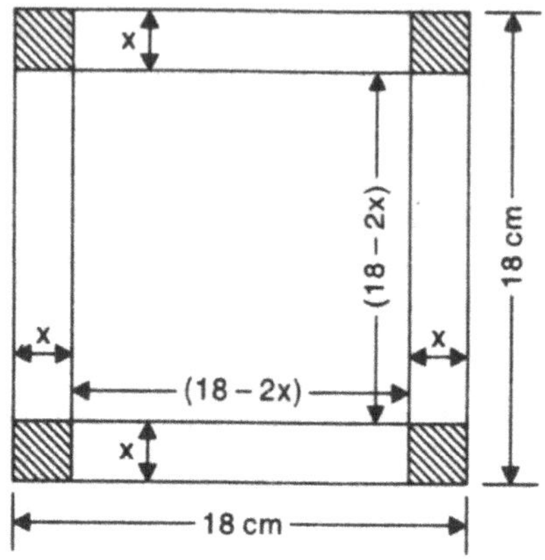

breadth of box = $18 - 2x$ cm

height of box = x cm

$\therefore$ Volume of the box = V

$$= \text{length} \times \text{breadth} \times \text{height}$$
$$= (18 - 2x)(18 - 2x)(x)$$
$$= x(18 - 2x)^2$$
$$= x(324 + 4x^2 - 72x)$$

$\Rightarrow$ $$V = 4x^3 - 72x^2 + 324x \quad ...(1)$$

$\Rightarrow$ $$\frac{dV}{dx} = 12x^2 - 144x + 324 \quad ...(2)$$

and $$\frac{d^2V}{dx^2} = 24x - 144 \quad ...(3)$$

$\therefore$ For a maximum or minimum, we must have

$$\frac{dV}{dx} = 0.$$

$\Rightarrow$ $12x^2 - 144x + 324 = 0$

$\Rightarrow$ $12(x^2 - 12x + 27) = 0$

$\Rightarrow$ $12(x - 9)(x - 3) = 0$

$\Rightarrow$ $x = 3, 9$ [Since, $x = 9$ cm is not possible, hence rejected]

$\therefore$ $x = 3$ cm

Now, $$\left.\frac{d^2V}{dx^2}\right]_{x=3} = 24(3) - 144 = 72 - 144 = -72 < 0$$

$\therefore$ Volume of the box is maximum, when $x = 3$.

i.e., the square of side 3 cm is cut off from each corner.

$\therefore$ Maximum volume $= V = 3(18 - 6)^2$

$$= 3(12)^2 = 3 \times 144 = 432 \text{ cm}^3.$$

Example 27. (*i*) *A square piece of tin of side 24 cm is to be made into a box without top by cutting a square from each corner and folding up the flaps to form a box. What should be the side of the square to be cut off so that the volume of the box is maximum. Also, find this maximum volume.*

(*ii*) *A rectangular sheet of tin 45 cm by 24 cm is to be made into a box without top, by cutting off squares from the corners and folding up the flaps. What should be the side of the square to be cut off so that the volume of the box is maximum.*

Solution. (*i*) Please try yourself.

[**Hint.** Volume of the box = $(24 - x)(24 - x) . x$

$$= (24 - x)^2 . x$$
$$V = 4x^3 - 96x^2 + 576x$$
$$\frac{dV}{dx} = 12x^2 - 192x + 576$$

and $$\frac{d^2V}{dx^2} = 24x - 192$$

Put $\frac{dV}{dx} = 0 \Rightarrow x = 4, 12.$

$\therefore$ $x = 4$ cm. [$\because$ $x = 12$ cm is not possible hence rejected]

Maximum volume $= (24-8)^2 . 4 = (16)^2 \times 4 = 256 \times 4 = 1024 \text{ cm}^3$]

(*ii*) Please try yourself.

[**Hint.** Volume of the box $= (45-2x)(24-2x) . x$

$$V = 4x^3 - 138x^2 + 1080x$$

$$\frac{dV}{dx} = 12x^2 - 276x + 1080$$

and $$\frac{d^2V}{dx^2} = 24x - 276$$

Put $\frac{dV}{dx} = 0 \Rightarrow x = 5, 18.$

[$\because$ $x = 18$ cm is not possible, hence rejected]

$\therefore$ $x = 5$ cm]

Example 28. (*i*) *A figure consists of a semi-circle, with a rectangle an its diameter. Give perimeter of the figure, find the dimensions in order that the area may be maximum.*

(*ii*) *A window is in the form of a rectangle surmounted by a semi-circular opening. The total perimeter of the window so as to admit maximum light through the whole opening.*

Solution. (*i*) Let $2x$ and $2y$ be the length and breadth of rectangle ABCD respectively.

$\therefore$ Radius of semi-circle $= x$

$\therefore$ Perimeter of the figure is given by

$$P = 2x + 4y + \pi x \quad ...(1)$$

Also, area of the figure = A = (Area of rectangle) + (Area of semi-circle)

$$A = (2x)(2y) + \frac{\pi x^2}{2}$$

$$A = 4xy + \frac{\pi x^2}{2}$$

[$\because$ $P = 2x + 4y + \pi x$

$\Rightarrow 2y = P - 2x - \pi x$]

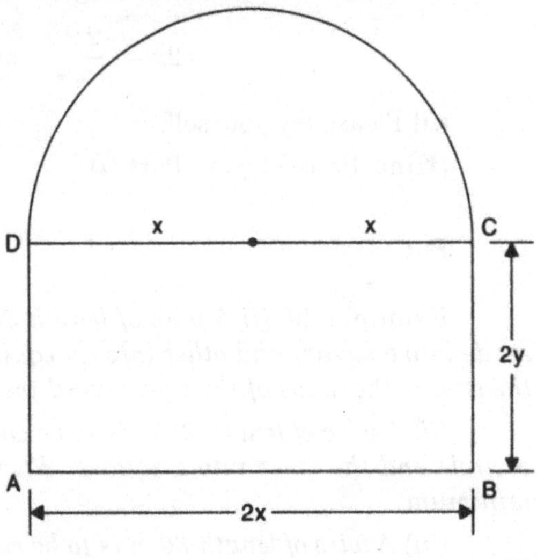

$$\Rightarrow \quad A = x(P - 2x - \pi x) + \frac{\pi x^2}{2}$$

$$\Rightarrow \quad A = Px - 2x^2 - \pi x^2 + \frac{\pi x^2}{2}$$

$$\Rightarrow \quad A = Px - 2x^2 - \frac{\pi x^2}{2} \quad ...(2)$$

$$\Rightarrow \quad \frac{dA}{dx} = P - 4x - \pi x \quad ...(3)$$

and $$\frac{d^2A}{dx^2} = -4 - \pi \qquad ...(4)$$

$\therefore$ For a maximum or minimum, we must have

$$\frac{dA}{dx} = 0 \quad \Rightarrow \quad P - 4x - \pi x = 0$$

$$\Rightarrow \quad 4x + \pi x = P$$

$$\Rightarrow \quad x(4 + \pi) = P$$

$$\Rightarrow \quad x = \frac{P}{4+\pi}$$

Now, $$\left.\frac{d^2A}{dx^2}\right]_{x=\left(\frac{P}{4+\pi}\right)} = -4 - \pi = -(4+\pi) < 0$$

$\therefore$ Area of the figure is maximum, when $x = \dfrac{P}{4+\pi}$

$\therefore$ From (1), we have

$$P = 2x + 4y + \pi x$$

$$\Rightarrow \quad 4y = P - 2x - \pi x$$

$$\Rightarrow \quad 4y = P - \frac{2P}{4+\pi} - \frac{\pi P}{4+\pi} = \frac{4P + \pi P - 2P - \pi P}{4+\pi}$$

$$\Rightarrow \quad 4y = \frac{2P}{4+\pi}$$

$$\Rightarrow \quad y = \frac{P}{2(4+\pi)}$$

$\therefore$ The dimensions of the figure are

$$2x = \frac{2P}{4+\pi} \quad \text{and} \quad 2y = \frac{P}{4+\pi}.$$

(*ii*) Please try yourself.

[**Hint.** Proceed as in Part (*i*).

Put P = 10 m] $\left[\textbf{Ans. } 2x = \dfrac{20}{4+\pi} \text{ and } 2y = \dfrac{10}{4+\pi}\right].$

Example 29. (*i*) *A wire of length 36 cm is cut into two pieces. One of the pieces is to be made into a square and other into an equilateral triangle. Find the length of each piece so that the sum of the areas of the square and the triangle is minimum.*

(*ii*) *A wire of length 28 m is to be cut into two pieces. One of the pieces is to be made into a circle and the other into a square. Where should the wire be cut. So that combined area is minimum.*

(*iii*) *A wire of length 36 m is to be cut into two pieces. One of the pieces is to be made into a square and the other into a circle. What should be the lengths of the two pieces, so that the combined area is minimum.*

Solution. (*i*) Let length of one part of wire = x cm

$\therefore$ The length other part of wire = $(36 - x)$ cm.

Let us suppose that, the part of length x cm is to be converted into an equilateral triangle.

Let side of equilateral triangle = a

$$\therefore \quad 3a = x$$

$$\Rightarrow \quad a = \frac{x}{3} \quad ...(1) \quad \begin{bmatrix} \because & \text{Perimeter of } \Delta = x \\ \Rightarrow & \text{sum of all sides of } \Delta = x \\ \Rightarrow & a + a + a = x \\ \Rightarrow & 3a = x \end{bmatrix}$$

$\therefore$ Area of equilateral triangle = $\frac{\sqrt{3}}{4}$ (side)^2

$$\Rightarrow \quad A_1 = \frac{\sqrt{3}}{4}\left(\frac{x}{3}\right)^2 \quad \left[\because a = \frac{x}{3}\right]$$

$$\Rightarrow \quad A_1 = \frac{\sqrt{3}}{36}x^2 \quad ...(2)$$

Let the part of length $(36 - x)$ cm is to be converted into a square.

$$\text{Let side of square} = b \quad \begin{bmatrix} \because & \text{Perimeter of square} = (36 - x) \\ \Rightarrow & 4\,(\text{side}) = 36 - x \\ \Rightarrow & 4b = 36 - x \end{bmatrix}$$

$$\Rightarrow \quad 4b = 36 - x$$

$$\Rightarrow \quad b = \left(\frac{36 - x}{4}\right) \quad ...(3)$$

$\therefore$ Area of the square = (side)^2

$$\Rightarrow \quad A_2 = \left(\frac{36 - x}{4}\right)^2 \quad ...(4)$$

$\therefore$ Combined area of two figures = A = $A_1 + A_2$

$$\Rightarrow \quad A = \frac{\sqrt{3}}{36}x^2 + \frac{1}{16}(36 - x)^2 \quad ...(5)$$

$$\Rightarrow \quad \frac{dA}{dx} = \frac{\sqrt{3}}{18}x + \frac{1}{8}(36 - x)(-1)$$

$$\Rightarrow \quad \frac{dA}{dx} = \frac{\sqrt{3}}{18}x - \frac{1}{8}(36 - x) \quad ...(6)$$

and

$$\frac{d^2A}{dx^2} = \frac{\sqrt{3}}{18} - \frac{1}{8}(-1)$$

$$\Rightarrow \quad \frac{d^2A}{dx^2} = \frac{\sqrt{3}}{18} + \frac{1}{8} \quad ...(7)$$

$\therefore$ For a maximum or minimum, we must have

$$\frac{dA}{dx} = 0$$

$$\Rightarrow \quad \frac{\sqrt{3}}{18}x - \frac{1}{8}(36 - x) = 0$$

$$\Rightarrow \quad \frac{\sqrt{3}}{18}x = \frac{36 - x}{8}$$

$$\Rightarrow \quad \left(\frac{\sqrt{3}}{18}x + \frac{x}{8}\right) = \frac{36}{8} \quad \Rightarrow \quad \left(\frac{4\sqrt{3}+9}{72}\right)x = \frac{9}{2}$$

$$\Rightarrow \quad x = \frac{9}{2} \times \frac{72}{(4\sqrt{3}+9)} = \frac{324}{4\sqrt{3}+9} \times \frac{4\sqrt{3}-9}{4\sqrt{3}-9}$$

$$= \frac{324(4\sqrt{3}-9)}{(4\sqrt{3})^2 - 9^2} = \frac{324(4\sqrt{3}-9)}{144-81} = \frac{324(4\sqrt{3}-9)}{63}$$

$$x = \frac{36}{7}(4\sqrt{3}-9)$$

Now, $$\left.\frac{d^2A}{dx^2}\right]_{x = \frac{36}{7}(4\sqrt{3}-9)} = \left(\frac{\sqrt{3}}{18} + \frac{1}{8}\right) > 0$$

$\therefore$ The area of the combined figure is minimum when $x = \frac{36}{7}(4\sqrt{3}-9)$.

$\therefore$ One part $= \frac{36}{7}(4\sqrt{3}-9)$

and other part $= 36 - x$

$$= 36 - \frac{36}{7}(4\sqrt{3}-9) = \frac{36}{7}(16 - 4\sqrt{3})$$

(*ii*) Let length of one part of wire = x m

$\therefore$ Length of other part of wire = $(28 - x)$ m

Let the part of length x m is to be converted into a circle of radius r.

$\therefore$ Perimeter of circle $= 2\pi r = x$

$$\Rightarrow \quad r = \frac{x}{2\pi} \qquad ...(1)$$

$\therefore$ Area of the circle $= \pi r^2$

$$\Rightarrow \quad A_1 = \pi\left(\frac{x}{2\pi}\right)^2 = \frac{\pi x^2}{4\pi^2}$$

$$\Rightarrow \quad A_1 = \frac{x^2}{4\pi} \qquad ...(2)$$

Let the part of length $(28 - x)$ to be converted into a square of side a.

$\therefore$ Perimeter of square = 4 (side) = (28 – x)

$$\Rightarrow \quad 4a = 28 - x$$

$$\Rightarrow \quad a = \left(\frac{28-x}{4}\right) \quad \text{...(3)}$$

$\therefore$ Area of the square = (side)^2

$$A_2 = \left(\frac{28-x}{4}\right)^2 \quad \text{...(4)}$$

$\therefore$ Combined area of two figures = A = $A_1 + A_2$

$$\Rightarrow \quad A = \frac{x^2}{4\pi} + \frac{1}{16}(28-x)^2 \quad \text{...(5)}$$

$$\Rightarrow \quad \frac{dA}{dx} = \frac{x}{2\pi} + \frac{1}{16}.2(28-x)(-1)$$

$$\Rightarrow \quad \frac{dA}{dx} = \frac{x}{2\pi} - \frac{1}{8}(28-x) \quad \text{...(6)}$$

and

$$\frac{d^2A}{dx^2} = \frac{1}{2\pi} + \frac{1}{8} \quad \text{...(7)}$$

$\therefore$ For a maximum or minimum, we must have

$$\frac{dA}{dx} = 0$$

$$\Rightarrow \quad \frac{x}{2\pi} - \frac{28-x}{8} = 0$$

$$\Rightarrow \quad \frac{x}{2\pi} = \frac{28-x}{8}$$

$$\Rightarrow \quad 4x = 28\pi - \pi x$$

$$\Rightarrow \quad (4+\pi)x = 28\pi$$

$$\Rightarrow \quad x = \frac{28\pi}{4+\pi}$$

Now,

$$\left.\frac{d^2A}{dx^2}\right]_{x=\frac{28\pi}{4+\pi}} = \left(\frac{1}{2\pi} + \frac{1}{8}\right) > 0$$

$\therefore$ The area of the combined figure is minimum, when

$$x = \frac{28\pi}{4+\pi}$$

$\therefore$ One part $= \dfrac{28\pi}{4+\pi}$

and other part $= 28 - x$

$$= 28 - \frac{28\pi}{4+\pi} = \frac{112 + 28\pi - 28\pi}{4+\pi} = \frac{112}{4+\pi}.$$

(*iii*) Please try yourself.

[**Hint.** Proceed as in part (*ii*).

Put 36 in place of 28.]

$$\left[\textbf{Ans. } \text{One part } (x) = \frac{36\pi}{\pi + 4} \quad \text{and} \quad \text{other part } (y) = \frac{144}{\pi + 4} \right].$$

Example 30. (*i*) *The cost of fuel for running a train is proportional to the square of the speed generated in kilometre per hour and costs Rs. 48 per hour, at 16 km per hour. What is the most economical speed, if the fixed charges are Rs. 300 per hour.*

(*ii*) *The cost of fuel for running a bus is proportional to the square of the speed generated in km/h, and costs Rs. 48 per hour, at the speed of 20 km per hour. What is the most economical speed if the fixed charges are Rs. 108 per hour.*

Solution. (*i*) Let the speed of train = x km/h

$\therefore$ Cost of fuel per hour is proportional to x^2 [Given]

$\Rightarrow$ Cost of fuel per hour α x^2

$\Rightarrow$ Cost of fuel per hour = kx^2, where $k \to$ constant.

It is given that

When speed x = 16 km/h.

Then, cost of fuel = Rs. 48 per hour

$\therefore$ $$48 = k\,(16)^2$$

$\Rightarrow$ $$48 = 256\,k$$

$\Rightarrow$ $$k = \frac{48}{256} = \frac{3}{16}$$

Let the total distance to be covered = s km.

$\therefore$ Time required to cover the distance = $\frac{s}{x}$ hours.

$\therefore$ Cost of running the train is given by

$$C = \frac{s}{x} \times \frac{3}{16}x^2 + \frac{300s}{x}$$

$\Rightarrow$ $$C = \frac{3}{16}sx + 300\,sx^{-1} \quad ...(1)$$

$\Rightarrow$ $$\frac{dC}{dx} = \frac{3}{16}s - 300\,sx^{-2} \quad ...(2)$$

and $$\frac{d^2C}{dx^2} = 0 + 600\,sx^{-3} \quad ...(3)$$

$\therefore$ For a maximum or minimum, we must have

$$\frac{dC}{dx} = 0$$

$\Rightarrow$ $$\frac{3s}{16} - \frac{300s}{x^2} = 0$$

$\Rightarrow$ $$\frac{300s}{x^2} = +\frac{3}{16}s = \frac{100}{x^2} = +\frac{1}{16}$$

$\Rightarrow$ $$+x^2 = 1600 \quad \Rightarrow \quad x = 40$$

Now, $$\left.\frac{d^2C}{dx^2}\right]_{x=40} = \frac{600s}{(40)^3} = \frac{600s}{6400} = \frac{3s}{32} > 0$$

∴ The cost is minimum, where $x = 40$ km/h

∴ The most economical speed is 40 km/h.

(*ii*) Please try yourself.

[**Hint.** Proceed as in part (*i*)] [**Ans.** 30 km/hr.]

Example 31. (*i*) *Show that the height of the cylinder of maximum volume that can be inscribed in a sphere of radius R is* $\frac{2R}{\sqrt{3}}$.

Or

Find the volume of the largest cylinder that can be inscribed in a sphere of radius r.

(*ii*) *Show that the height of a closed right circular cylinder of given surface and maximum volume is equal to the diameter of its base.*

Solution. (*i*) Let r and h be the radius and height of the cylinder inscribed in a circle of radius R, respectively.

∴ Volume of cylinder $= V = \pi r^2 h$ ...(1)

∴ From the figure

In Δ BCD ; By Pythagoras theorem, we have

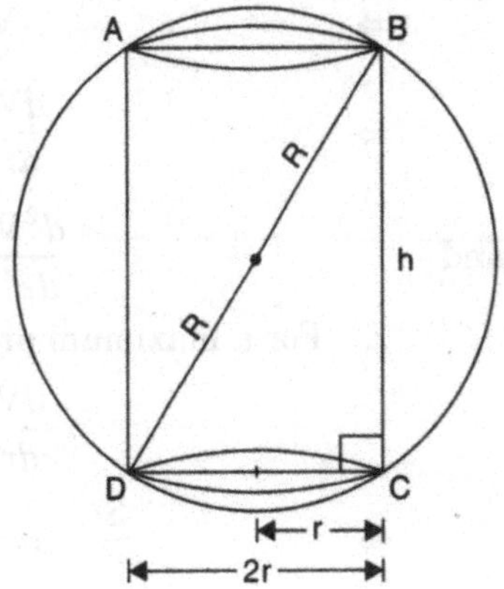

$$(BD)^2 = (BC)^2 + (CD)^2$$

$$\Rightarrow \quad (2R)^2 = h^2 + (2r)^2$$

$$\Rightarrow \quad 4R^2 = h^2 + 4r^2$$

$$\Rightarrow \quad 4r^2 = 4R^2 - h^2$$

$$\Rightarrow \quad r^2 = \left(\frac{4R^2 - h^2}{4}\right) \quad ...(2)$$

Putting this value of r^2 in (1), we get

$$V = \pi\left(\frac{4R^2 - h^2}{4}\right)h$$

$$\Rightarrow \quad V = \pi R^2 h - \frac{\pi}{4}h^3 \quad ...(3)$$

$$\Rightarrow \quad \frac{dV}{dh} = \pi R^2 - \frac{3\pi}{4}h^2 \quad ...(4)$$

and $$\frac{d^2V}{dh^2} = 0 - \frac{6\pi h}{4} = -\frac{3\pi}{2}h \quad ...(5)$$

∴ For a maximum or minimum, we have

$$\frac{dV}{dh} = 0$$

$$\Rightarrow \quad \pi R^2 - \frac{3\pi}{4}h^2 = 0$$

$$\Rightarrow \quad \frac{3\pi}{4}h^2 = \pi R^2$$

$$\Rightarrow \qquad h^2 = \frac{4R^2}{3} \qquad \Rightarrow \qquad h = \frac{2R}{\sqrt{3}}$$

Now, $$\left.\frac{d^2V}{dh^2}\right]_{h=\frac{2R}{\sqrt{3}}} = -\frac{3\pi}{2}\left(\frac{2R}{\sqrt{3}}\right) = -\sqrt{3}\pi R < 0$$

$\therefore$ Volume of cylinder is maximum, when $h = \dfrac{2R}{\sqrt{3}}$.

(*ii*) Let r and h be the radius and height of the closed cylinder respectively.

$\therefore$ Surface area of the closed cylinder $= S = \pi r^2 + \pi r^2 + 2\pi rh \Rightarrow S = 2\pi r^2 + 2\pi rh$ ...(1)

and volume of the closed cylinder $= V = \pi r^2 h$

$$\Rightarrow \qquad V = \pi r^2\left(\frac{S - 2\pi r^2}{2\pi r}\right) \qquad \left[\begin{array}{l} \because \ S = 2\pi r^2 + 2\pi rh \\ \Rightarrow \ S - 2\pi r^2 = 2\pi rh \\ \Rightarrow \ h = \dfrac{S - 2\pi r^2}{2\pi r} \end{array}\right]$$

$$\Rightarrow \qquad V = \frac{Sr}{2} - \pi r^3 \qquad ...(2)$$

$$\Rightarrow \qquad \frac{dV}{dr} = \frac{S}{2} - 3\pi r^2 \qquad ...(3)$$

and $$\frac{d^2V}{dr^2} = 0 - 6\pi r = -6\pi r \qquad ...(4)$$

$\therefore$ For a maximum or minimum, we must have

$$\frac{dV}{dr} = 0$$

$$\Rightarrow \qquad \frac{S}{2} - 3\pi r^2 = 0$$

$$\Rightarrow \qquad 3\pi r^2 = \frac{S}{2} \qquad [\because \ S = 2\pi r^2 + 2\pi rh]$$

$$\Rightarrow \qquad S = 6\pi r^2$$

$$\Rightarrow \qquad 2\pi r^2 + 2\pi rh = 6\pi r^2$$

$$\Rightarrow \qquad 2\pi rh = 4\pi r^2 \qquad \Rightarrow \qquad h = 2r$$

$$\Rightarrow \qquad r = \frac{h}{2}$$

Now, $$\left.\frac{d^2V}{dr^2}\right]_{r=\frac{h}{2}} = -6\pi\left(\frac{h}{2}\right) = -3\pi h < 0$$

$\therefore$ Volume of cylinder is maximum, when $r = \dfrac{h}{2}$

i.e., $h = 2r$ *i.e.*, when height of the cylinder is equal to the diameter of its base.

Example 32. (*i*) *An open box with a square base is to be made out of a given quantity of sheet of area a^2. Show that the maximum volume of the box is $\frac{a^3}{6\sqrt{3}}$.*

(*ii*) *Show that a cylinder of a given volume which is open at the top, has minimum total surface area, provided its height is equal to the radius of its base.*

(*iii*) *A closed right circular cylinder has volume 2156 cm³. What should be the radius of its base, so that its total surface area may be minimum.*

Solution. (*i*) Let x be the side of square base of the box and h be the height of the box.

$\therefore$ Surface area of the box = S = $x^2 + 4xh$ ...(1)

$$\Rightarrow \quad x^2 + 4xh = a^2 \qquad [\because \text{ Surface area} = a^2 \text{ (Given)}]$$

$$\Rightarrow \quad 4xh = a^2 - x^2$$

$$\Rightarrow \quad h = \left(\frac{a^2 - x^2}{4x}\right) \qquad ...(2)$$

Also, volume of the box = V = $x \times x \times h$

$$\Rightarrow \quad V = x^2 h$$

$$\Rightarrow \quad V = x^2\left(\frac{a^2 - x^2}{4x}\right) \qquad [\because \text{ By using (2)}]$$

$$\Rightarrow \quad V = \frac{a^2 x}{4} - \frac{1}{4}x^3 \qquad ...(3)$$

$$\Rightarrow \quad \frac{dV}{dx} = \frac{a^2}{4} - \frac{3}{4}x^2$$

and
$$\frac{d^2V}{dx^2} = 0 - \frac{6}{4}x = -\frac{3}{2}x$$

$\therefore$ For a maximum or minimum, we must have

$$\frac{dV}{dx} = 0$$

$$\Rightarrow \quad \frac{a^2}{4} - \frac{3}{4}x^2 = 0$$

$$\Rightarrow \quad \frac{3}{4}x^2 = \frac{a^2}{4}$$

$$\Rightarrow \quad x^2 = \frac{a^2}{3} \Rightarrow x = \frac{a}{\sqrt{3}}$$

Now,
$$\left.\frac{d^2V}{dx^2}\right]_{x=\frac{a}{\sqrt{3}}} = -\frac{3}{2}\left(\frac{a}{\sqrt{3}}\right) = -\frac{\sqrt{3}\,a}{2} < 0$$

$\therefore$ Volume of the box is maximum, when $x = \frac{a}{\sqrt{3}}$.

$\therefore$ From (3), we have

Maximum volume of the box

$$= \frac{a^2}{4}(x) - \frac{1}{4}x^3 = \frac{a^2}{4}\left(\frac{a}{\sqrt{3}}\right) - \frac{1}{4}\left(\frac{a}{\sqrt{3}}\right)^3$$

$$= \frac{a^3}{4\sqrt{3}} - \frac{a^3}{12\sqrt{3}} = \frac{3a^3 - a^3}{12\sqrt{3}} = \frac{2a^3}{12\sqrt{3}} = \frac{a^3}{6\sqrt{3}}.$$

(*ii*) Let r and h be the height of the cylinder respectively.

$\therefore$ Volume of the cylinder $= V = \pi r^2 h$ ...(1)

and total surface area of cylinder $= S = \pi r^2 + 2\pi rh$

$\Rightarrow$ $$S = \pi r^2 + 2\pi r\left(\frac{V}{\pi r^2}\right) \qquad \left[\because V = \pi r^2 h \Rightarrow h = \frac{V}{\pi r^2}\right]$$

$\Rightarrow$ $$S = \pi r^2 + \frac{2V}{r} \qquad ...(2)$$

$\Rightarrow$ $$\frac{dS}{dr} = 2\pi r - \frac{2V}{r^2} \qquad ...(3)$$

and $$\frac{d^2S}{dr^2} = 2\pi + \frac{4V}{r^3} \qquad ...(4)$$

$\therefore$ For a maximum or minimum, we must have

$$\frac{dS}{dr} = 0$$

$\Rightarrow$ $$2\pi r - \frac{2V}{r^2} = 0$$

$\Rightarrow$ $$\frac{2V}{r^2} = 2\pi r$$

$\Rightarrow$ $$\pi r^3 = V$$

$\Rightarrow$ $$\pi r^3 = \pi r^2 h \qquad [\because V = \pi r^2 h]$$

$\Rightarrow$ $$h = r.$$

Now, $$\left.\frac{d^2S}{dr^2}\right]_{h=r} = \left(\frac{4V}{h^3} + 2\pi\right) > 0$$

$\therefore$ Surface area of cylinder is minimum, when $h = r$

i.e., height of the cylinder is equal to the radius of its base.

(*iii*) Let r and h be the radius and height of the cylinder respectively.

$\therefore$ Volume of cylinder $= \pi r^2 h$

$\Rightarrow$ $$\pi r^2 h = 2156 \qquad [\because \text{Volume of cylinder} = 2156 \text{ cm}^3 \text{ (Given)}]$$

$\Rightarrow$ $$h = \frac{2156}{\pi r^2} \qquad ...(1)$$

Also, Total surface area of the cylinder = S = $2\pi r^2 + 2\pi rh$

$$\Rightarrow \quad S = 2\pi r^2 + 2\pi r\left(\frac{2156}{\pi r^2}\right) \qquad [\because \text{ By using (1)}]$$

$$\Rightarrow \quad S = 2\pi r^2 + \frac{4312}{r} \qquad ...(2)$$

$$\Rightarrow \quad \frac{dS}{dr} = 4\pi r - \frac{4312}{r^2} \qquad ...(3)$$

and
$$\frac{d^2S}{dr^2} = 4\pi + \frac{8624}{r^3} \qquad ...(4)$$

$\therefore$ For a maximum or minimum, we must have

$$\frac{dS}{dr} = 0$$

$$\Rightarrow \quad 4\pi r - \frac{4312}{r^2} = 0 \Rightarrow \frac{4312}{r^2} = 4\pi r$$

$$\Rightarrow \quad 1078 = \pi r^3$$

$$\Rightarrow \quad \frac{1078 \times 7}{22} = r^3$$

$$\Rightarrow \quad r^3 = 49 \times 7$$

$$\Rightarrow \quad r^3 = 7^3 \Rightarrow r = 7$$

Now,
$$\left.\frac{d^2S}{dr^2}\right]_{r=7} = \left[4\pi + \frac{8624}{(7)^3}\right] = \left[4\pi + \frac{8624}{343}\right] > 0$$

$\therefore$ The total surface area of cylinder is minimum, when $r = 7$ cm.

Example 33. (*i*) *Show that the semi-vertical of the cone of maximum volume of given slant height is* $\tan^{-1}\sqrt{2}$.

(ii) Show that the semi-vertical angle of a right circular cone of given surface area and maximum volume is $\sin^{-1}\left(\frac{1}{3}\right)$.

Solution. (*i*) Let r and h be the radius and height of the right circular cone respectively.

Let l be the slant height of cone, and Q be the semi-vertical angle.

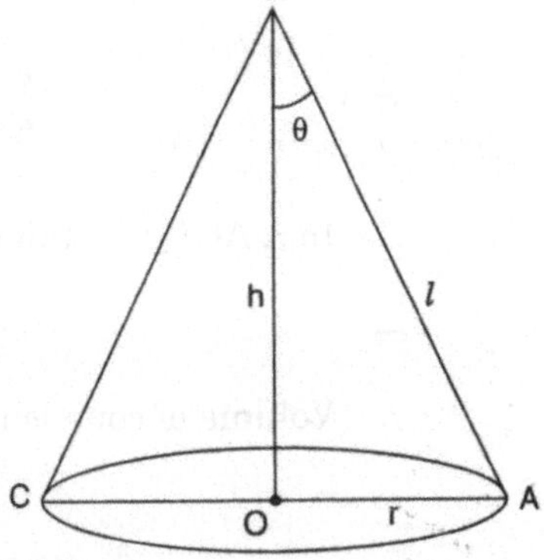

$\therefore$ Volume of the cone $= V = \frac{1}{3}\pi r^2 h$ ...(1)

$\therefore$ In Δ AOB, by Pythagoras theorem,

We have $(AB)^2 = (AO)^2 + (BO)^2$

$$\Rightarrow \quad l^2 = r^2 + h^2$$

$$\Rightarrow \quad r^2 = l^2 - h^2 \qquad ...(2)$$

Putting this value of r^2 in (1), we have

$$V = \frac{1}{3}\pi (l^2 - h^2)\, h$$

$$\Rightarrow \quad V = \frac{1}{3}\pi (l^2 h - h^3) \qquad ...(3)$$

$$\Rightarrow \quad \left(\frac{dV}{dh}\right) = \frac{1}{3}\pi (l^2 - 3h^2) \qquad ...(4)$$

and

$$\frac{d^2V}{dh^2} = \frac{\pi}{3}(0 - 6h) = -2\pi h \qquad ...(5)$$

∴ For a maximum or minimum, we must have

$$\frac{dV}{dh} = 0$$

$$\Rightarrow \quad \frac{1}{3}\pi (l^2 - 3h^2) = 0$$

$$\Rightarrow \quad l^2 - 3h^2 = 0 \Rightarrow 3h^2 = l^2 \Rightarrow h^2 = \frac{l^2}{3}$$

$$\Rightarrow \quad h = \frac{l}{\sqrt{3}}$$

Now,

$$\left.\frac{d^2V}{dh^2}\right]_{h=\frac{l}{\sqrt{3}}} = -2\pi\left(\frac{l}{\sqrt{3}}\right) < 0$$

∴ The volume of cone is maximum, when

$$h = \frac{l}{\sqrt{3}}.$$

∴ **From (2)**, we have $r^2 = l^2 - h^2$

$$\Rightarrow \quad r^2 = l^2 - \frac{l^2}{3} = \frac{2l^2}{3} = 2\left(\frac{l}{\sqrt{3}}\right)^2 = 2h^2$$

$$\Rightarrow \quad \frac{r^2}{h^2} = 2 \Rightarrow \frac{r}{h} = \sqrt{2}$$

∴ In Δ AOB, $\tan\theta = \frac{r}{h} = \sqrt{2}$

$$\Rightarrow \quad \theta = \tan^{-1}\sqrt{2}$$

∴ Volume of cone is **maximum**, when semi-vertical angle $\theta = \tan^{-1}\sqrt{2}$.

(*ii*) Let r and h be the radius and height of the right circular cone respectively. Let l be the slant height of the cone and θ be the semi-vertical angle.

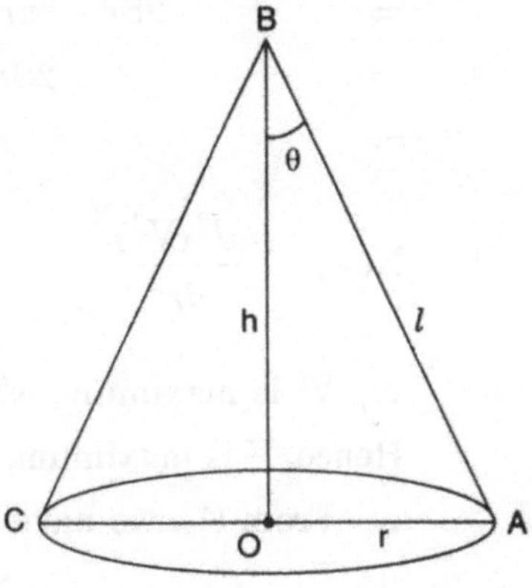

$\therefore$ Total surface area of the cone = S

$$S = \pi r^2 + \pi rl \quad ...(1)$$

and the volume of cone = V

$$V = \frac{1}{3}\pi r^2 h$$

$$\Rightarrow \quad V^2 = \frac{1}{9}\pi^2 r^4 h^2 \quad ...(2)$$

$$\Rightarrow \quad V^2 = \frac{1}{9}\pi^2 r^4 (l^2 - r^2) \quad ...(3)$$

$\left[\because \text{ Slant height of cone} = l = \sqrt{r^2 + h^2} \Rightarrow l^2 = r^2 + h^2 \Rightarrow h^2 = l^2 - r^2\right]$

$$\Rightarrow \quad V^2 = \frac{1}{9}\pi^2 r^4 \left[\left(\frac{S - \pi r^2}{\pi r}\right)^2 - r^2\right]$$

$\left[\because S = \pi r^2 + \pi rl \Rightarrow \pi rl = S - \pi r^2 \Rightarrow l = \frac{S - \pi r^2}{\pi r} \Rightarrow l^2 = \left(\frac{S - \pi r^2}{\pi r}\right)^2\right]$

$$= \frac{\pi^2 r^4}{9}\left[\frac{(S - \pi r^2)^2}{\pi^2 r^2} - r^2\right]$$

$$= \frac{\pi^2 r^4}{9}\left[\frac{S^2 + \pi^2 r^4 - 2S\pi r^2 - \pi^2 r^4}{\pi^2 r^2}\right] = \frac{r^2}{9}[S^2 - 2S\pi r^2]$$

$$V^2 = \frac{S}{9}(Sr^2 - 2\pi r^4) \quad ...(4)$$

$$\Rightarrow \quad \frac{d(V^2)}{dr} = \frac{S}{9}(2Sr - 8\pi r^3) \quad ...(5)$$

and

$$\frac{d^2(V^2)}{dr^2} = \frac{S}{9}(2S - 24\pi r^2) \quad ...(6)$$

$\therefore$ For a maximum or minimum, we must have

$$\frac{d(V^2)}{dr} = 0$$

$$\Rightarrow \quad \frac{S}{9}(2Sr - 8\pi r^3) = 0$$

$$\Rightarrow \qquad 2Sr - 8\pi r^3 = 0$$

$$\Rightarrow \qquad 2Sr = 8\pi r^3$$

$$\Rightarrow \qquad S = 4\pi r^2$$

Now, $$\left.\frac{d^2(V^2)}{dr^2}\right]_{S = 4\pi r^2} = \frac{S}{9}(2S - 6S) = \frac{S}{9}(-4S) = -\frac{4S^2}{9} < 0$$

$\therefore$ V^2 is maximum, when $S = 4\pi r^2$

Hence, V is maximum, when $S = 4\pi r^2$.

$\therefore$ From (1), we have

$$S = \pi r^2 + \pi r l$$

$$\Rightarrow \qquad 4\pi r^2 = \pi r^2 + \pi r l$$

$$\Rightarrow \qquad 3\pi r^2 = \pi r l$$

$$\Rightarrow \qquad 3r = l \Rightarrow l = 3r.$$

$\therefore$ In Δ AOE, $\sin\theta = \dfrac{r}{l} = \dfrac{r}{3r} = \dfrac{1}{3}$

$$\Rightarrow \qquad \theta = \sin^{-1}\left(\frac{1}{3}\right)$$

Hence, the volume of the cone is maximum, when the semi-vertical angle

$$\theta = \sin^{-1}\left(\frac{1}{3}\right).$$

Example 34. *Show that the right circular cone of least curved surface and given volume has an altitude equal to $\sqrt{2}$ times the radius of the base.*

Solution. Let r and h be radius and height of the right circular cone respectively. Let l be the slant height of cone.

$\therefore$ Volume of the cone $= V = \dfrac{1}{3}\pi r^2 h$

$= \text{a constant}$

$= \dfrac{1}{3}\pi k$ (say)

$$\therefore \qquad k = r^2 h \Rightarrow h = \frac{k}{r^2} \qquad \text{...(1)}$$

$\therefore$ Curved surface area of cone

$$= S = \pi r l \qquad \left[\begin{array}{l}\because \text{ Slant height of cone} = l \\ \Rightarrow l = \sqrt{r^2 + h^2}\end{array}\right]$$

$$\Rightarrow \qquad S = \pi r\sqrt{r^2 + h^2}$$

$$\Rightarrow \qquad S = \pi r \sqrt{r^2 + \frac{k^2}{r^4}} \qquad [\because \text{ By using (1)}]$$

$$\Rightarrow \qquad S = \pi r \frac{\sqrt{r^6 + k^2}}{r^2} = \frac{\pi}{r} \sqrt{r^6 + k^2}$$

$$\Rightarrow \qquad \frac{dS}{dr} = \pi \left[\frac{r \cdot \frac{1}{2\sqrt{r^6 + k^2}} \cdot (6r^5) - \sqrt{r^6 + k^2} \cdot 1}{r^2} \right]$$

$$= \pi \left[\frac{\frac{3r^6}{\sqrt{r^6 + k^2}} - \sqrt{r^6 + k^2}}{r^2} \right] = \pi \left[\frac{3r^6 - r^6 - k^2}{r^2 \sqrt{r^6 + k^2}} \right]$$

$$\frac{dS}{dr} = \frac{\pi(2r^6 - k^2)}{r^2 \sqrt{r^6 + k^2}} \qquad ...(2)$$

$\therefore$ For a maximum or minimum, we must have

$$\frac{dS}{dr} = 0$$

$$\Rightarrow \qquad \frac{\pi(2r^6 - k^2)}{r^2 \sqrt{r^6 + k^2}} = 0 \Rightarrow \pi(2r^6 - k^2) = 0$$

$$\Rightarrow \qquad 2r^6 - k^2 = 0$$

$$\Rightarrow \qquad 2r^6 = k^2$$

$$\Rightarrow \qquad r^6 = \frac{k^2}{2} = \frac{(r^2 h)^2}{2}$$

$$\Rightarrow \qquad r^6 = \frac{r^4 h^2}{2}$$

$$\Rightarrow \qquad 2r^2 = h^2$$

$$\Rightarrow \qquad h = \sqrt{2}\, r$$

Now, when $h = \sqrt{2}\, r$, $\frac{dS}{dr}$ changes sign from (–ve) to (+ve) as r increases through $\left(\frac{h}{\sqrt{2}}\right)$.

$\therefore$ S is minimum, when $h = \sqrt{2}\, r$.

Hence, a right circular cone has least curved surface area when altitude of cone is equal to $\sqrt{2}$ times the radius the base.

Example 35. *(i) Show that the height of the cone of maximum volume that can be inscribed in a sphere of radius R is* $\frac{4R}{3}$.

(ii) Show that the volume of the largest cone that can be inscribed a sphere is $\frac{8}{27}$ *of the volume of the sphere.*

Solution. (*i*) Let *r* and *h* be the radius and height of the right circular cone respectively. Let R be the radius of the sphere.

$\therefore$ In Δ ODC

By using Pythagoras theorem, we have

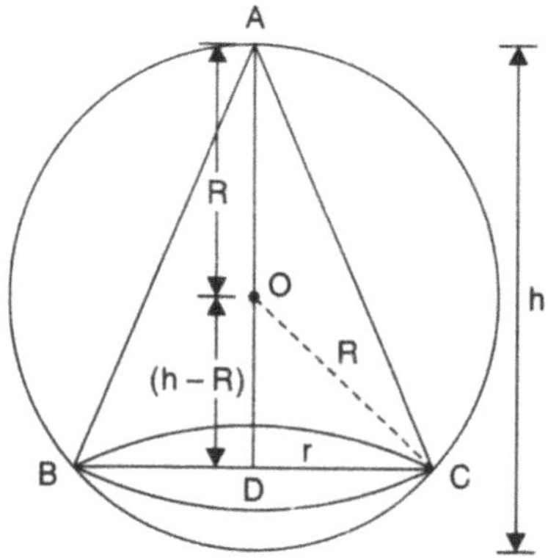

$$(OC)^2 = OD^2 + CD^2$$

$$\Rightarrow \quad R^2 = (h - R)^2 + r^2$$

$$\Rightarrow \quad R^2 = h^2 + R^2 - 2hR + r^2$$

$$\Rightarrow \quad r^2 = 2hR - h^2 \qquad ...(1)$$

Now, volume of the cone

$$= V = \frac{1}{3}\pi r^2 h$$

[$\because$ By using (1)]

$$\Rightarrow \quad V = \frac{1}{3}\pi(2hR - h^2)h$$

$$\Rightarrow \quad V = \frac{2}{3}\pi Rh^2 - \frac{1}{3}\pi h^3 \qquad ...(2)$$

$$\Rightarrow \quad \frac{dV}{dh} = \frac{4}{3}\pi Rh - \pi h^2 \qquad ...(3)$$

and $$\frac{d^2V}{dh^2} = \frac{4}{3}\pi R - 2\pi h \qquad ...(4)$$

$\therefore$ For a maximum or minimum, we must have

$$\frac{dV}{dh} = 0$$

$$\Rightarrow \quad \frac{4}{3}\pi Rh - \pi h^2 = 0$$

$$\Rightarrow \quad \frac{4}{3}\pi Rh = \pi h^2$$

$$\Rightarrow \quad h = \frac{4}{3}R$$

Now, $$\left.\frac{d^2V}{dh^2}\right]_{h=\frac{4}{3}R} = \frac{4}{3}\pi R - 2\pi\left(\frac{4}{3}R\right) = \frac{4}{3}\pi R - \frac{8\pi R}{3} = -\frac{4}{3}\pi R < 0$$

$\therefore$ The volume of the cone is maximum, when $h = \frac{4}{3}R$.

(*ii*) Let r and h be the radius and height of the right circular cylinder respectively. Let R be the radius of the sphere.

$\therefore$ In Δ ODC

By using Pythagoras theorem,

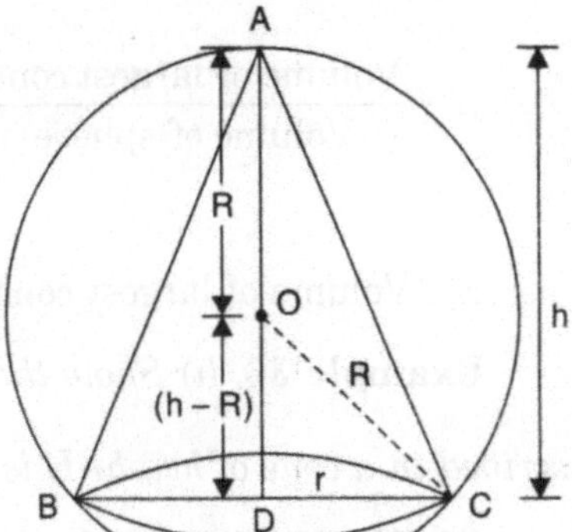

we have $(OC)^2 = (OD)^2 + (CD)^2$

$\Rightarrow \quad R^2 = (h - R)^2 + r^2$

$\Rightarrow \quad R^2 = h^2 + R^2 - 2hR + r^2$

$\Rightarrow \quad r^2 = 2hR - h^2 \quad \text{...(1)}$

Now, volume of the cone = $V = \frac{1}{3}\pi r^2 h$ [$\because$ By using (1)]

$\Rightarrow \quad V = \frac{1}{3}\pi(2hR - h^2)\,h$

$\Rightarrow \quad V = \frac{2}{3}\pi R h^2 - \frac{1}{3}\pi h^3 \quad \text{...(2)}$

$\Rightarrow \quad \frac{dV}{dh} = \frac{4}{3}\pi R h - \pi h^2 \quad \text{...(3)}$

and $\quad \frac{d^2V}{dh^2} = \frac{4}{3}\pi R - 2\pi h$

$\therefore$ For a maximum or minimum, we must have

$$\frac{dV}{dh} = 0$$

$\Rightarrow \quad \frac{4}{3}\pi R h - \pi h^2 = 0$

$\Rightarrow \quad \frac{4}{3}\pi R h = \pi h^2$

$\Rightarrow \quad h = \frac{4}{3}R$

Now, $\left.\frac{d^2V}{dh^2}\right]_{h=\frac{4}{3}R} = \frac{4}{3}\pi R - 2\pi\left(\frac{4}{3}R\right) = \frac{4}{3}\pi R - \frac{8\pi R}{3} = \frac{-4}{3}\pi R < 0$

$\therefore$ The volume of cone is maximum, when $h = \frac{4}{3}$ R.

$\therefore$ From (2), we have

Volume of largest cone $= V = \frac{2}{3}\pi R h^2 - \frac{1}{3}\pi h^3 \quad \left[\because h = \frac{4}{3}R\right]$

$$= \frac{2}{3}\pi R\left(\frac{4}{3}R\right)^2 - \frac{1}{3}\pi\left(\frac{4}{3}R\right)^3 = \frac{32\pi R^3}{27} - \frac{64\pi R^3}{81}$$

$$= \frac{96\pi R^3 - 64\pi R^3}{81} = \frac{32\pi R^3}{81}$$

Also, the volume of sphere $= \frac{4}{3}\pi R^3$.

$$\therefore \quad \frac{\text{Volume of largest cone}}{\text{Volume of sphere}} = \frac{\frac{32}{81}\pi R^3}{\frac{4}{3}\pi R^3} = \frac{32}{81} \times \frac{3}{4} = \frac{8}{27}$$

$\therefore$ Volume of largest cone $= \frac{8}{27}$. (volume of sphere).

Example 36. (*i*) *Show that the height of the cylinder of maximum volume that can be inscribed in a cone of height H is* $\frac{H}{3}$.

(*ii*) *Show that the volume of the greatest cylinder which can be inscribed in a cone of height h and semi-vertical angle 'θ' is* $\frac{4}{27}\pi H^3 \tan^2\theta$.

(*iii*) *Prove that the radius of the right circular cylinder of greatest curved surface which can be inscribed in a given cone is half of that of the cone.*

Or

A right circular cylinder is inscribed in a cone. Show that the curved surface area of the cylinder is maximum when the diameter of the cylinder is equal to the radius of the base of the cone.

Solution. (*i*) Let r and h be the radius and height of cylinder inscribed in a cone respectively.

Let R and H be the radius and height of the cone respectively.

$\therefore$ From the figure,

Δ AOD and Δ AO′C are similar.

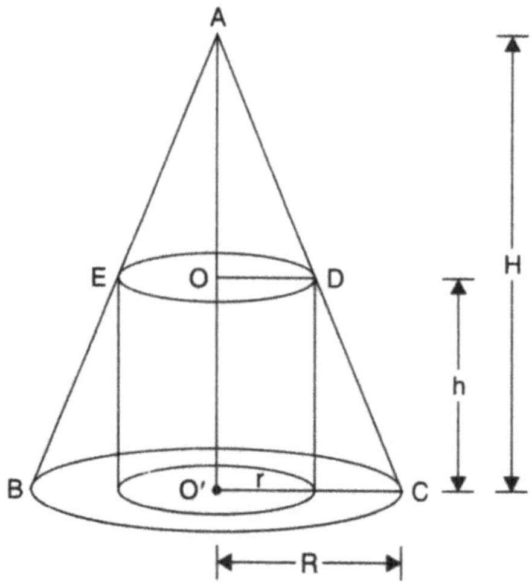

$$\Rightarrow \quad \frac{AO}{AO'} = \frac{OD}{O'C} \quad \Rightarrow \quad \frac{(H-h)}{H} = \frac{r}{R}$$

$$\Rightarrow \quad r = \frac{(H-h)R}{H} \quad \ldots(1)$$

$\therefore$ Volume of the cylinder $= V = \pi r^2 h$

[$\because$ By using (1)]

$$\Rightarrow \quad V = \pi\left(\frac{(H-h)R}{H}\right)^2 h$$

$$\Rightarrow \quad V = \frac{\pi R^2}{H^2}(H^2 + h^2 - 2hH)\, h$$

$$\Rightarrow \quad V = \frac{\pi R^2}{H^2}(H^2 h + h^3 - 2h^2 H) \quad \ldots(2)$$

$$\Rightarrow \quad \frac{dV}{dh} = \frac{\pi R^2}{H^2}(H^2 + 3h^2 - 4hH) \quad \ldots(3)$$

and $$\frac{d^2V}{dh^2} = \frac{\pi R^2}{H^2}(6h - 4H) \quad ...(4)$$

$\therefore$ For a maximum or minimum, we must have

$$\frac{dV}{dh} = 0$$

$$\Rightarrow \quad \frac{\pi R^2}{H^2}(H^2 + 3h^2 - 4hH) = 0$$

$$\Rightarrow \quad H^2 + 3h^2 - 4hH = 0$$

$$\Rightarrow \quad 3h^2 - 4hH + H^2 = 0$$

$$\Rightarrow \quad 3h^2 - 3hH - hH + H^2 = 0$$

$$\Rightarrow \quad 3h(h - H) - H(h - H) = 0$$

$$\Rightarrow \quad (h - H)(3h - H) = 0$$

$$\Rightarrow \quad h = H, \frac{H}{3} \qquad [\because \ h \neq H]$$

$$\Rightarrow \quad h = \frac{H}{3}$$

Now, $$\left.\frac{d^2V}{dh^2}\right]_{h=\frac{H}{3}} = \frac{\pi R^2}{H^2}\left(\frac{6 . H}{3} - 4H\right) = \frac{\pi R^2}{H^2}(-2H) = -\frac{2\pi R^2}{H} < 0$$

$\therefore$ Volume of the cylinder is maximum, when $h = \frac{H}{3}$.

(ii) Let r and h be the radius and height of cylinder inscribed in a cone respectively. Let R and H be the radius and height of cone respectively.

$\therefore$ From the figure,

Δ AOD and Δ AO′C are similar.

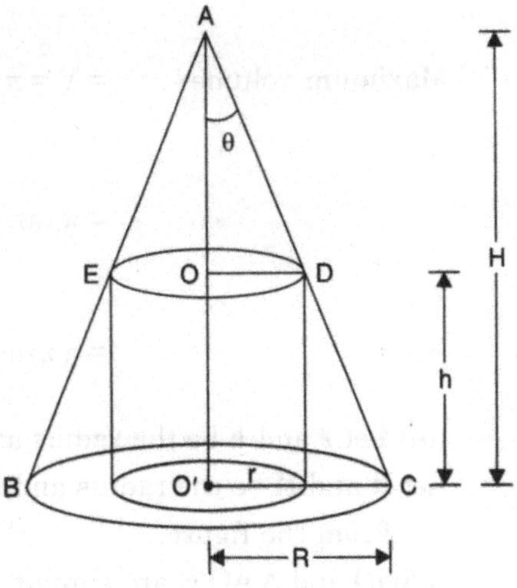

$$\Rightarrow \quad \frac{AO}{AO'} = \frac{OD}{O'C} \Rightarrow \frac{(H - h)}{H} = \frac{r}{R}$$

$$\Rightarrow \quad r = \frac{(H - h)R}{H} \quad ...(1)$$

Also, in ΔAO′C

$$\tan\theta = \frac{O'C}{AO'} = \frac{R}{H}$$

$$\Rightarrow \quad R = H\tan\theta \quad ...(2)$$

$\therefore$ From (1) and (2), we have

$$r = \frac{(H - h)H\tan\theta}{H}$$

$$= (H - h)\tan\theta \quad ...(3)$$

$\therefore$ Volume of the cylinder = $V = \pi r^2 h$ [$\because$ By using (3)]

$$\Rightarrow \quad V = \pi((H - h)\tan\theta)^2 . h$$

$\Rightarrow \quad V = \pi (H - h)^2 \tan^2 \theta \,.\, h$

$\Rightarrow \quad V = \pi \tan^2 \theta (H^2 + h^2 - 2hH) h$

$\Rightarrow \quad V = \pi \tan^2 \theta (H^2 h + h^3 - 2h^2 H) \quad ...(4)$

$\Rightarrow \quad \frac{dV}{dh} = \pi \tan^2 \theta (H^2 + 3h^2 - 4hH) \quad ...(5)$

and $\quad \frac{d^2V}{dh^2} = \pi \tan^2 \theta (\because \quad 6h - 4H) \quad ..(6)$

$\therefore$ For a maximum or minimum, we must have

$$\frac{dV}{dh} = 0$$

$\Rightarrow \quad \pi \tan^2 \theta (H^2 + 3h^2 - 4hH) = 0$

$\Rightarrow \quad 3h^2 - 4hH + H^2 = 0$

$\Rightarrow \quad (h - H)(3h - H) = 0$

$\Rightarrow \quad h = H, \frac{H}{3} \quad [\because \quad h \neq H]$

$\Rightarrow \quad h = \frac{H}{3}$

Now, $\quad \left.\frac{d^2V}{dh^2}\right]_{h = \frac{H}{3}} = \pi \tan^2 \theta \left(\frac{6 . H}{3} - 4H\right) = \pi \tan^2 \theta (-2H) = -2H\pi \tan^2 \theta < 0$

$\therefore$ Volume of the cylinder is maximum, when $h = \frac{H}{3}$.

$\therefore$ From equation (4), we have

Maximum volume $\quad = V = \pi \tan^2 \theta \left[H^2\left(\frac{H}{3}\right) + \left(\frac{H}{3}\right)^3 - 2\left(\frac{H}{3}\right)^2 H\right]$

$$= \pi \tan^2 \theta \left[\frac{H^3}{3} + \frac{H^3}{27} - \frac{2H^3}{9}\right]$$

$$= \pi \tan^2 \theta \left[\frac{9H^3 + H^3 - 6H^3}{27}\right] = \frac{4}{27}\pi H^3 \tan^2 \theta.$$

(*iii*) Let r and h be the radius and height of cylinder inscribed in a cone respectively. Let R and H be the radius and height of cone respectively.

$\therefore$ From the figure,

Δ AOD and Δ AO′C are similar.

$\Rightarrow \quad \frac{AO}{AO'} = \frac{OD}{O'C}$

$$\Rightarrow \quad \frac{H-h}{H} = \frac{r}{R}$$

$$\Rightarrow \quad H - h = \frac{Hr}{R}$$

$$\Rightarrow \quad H - \frac{Hr}{R} = h \qquad ...(1)$$

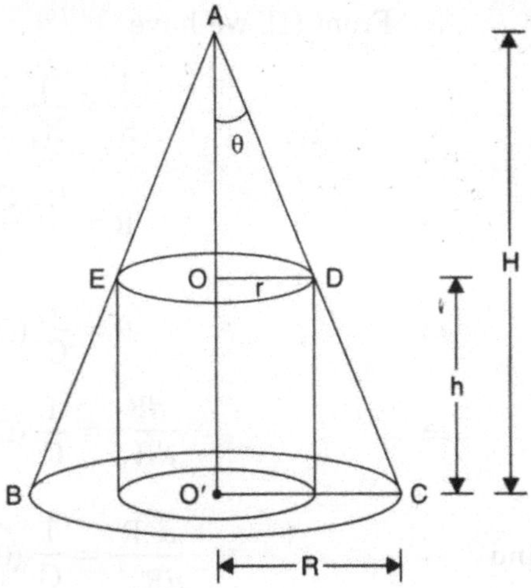

$\therefore$ The curved surface area of cylinder = S = $2\pi rh$

$$\Rightarrow \quad S = 2\pi r\left(H - \frac{Hr}{R}\right)$$

$$\Rightarrow \quad S = \frac{2\pi H}{R}(rR - r^2) \qquad ...(2)$$

$$\Rightarrow \quad \frac{dS}{dr} = \frac{2\pi H}{R}(R - 2r) \qquad ...(3)$$

and
$$\frac{d^2S}{dr^2} = \frac{2\pi H}{R}(0-2) = -\frac{4\pi H}{R} \qquad ...(4)$$

$\therefore$ For a maximum or minimum, we must have

$$\frac{dS}{dr} = 0$$

$$\Rightarrow \quad \frac{2\pi H}{R}(R - 2r) = 0$$

$$\Rightarrow \quad R - 2r = 0$$

$$\Rightarrow \quad 2r = R$$

$$\Rightarrow \quad r = \frac{R}{2}$$

Now,
$$\left.\frac{d^2S}{dr^2}\right]_{r=\frac{R}{2}} = -\frac{4\pi H}{R} < 0$$

$\therefore$ The curved surface area is maximum, when $r = \frac{R}{2}$.

Example 37. *If* $\frac{1}{R} = \frac{1}{R_1} + \frac{1}{R_2}$ *, where* $R_1 + R_2$ *= constant,* $R_1 > 0, R_2 > 0$*, show that for maximum value of R,* $R_1 = R_2$.

Solution. We have , $\frac{1}{R} = \frac{1}{R_1} + \frac{1}{R_2}$ ...(1)

$R_1 + R_2$ = constant (C) say

$$\Rightarrow \quad R_1 + R_2 = C \;\Rightarrow\; R_2 = C - R_1$$

$\therefore$ From (1), we have

$$\frac{1}{R} = \frac{1}{R_1} + \frac{1}{C - R_1} = \frac{C - R_1 + R_1}{R_1(C - R_1)} = \frac{C}{R_1(C - R_1)}$$

$$\Rightarrow \quad R = \frac{R_1(C - R_1)}{C} = \frac{CR_1 - R_1^2}{C}$$

$$\Rightarrow \quad R = \frac{1}{C}(CR_1 - R_1^2) \quad \text{...(2)}$$

$$\Rightarrow \quad \frac{dR}{dR_1} = \frac{1}{C}(C - 2R_1) \quad \text{...(3)}$$

and

$$\frac{d^2R}{dR_1^2} = \frac{1}{C}(0 - 2) = -\frac{2}{C} \quad \text{...(4)}$$

$\therefore$ For a maximum or minimum, we must have

$$\frac{dR}{dR_1} = 0$$

$$\Rightarrow \quad \frac{1}{C}(C - 2R_1) = 0$$

$$\Rightarrow \quad C - 2R_1 = 0 \Rightarrow C = 2R_1$$

$$\Rightarrow \quad R_1 = \frac{C}{2}$$

Now,

$$\left.\frac{d^2R}{dR_1^2}\right]_{R_1 = \frac{C}{2}} = -\frac{2}{C} < 0$$

$\therefore$ Value of R is maximum, when $R_1 = \frac{C}{2}$.

Since,

$$R_1 + R_2 = C$$

$$\Rightarrow \quad \frac{C}{2} + R_2 = C \Rightarrow R_2 = C - \frac{C}{2} = \frac{C}{2}$$

$$\Rightarrow \quad R_1 = R_2 = \frac{C}{2}$$

$\therefore$ Value of R is maximum, when $R_1 = R_2$.

Example 38. *Of all the closed right circular cylindrical cans which encloses a given volume of 100 cm³, find the dimensions of the can which has the minimum surface area.*

Solution. Let r and h be the radius and height of the cylindrical can respectively.

$\therefore$ Volume of the cylindrical can = 100 cm³

$$\Rightarrow \quad \pi r^2 h = 100$$

$$\Rightarrow \quad h = \frac{100}{\pi r^2} \quad \text{...(1)}$$

Now, the total surface area of the cylindrical can = S = $2\pi rh + 2\pi r^2$

$$\Rightarrow \quad S = 2\pi r\left(\frac{100}{\pi r^2}\right) + 2\pi r^2 \qquad [\because \text{ By using (1)}]$$

$$\Rightarrow \quad S = \frac{200}{r} + 2\pi r^2 \qquad ...(2)$$

$$\Rightarrow \quad \frac{dS}{dr} = -\frac{200}{r^2} + 4\pi r \qquad ...(3)$$

and
$$\frac{d^2S}{dr^2} = \frac{400}{r^3} + 4\pi \qquad ...(4)$$

$\therefore$ For a maximum or minimum, we must have

$$\frac{dS}{dr} = 0$$

$$\Rightarrow \quad \frac{-200}{r^2} + 4\pi r = 0 \quad \Rightarrow \quad \frac{-200 + 4\pi r^3}{r^2} = 0$$

$$\Rightarrow \quad -200 + 4\pi r^3 = 0 \quad \Rightarrow \quad 4\pi r^3 = 200$$

$$\Rightarrow \quad r^3 = \frac{50}{\pi} \quad \Rightarrow \quad r = \left(\frac{50}{\pi}\right)^{1/3}$$

Now,
$$\left.\frac{d^2S}{dr^2}\right]_{r=\left(\frac{50}{\pi}\right)^{1/3}} = \frac{400}{\left[\left(\frac{50}{\pi}\right)^{1/3}\right]^3} + 4\pi = \frac{400}{\left(\frac{50}{\pi}\right)} + 4\pi$$

$$= 8\pi + 4\pi = 12\pi > 0$$

$\therefore$ The surface area of the cylinder is minimum, when $r = \left(\frac{50}{\pi}\right)^{1/3}$.

$\therefore$ From (1), we have

$$h = \frac{100}{\pi\left(\left(\frac{50}{\pi}\right)^{1/3}\right)^2} = \frac{100}{\pi\left(\frac{50}{\pi}\right)^{2/3}}.$$

Example 39. *Find the point on the curve $y^2 = 4x$ which is nearest to the point (2, 1).*

Solution. We have $y^2 = 4x$ ...(1)

Let P(x, y) be any point on the parabola and A (2, 1) be the given point on the parabola. Then, By distance formula, we have

$$(AP) = \sqrt{(x-2)^2 + (y-1)^2} \qquad \left[\because \; y^2 = 4x \;\Rightarrow\; x = \frac{y^2}{4}\right]$$

$$\Rightarrow \qquad AP = \sqrt{\left(\frac{y^2}{4} - 2\right)^2 + (y-1)^2}$$

$$\Rightarrow \qquad (AP)^2 = \left(\frac{y^2}{4} - 2\right)^2 + (y-1)^2 \qquad ...(2)$$

$$\Rightarrow \qquad \frac{d(AP)^2}{dy} = 2\left(\frac{y^2}{4} - 2\right)\left(\frac{2y}{4}\right) + 2(y-1) . 1$$

$$\Rightarrow \qquad \frac{d(AP)^2}{dy} = \frac{y^3}{4} - 2y + 2y - 2$$

$$\Rightarrow \qquad \frac{d(AP)^2}{dy} = \frac{y^3}{4} - 2 \qquad ...(3)$$

and
$$\frac{d^2(AP)^2}{dy^2} = \frac{3y^2}{4} \qquad ...(4)$$

$\therefore$ For a maximum or minimum, we must have

$$\frac{d(AP)^2}{dy} = 0$$

$$\Rightarrow \qquad \frac{y^3}{4} - 2 = 0 \Rightarrow \frac{y^3 - 8}{4} = 0$$

$$\Rightarrow \qquad y^3 - 8 = 0 \Rightarrow y^3 = 8 \Rightarrow y = 2$$

Now, $\left.\dfrac{d^2(AP)^2}{dy^2}\right]_{y=2} = \dfrac{3(2)^2}{4} = 3 > 0$

$\therefore$ $(AP)^2$ is minimum, when $y = 2$

Hence, the distance (AP) is minimum, when $y = 2$.

$\therefore$ From (1), we have

$$y^2 = 4x$$

$$\Rightarrow \qquad (2)^2 = 4x \Rightarrow 4 = 4x \Rightarrow x = 1.$$

$\therefore$ The point (1, 2) is nearest to the point (2, 1).

Example 40. *A square tank of capacity 250 m^3 has to be dug out. The cost of the land is Rs. 50 per m^2. The cost of digging increases with depth and for the whole tank, it is Rs 400 h^2, where h be the depth of the tank. What should be the dimensions of the tank, so that the cost be minimum.*

Solution. Let the length of side of the square base = x m and depth of the tank = h m.

$\therefore$ Volume (capacity) of tank = 250 m^3

$$\Rightarrow \qquad x^2h = 250 \qquad ...(1)$$

The area of the base of the tank = x^2 m^2.

$\therefore$ Cost of land $= \text{Rs. } 50\, x^2$

and cost of digging $= \text{Rs. } 400\, h^2$

$\therefore$ Total cost $= C = 50\, x^2 + 400\, h^2$ ...(2)

$$\Rightarrow \quad C = 50\left(\frac{250}{h}\right) + 400h^2 \qquad \left[\because x^2h = 250 \Rightarrow x^2 = \frac{250}{h}\right]$$

$$\Rightarrow \quad C = \frac{12500}{h} + 400h^2 \qquad ...(3)$$

$$\therefore \quad \frac{dC}{dh} = \frac{-12500}{h^2} + 800h \qquad ...(4)$$

and

$$\frac{d^2C}{dh^2} = \frac{25000}{h^3} + 800 \qquad ...(5)$$

$\therefore$ For a maximum or minimum, we must have

$$\frac{dC}{dh} = 0$$

$$\Rightarrow \quad \frac{-12500}{h^2} + 800h = 0 \quad \Rightarrow \quad \frac{-12500 + 800h^3}{h^2} = 0$$

$$\Rightarrow \quad 800h^3 - 12500 = 0 \quad \Rightarrow \quad 800h^3 = 12500$$

$$\Rightarrow \quad h^3 = \frac{12500}{800} = \frac{125}{8} = \left(\frac{5}{2}\right)^3$$

$$\Rightarrow \quad h = \frac{5}{2}$$

Now,

$$\left.\frac{d^2C}{dh^2}\right]_{h=\frac{5}{2}} = \frac{25000}{\left(\frac{5}{2}\right)^3} + 800 = \frac{25000 \times 8}{125} + 800 = 1600 + 800 = 2400 > 0$$

$\therefore$ The cost C is minimum, when $h = \frac{5}{2}$ m

$\therefore$ From (1), we have $x^2h = 250$

$$\Rightarrow \quad x^2\left(\frac{5}{2}\right) = 250$$

$$\Rightarrow \quad x^2 = \frac{250 \times 2}{5} = 100 = (10)^2$$

$$\Rightarrow \quad x = 10 \text{ m.}$$

EXERCISE FOR PRACTICE

Find the local maxima and local minima, if any, of the following functions. Find also the local maximum and local minimum values : (Q. No. 1—5)

1. $2x^3 - 3x^2 - 12x + 4$

2. $2x^3 - 24x + 107$.

3. $(x-1)^3 (x+1)^2$

4. $\sin 2x, 0 < x < \pi$.

5. $2 \sin x + \sin 2x , 0 \le x \le 2\pi$.

Find the absolute maximum and absolute minimum values of the following functions on the given intervals : (Q. No. 6—10)

6. $3x^4 - 8x^3 + 12x^2 - 48x + 25 , x \in [0, 3]$.

7. $\dfrac{x+1}{\sqrt{x^2+1}} , x \in [0, 2]$.

8. $x + \sin x , x \in [0, 2\pi]$.

9. $\sqrt{5} (\sin x + \cos 2x) , x \in \left[0, \dfrac{\pi}{2}\right]$.

10. $2 \sin x + \sin 2x , x \in [0, 2\pi]$.

11. Find the largest possible area of a right angled triangle whose hypotenuse is 5 cm long.

12. Find the point on the curve $y^2 = 2x$ which is nearest to the point $(1, -4)$.

13. A window is in the form of a rectangle surmounted by a semi-circular opening. The total perimeter of the window is 20 m. Find the dimensions of the window so as to admit maximum light through the whole opening.

14. Show that the height of the cylinder of maximum volume that can be inscribed in a sphere of radius 10 cm is $\dfrac{20}{\sqrt{3}}$ cm.

15. A wire of length 36 cm is to be cut into two pieces. One of the pieces is to be made into a square and the other into an equilateral triangle. Where should the wire be cut so that the combined area is minimum.

16. The cost of fuel for running a bus is proportional to the square of the speed generated in km/hr. It costs Rs. 48 per hour. When the bus is moving at a speed of 20 km/hr. What is the most economical speed if the fixed charges are Rs. 100 per hour.

Answers

1. Local maximum value = 11 at $x = -1$, Local minimum value = -16 at $x = 2$.

2. Local maximum value = 139 at $x = -2$, Local minimum value = 75 at $x = 2$.

3. Local maximum value = 0 at $x = -1$, Local minimum value = $\dfrac{-3456}{3125}$ at $x = -\dfrac{1}{5}$

4. Local maximum value = 1 at $x = \dfrac{\pi}{4}$, Local minimum value = -1 at $x = \dfrac{3\pi}{4}$.

5. Local maximum value = $\dfrac{3\sqrt{3}}{2}$ at $x = \dfrac{\pi}{3}$, Local minimum value = $\dfrac{-3\sqrt{3}}{2}$ at $x = \dfrac{5\pi}{3}$.

6. Absolute maximum value = 25 at $x = 0$, Absolute minimum value = -39 at $x = 2$.

7. Absolute maximum value = $\sqrt{2}$ at $x = 1$, Absolute minimum value = 1 at $x = 0$.

8. Absolute maximum value = 2π at $x = 2\pi$, Absolute minimum value = 0 at $x = 0$.

9. Absolute maximum value = $\frac{9\sqrt{5}}{8}$ at $x = \sin^{-1}\left(\frac{1}{4}\right)$, Absolute minimum value = 0 at $x = \frac{\pi}{2}$.

10. Absolute maximum value = $\frac{3\sqrt{3}}{2}$ at $x = \frac{\pi}{3}$, Absolute minimum value = $\frac{-3\sqrt{3}}{2}$ at $x = \frac{5\pi}{3}$.

11. $\frac{25}{4}$ sq. cm.

12. (2, – 2).

13. Radius of semi-circle = $\frac{20}{\pi + 4}$ m.

Side of rectangle = $\frac{20}{\pi + 4}$ m.

15. $\frac{144}{4 + 3\sqrt{3}}$ cm, $\frac{108\sqrt{3}}{4 + 3\sqrt{3}}$ cm.

16. $\frac{50}{\sqrt{3}}$ km/hr.

7

Curve Sketching

7.1 INTRODUCTION

We have already learnt how to draw graphs of linear functions, polynomial functions and trigonometric functions. In this chapter, we will discuss how to use methods of calculus to analyse various types of graphs that occur in applications.

The process of drawing the graph of functions is called curve sketching. We shall now sketch some curves using the results of differential calculus with the help of derivatives.

7.2 PROCEDURE OF CURVE SKETCHING

The following steps are taken in the process of curve sketching.

1. Symmetry :

(*a*) *Symmetry about x-axis*. The graph of the curve is symmetrical about x-axis, if on changing y to $-y$, the equation of the curve remains unchanged.

(*b*) *Symmetry about y-axis*. The graph of the curve is symmetrical about y-axis, if on changing x to $-x$, the equation of the curve remains unchanged.

(*c*) *Symmetry in opposite quadrants*. The graph of the curve is symmetrical about the origin or symmetric in opposite quadrants, if on changing x to $-x$ and y to $-y$, the equation of the curve remains unchanged.

(*d*) *Symmetry about the line $y = x$*. The graph of the curve is symmetrical about the line $y = x$ if, x is changed to y and y is changed to x, the equation of the curve remains unchanged.

2. Origin. Find whether the curve passes through the origin or not.

3. Points of Intersection of the curve with the co-ordinate axes :

Find points where the curve meets the co-ordinate axes.

Putting $y = 0$, in the equation of the curve, we can find the points of intersection of the curve with the x-axis. Similarly, putting $x = 0$, we can find the points of intersection of the curve with the y-axis.

4. Monotonocity or (Intervals of increase or decrease). The intervals/regions where the function is strictly increasing or strictly decreasing are found out by studying the sign of the derivative.

5. Extreme values/Local maximum and Local minimum values :

The extreme points are the turning points of the curve. These points are very useful for curve sketching.

6. Table of values of the function. Giving some values to x, find the corresponding values of y which gives us some points on the curve. The graph of the function is drawn by free hand.

SOME SOLVED EXAMPLES

Example 1. *Sketch the curve :*

$$y = -x^2 + 5x - 4$$

Solution. We have

$$y = -x^2 + 5x - 4 \quad \text{...(1)}$$

1. *Symmetry.* The curve is not symmetrical about any axis.
2. *Origin.* The curve does not passes through the origin.
3. *Points of intersection with axes*

Put $x = 0$ in (1), we get

$$y = 0 + 0 - 4 \Rightarrow y = -4$$

$\therefore$ The curve intersects the y-axis at $(0, -4)$.

Put $y = 0$ in (1), we get

$$0 = -x^2 + 5x - 4$$

$$\Rightarrow \quad x^2 - 5x + 4 = 0 \Rightarrow (x-1)(x-4) = 0$$

$$\Rightarrow \quad x = 1, 4.$$

$\therefore$ The curve intersects the x-axis at $(1, 0)$ and $(4, 0)$.

4. *Monotonocity*

Differentiating (1) w.r.t. x, we get

$$\frac{dy}{dx} = -2x + 5 = -2\left(x - \frac{5}{2}\right) \quad \text{...(2)}$$

$$\therefore \quad \frac{dy}{dx} = 0 \Rightarrow -2x + 5 = 0$$

$$\Rightarrow \quad -2x = -5 \Rightarrow x = \frac{5}{2} = 2.5.$$

When $x < 2.5$

$$\Rightarrow \quad \frac{dy}{dx} = (-2)(-) = +\text{ve}$$

$\Rightarrow$ y is strictly increasing on $(-\infty, 2.5)$

When $x > 2.5$

$$\Rightarrow \quad \frac{dy}{dx} = (-2)(+) = -\text{ ve}$$

$\Rightarrow$ y is strictly decreasing on $(2.5, \infty)$

5. *Extreme values*

$$\frac{d^2y}{dx^2} = -2$$

$$\Rightarrow \quad \left.\frac{d^2y}{dx^2}\right]_{x=2.5} = -2 < 0$$

∴ y is a maximum, when $x = 2.5$

∴ Local maximum value

$$y = -(2.5)^2 + 5(2.5) - 4 = -6.25 + 12.5 - 4$$

$$\Rightarrow \qquad y = 2.25$$

6. *Table of values :*

x	0	1	4	5
y	−4	0	0	−4

The graph of the given function is as shown in the figure.

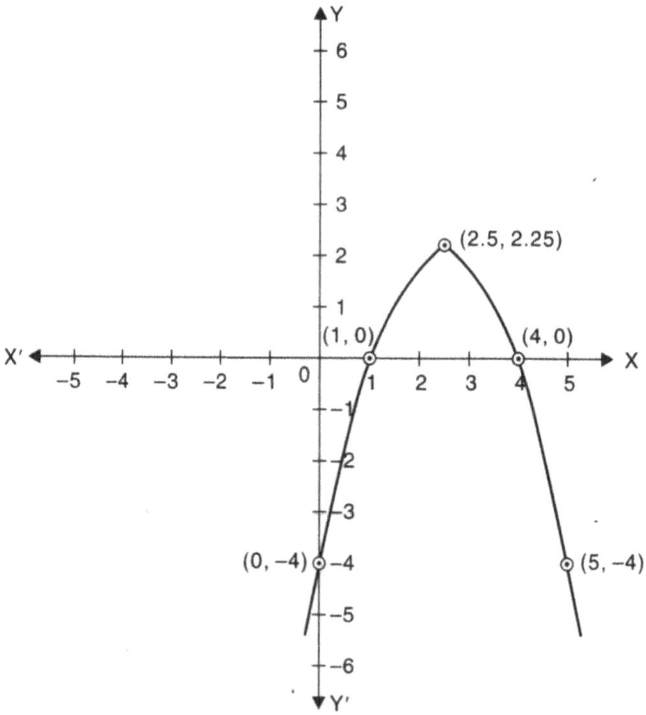

Example 2. *Sketch the curve* $y = x^3 + 1$.

Solution. We have

$$y = x^3 + 1 \qquad \text{...(1)}$$

1. *Symmetry.* The curve is not symmetrical about any axes.
2. *Origin.* The curve does not passes through the origin.
3. *Points of intersection with axes :*

Put $x = 0$ in (1), we get

$$y = 0 + 1 \Rightarrow y = 1$$

∴ The curve intersects the y-axis at (0, 1)

Put $y = 0$ in (1), we get

$$0 = x^3 + 1 \Rightarrow x^3 = -1 \Rightarrow x = -1$$

∴ The curve intersects the x-axis at (− 1, 0).

4. *Monotonocity*

Differentiating (1) w.r.t. x, we get

$$\frac{dy}{dx} = 3x^2 \qquad ...(2)$$

$\therefore$ $\frac{dy}{dx} = 0 \Rightarrow 3x^2 = 0$

$\Rightarrow$ $x = 0$

When $x < 0$, $\frac{dy}{dx} = 3(+) = +ve$

$\Rightarrow$ y is strictly increasing on $(-\infty, 0)$.

When $x > 0$, $\frac{dy}{dx} = 3(+) = +ve$

$\Rightarrow$ y is strictly increasing on $(0, \infty)$.

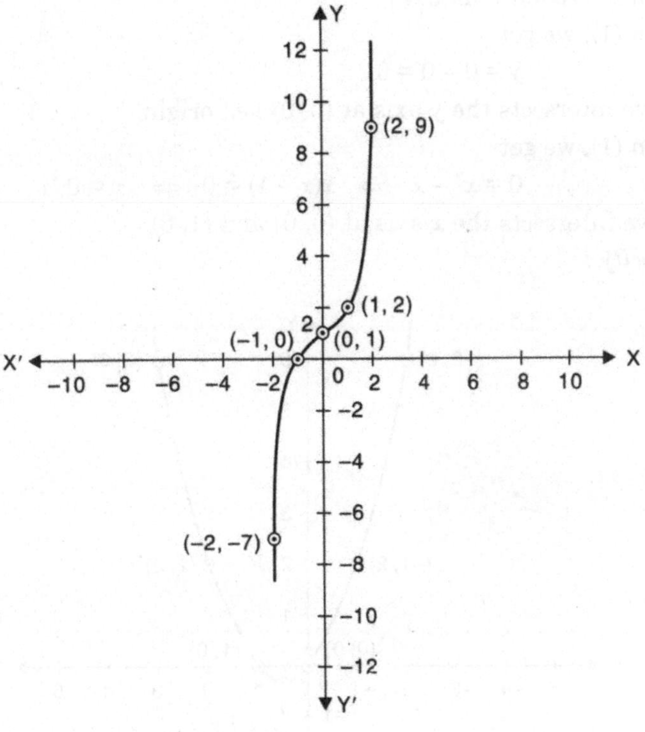

5. *Extreme values :*

$$\frac{d^2y}{dx^2} = 6x$$

$\Rightarrow$ $\left.\frac{d^2y}{dx^2}\right]_{x=0} = 6(0) = 0$

i.e., $x = 0$ is a point of inflexion.

$\therefore$ At $x = 0$, $y = (0)^3 + 1 = 1$

$\therefore$ (0, 1) is a point of inflexion.

6. *Table of values :*

x	-1	0	1	2	-2
y	0	1	2	9	-7

The graph of the given function is as shown in the figure.

Example 3. *Sketch the curve* $y = x^3 + 8$.

Solution. Please try yourself.

[**Hint.** Proceed as in example 2].

Example 4. *Sketch the curve* $y = x^2 - x$

Solution. We have

$$y = x^2 - x \qquad \text{...(1)}$$

1. *Symmetry.* The curve is not symmetrical about any axis.
2. *Origin.* The curve passes through the origin.
3. *Points of intersection with axes :*

Put $x = 0$ in (1), we get

$$y = 0 - 0 = 0$$

$\therefore$ The curve intersects the y-axis at (0, 0) *i.e.,* origin.

Put $y = 0$ in (1), we get

$$0 = x^2 - x \Rightarrow x(x-1) = 0 \Rightarrow x = 0, 1$$

$\therefore$ The curve intersects the x-axis at (0, 0) and (1, 0).

4. *Monotonocity :*

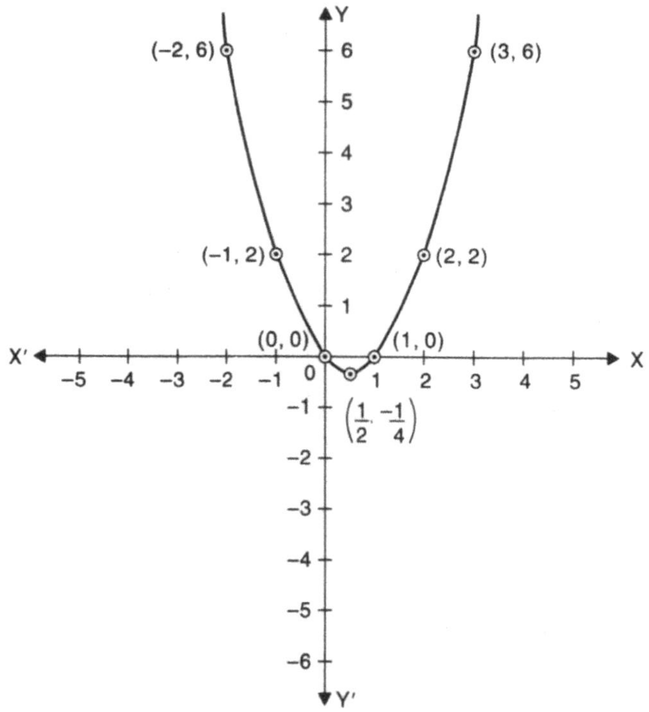

Differentiating (1) w.r.t. x, we get

$$\frac{dy}{dx} = 2x - 1$$

$\therefore$ $$\frac{dy}{dx} = 0, \quad \Rightarrow \quad 2x - 1 = 0$$

$\Rightarrow$ $$x = \frac{1}{2}$$

When $x < \frac{1}{2}$, $\frac{dy}{dx} = -\text{ve}$

$\Rightarrow$ y is strictly decreasing on $\left(-\infty, \frac{1}{2}\right)$

When $x > \frac{1}{2}$, $\frac{dy}{dx} = +\text{ve}$

$\Rightarrow$ y is strictly increasing on $\left(\frac{1}{2}, \infty\right)$

5. *Extreme values :*

$$\frac{d^2y}{dx^2} = 2$$

$\Rightarrow$ $$\left.\frac{d^2y}{dx^2}\right]_{x=1/2} = 2 > 0$$

$\Rightarrow$ y is a minimum, when $x = \frac{1}{2}$

$\therefore$ Local minimum value

$$y = \left(\frac{1}{2}\right)^2 - \frac{1}{2} = \frac{1}{4} - \frac{1}{2} = \left(\frac{1-2}{4}\right) = -\frac{1}{4}$$

$\Rightarrow$ $$y = -0.25$$

6. *Table of values :*

x	-2	-1	0	1	2	3
y	6	2	0	0	2	6

The graph of the given function is as shown in the figure.

Example 5. *Sketch the curve :* $y = x^3 - 4x$

Solution. We have

$$y = x^3 - 4x \qquad \text{...(1)}$$

1. *Symmetry :*

Put $x = -x$ and $y = -y$

$\Rightarrow$ $$-y = (-x)^3 - 4(-x) \quad \Rightarrow \quad -y = -x^3 + 4x$$

$\Rightarrow \qquad y = x^3 - 4x$, which is same as (1)

$\therefore$ The given curve is symmetric in opposite quadrants.

2. *Origin.* The curve passes through the origin.

3. *Points of intersection with axes :*

Put $x = 0$ in (1), we get

$$y = 0 - 0 \Rightarrow y = 0$$

$\therefore$ The curve intersects the y-axis at (0, 0) *i.e.,* origin.

Put $y = 0$ in (1), we get

$$0 = x^3 - 4x \Rightarrow x(x^2 - 4) = 0$$

$$\Rightarrow \qquad x(x + 2)(x - 2) = 0 \Rightarrow x = 0, -2, 2.$$

$\therefore$ The curve intersects the x-axis at (0, 0) (– 2, 0) and (2, 0).

4. *Monotonocity :*

Differentiating (1) w.r.t. x, we get

$$\frac{dy}{dx} = 3x^2 - 4 \Rightarrow 3\left(x^2 - \frac{4}{3}\right) \qquad ...(2)$$

$$\therefore \qquad \frac{dy}{dx} = 0 \Rightarrow 3x^2 - 4 = 0$$

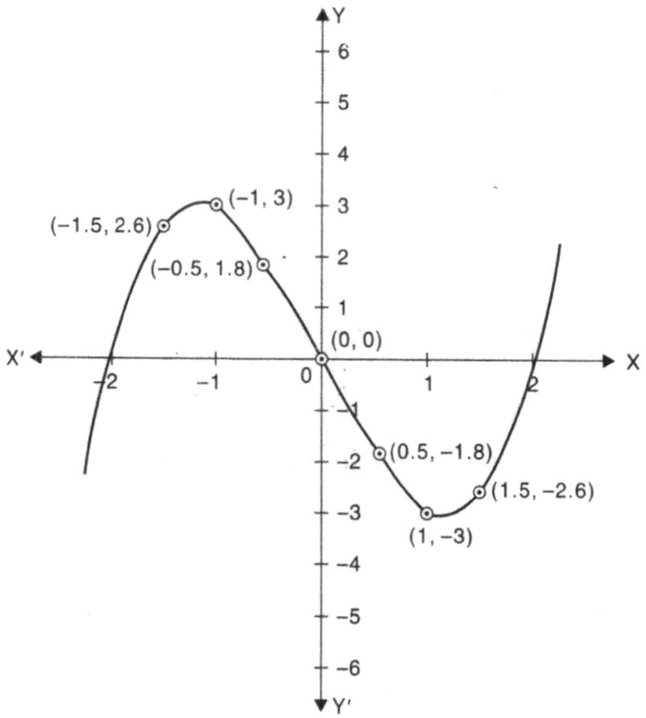

$$\Rightarrow \qquad x^2 = \frac{4}{3} \quad \Rightarrow \quad x = \pm\frac{2}{\sqrt{3}} \quad \Rightarrow \quad x = \frac{-2}{\sqrt{3}}, \frac{+2}{\sqrt{3}}$$

$\therefore$ When $x < \frac{-2}{\sqrt{3}} \Rightarrow \frac{dy}{dx} = 3(-)(-) = +\text{ve}$

$\Rightarrow$ y is strictly increasing on $\left(-\infty, \frac{-2}{\sqrt{3}}\right)$

When $\frac{-2}{\sqrt{3}} < x < \frac{2}{\sqrt{3}} \Rightarrow \frac{dy}{dx} = 3\,(+)\,(-) = -\text{ve}$

$\Rightarrow$ y is strictly decreasing on $\left(\frac{-2}{\sqrt{3}}, \frac{2}{\sqrt{3}}\right)$.

When $x > \frac{2}{\sqrt{3}} \Rightarrow \frac{dy}{dx} = 3\,(+)\,(+) = +\text{ve}$

$\Rightarrow$ y is strictly increasing on $\left(\frac{2}{\sqrt{3}}, \infty\right)$

5. *Extreme values :*

$$\Rightarrow \qquad \frac{d^2y}{dx^2} = 6x$$

$$\Rightarrow \qquad \left.\frac{d^2y}{dx^2}\right]_{x=\frac{-2}{\sqrt{3}}} = 6\left(\frac{-2}{\sqrt{3}}\right) < 0$$

$\therefore$ y is a maximum at $x = \frac{-2}{\sqrt{3}}$.

$\therefore$ Local maximum value

$$y = \left(\frac{-2}{\sqrt{3}}\right)^3 - 4\left(\frac{-2}{\sqrt{3}}\right) = \frac{-8}{3\sqrt{3}} + \frac{8}{\sqrt{3}} = \frac{16}{3\sqrt{3}}$$

$$\left.\frac{d^2y}{dx^2}\right]_{x=\frac{2}{\sqrt{3}}} = 6\left(\frac{2}{\sqrt{3}}\right) > 0$$

$\therefore$ y is a minimum at $x = \frac{2}{\sqrt{3}}$

$\therefore$ Local minimum value

$$y = \left(\frac{2}{\sqrt{3}}\right)^3 - 4\left(\frac{2}{\sqrt{3}}\right) = \frac{8}{3\sqrt{3}} - \frac{8}{\sqrt{3}} = \frac{-16}{3\sqrt{3}}.$$

6. *Table of values :*

x	− 1.5	− 1	− 0.5	0	0.5	1	1.5
y	2.6	3	1.8	0	− 1.8	− 3	− 2.6

The graph of the given function is as shown in the figure.

Example 6. *Sketch the curve* $y = x^3 - 3x^2 - 9x + 7$

Solution. We have

$$y = x^3 - 3x^2 - 9x + 7 \qquad \text{...(1)}$$

1. *Symmetry.* The curve is not symmetrical about any axis.

2. *Origin.* The curve does not passes through the origin.

3. *Points of intersection with axes :*

Put $x = 0$ in (1), we get

$$y = 0 - 0 - 0 + 7 \Rightarrow y = 7$$

∴ The curve intersects the y-axis at (0, 7).

4. *Monotonocity :*

Differentiating (1) w.r.t. x, we get

$$\frac{dy}{dx} = 3x^2 - 6x - 9 = 3(x^2 - 2x - 3) = 3(x + 1)(x - 3)$$

$$\therefore \quad \frac{dy}{dx} = 0,$$

$$\Rightarrow \quad 3(x + 1)(x - 3) = 0 \Rightarrow x = -1, 3$$

When $x < -1$

$$\Rightarrow \quad \frac{dy}{dx} = 3\,(-)\,(-) = +\text{ve}$$

∴ y is strictly increasing on $(-\infty, -1)$.

When $-1 < x < 3$,

$$\Rightarrow \quad \frac{dy}{dx} = 3\,(+)(-) = -\text{ve}$$

⇒ y is strictly decreasing on $(-1, 3)$.

When $x > 3$

$$\Rightarrow \quad \frac{dy}{dx} = 3\,(+)\,(+) = +\text{ve}$$

⇒ y is strictly increasing on $(3, \infty)$.

5. *Extreme values :*

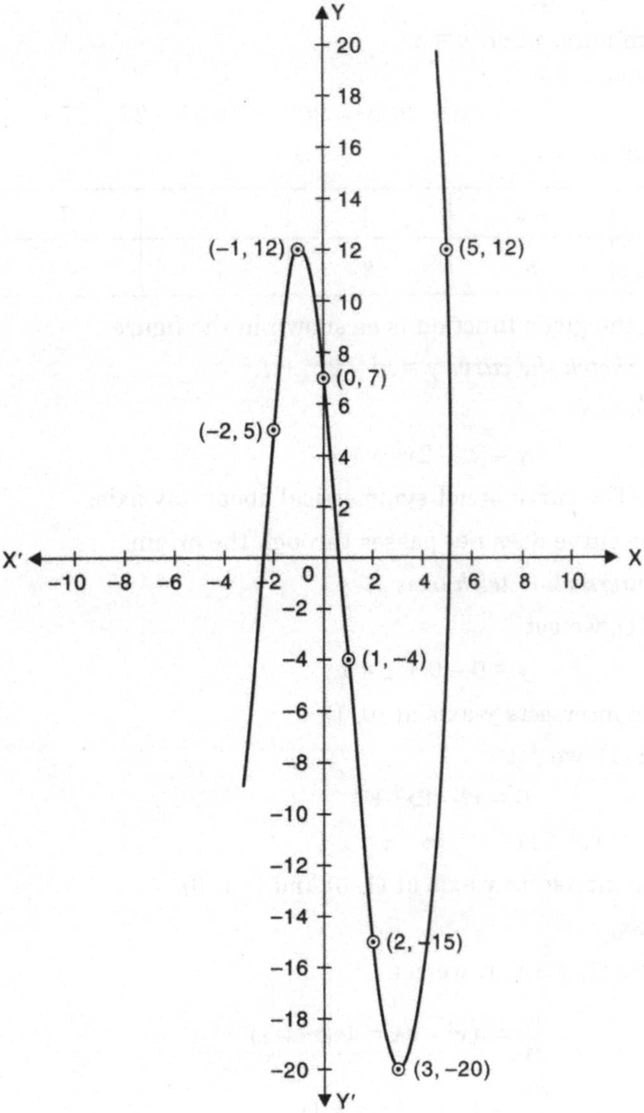

$$\Rightarrow \qquad \frac{d^2y}{dx^2} = 6x - 6$$

$$\Rightarrow \qquad \left.\frac{d^2y}{dx^2}\right]_{x=-1} = 6(-1) - 6 = -6 - 6 = -12 < 0$$

$\therefore$ y is a maximum, when $x = -1$.

Local maximum value

$$y = (-1)^3 - 3(-1)^2 - 9(-1) + 7 = -1 - 3 + 9 + 7 = 12$$

$$\Rightarrow \qquad \left.\frac{d^2y}{dx^2}\right]_{x=3} = 6(3) - 6 = 18 - 6 = 12 > 0$$

$\therefore$ y is a minimum, when $x = 3$.

Local minimum value

$$y = (3)^3 - 3(3)^2 - 9(3) + 7 = 27 - 27 - 27 + 7 = -20.$$

6. *Table of values :*

x	-2	-1	0	1	5
y	5	12	7	-4	12

The graph of the given function is as shown in the figure.

Example 7. *Sketch the curve $y = x^4 - 2x^2 + 1$.*

Solution. We have

$$y = x^4 - 2x^2 + 1 \qquad \ldots(1)$$

1. *Symmetry.* The curve is not symmetrical about any axis.
2. *Origin.* The curve does not passes through the origin.
3. *Points of intersection with axes :*

Put $x = 0$ in (1), we get

$$y = 0 - 0 + 1 = 1$$

$\therefore$ The curve intersects y-axis at (0, 1)

Put $y = 0$ in (1), we get

$$0 = x^4 - 2x^2 + 1$$

$$\Rightarrow \qquad (x^2 - 1)(x^2 - 1) = 0 \quad \Rightarrow \quad x = \pm 1$$

$\therefore$ The curve intersects x-axis at (1, 0) and (– 1, 0).

4. *Monotonocity :*

Differentiating (1) w.r.t. x, we get

$$\Rightarrow \qquad \frac{dy}{dx} = 4x^3 - 4x = 4x(x^2 - 1)$$

$$\Rightarrow \qquad \frac{dy}{dx} = 4x(x-1)(x+1) \qquad \ldots(2)$$

$$\therefore \qquad \frac{dy}{dx} = 0$$

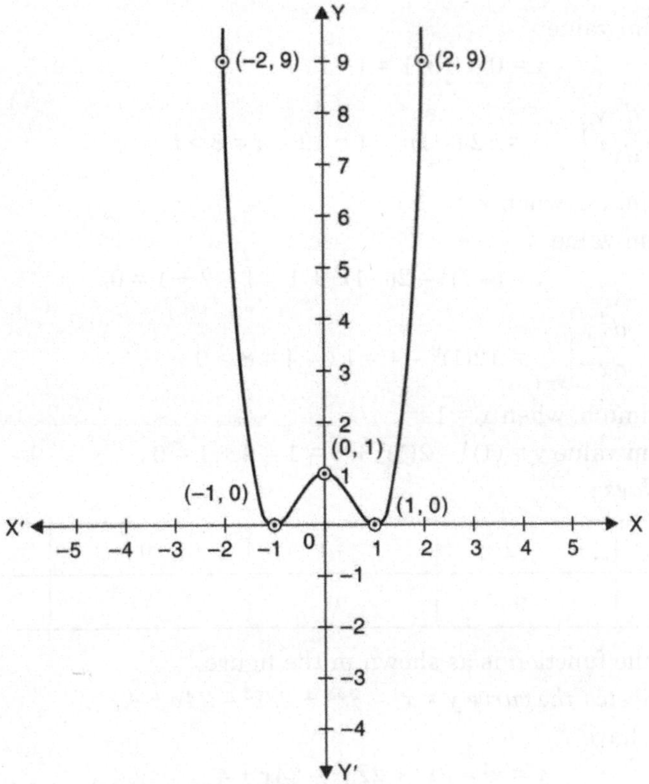

$\Rightarrow \quad 4x(x-1)(x+1) = 0 \quad \Rightarrow \quad x = 0, 1, -1$

When $x < -1$, $\quad \dfrac{dy}{dx} = 4(-)(-)(-) = (-ve)$

$\Rightarrow$ y is strictly decreasing on $(-\infty, -1)$

When $-1 < x < 0$, $\quad \dfrac{dy}{dx} = 4(-)(-)(+) = +ve$

$\Rightarrow$ y is strictly increasing on $(0, -1)$

When $0 < x < 1$, $\quad \dfrac{dy}{dx} = 4(+)(-)(+) = -ve$

$\Rightarrow$ y is strictly decreasing on $(0, 1)$

When $x > 1$, $\quad \dfrac{dy}{dx} = 4(+)(+)(+) = +ve$

$\Rightarrow$ y is strictly increasing on $(1, \infty)$.

5. *Extreme values* :

$$\Rightarrow \quad \frac{d^2y}{dx^2} = 12x^2 - 4$$

$$\therefore \quad \left.\frac{d^2y}{dx^2}\right]_{x=0} = 12(0) - 4 = -4 < 0$$

$\therefore$ y is a maximum, when $x = 0$

Local maximum value

$$y = 0 - 0 + 1 = 1$$

$$\left.\frac{d^2y}{dx^2}\right]_{x=-1} = 12\,(-1)^2 - 4 = 12 - 4 = 8 > 0$$

$\therefore$ y is a minimum, when $x = -1$

Local minimum value

$$y = (-1)^4 - 2(-1)^2 + 1 = 1 - 2 + 1 = 0$$

$$\left.\frac{d^2y}{dx^2}\right]_{x=1} = 12(1)^2 - 4 = 12 - 4 = 8 > 0$$

$\therefore$ y is a minimum, when $x = 1$

Local minimum value $y = (1)^4 - 2(1)^2 + 1 = 1 - 2 + 1 = 0$

6. *Table of values* :

x	-2	-1	0	2
y	9	0	1	9

The graph of the function is as shown in the figure.

Example 8. *Sketch the curve* $y = x^4 - 8x^3 + 22x^2 - 24x + 4$.

Solution. We have

$$y = x^4 - 8x^3 + 22x^2 - 24x + 4 \quad \text{...(1)}$$

1. *Symmetry* : The curve is not symmetrical about any axis.
2. *Origin* : The curve does not passes through the origin.
3. *Points of intersection with axes* :

Put $x = 0$ in (1), we get

$$y = 0 - 0 + 0 - 0 + 4 = 4$$

$\therefore$ The curve intersects y-axis at (0, 4).

4. *Monotonocity* :

Differentiating (1) w.r.t. x, we get

$$\frac{dy}{dx} = 4x^3 - 24x^2 + 44x - 24$$

$$= 4(x^3 - 6x^2 + 11x - 6) = 4(x - 1)(x - 2)(x - 3)$$

$$\therefore \quad \frac{dy}{dx} = 0$$

$\Rightarrow$ $4(x - 1)(x - 2)(x - 3) = 0$

$\Rightarrow$ $x = 1, 2, 3.$

When $x < 1$, $\dfrac{dy}{dx} = 4(-)(-)(-) = -\text{ ve}$

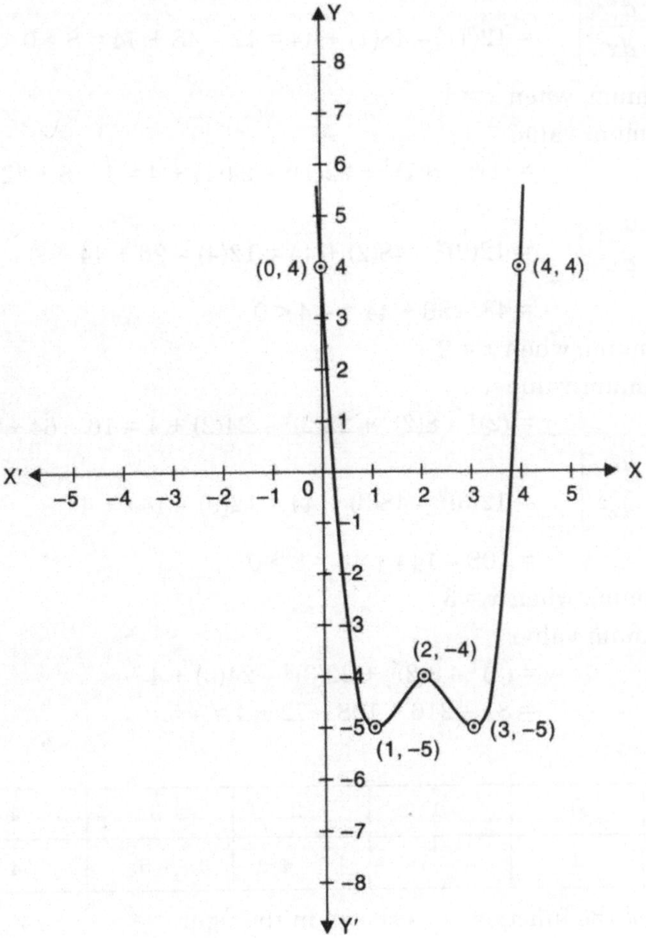

$\Rightarrow$ y is strictly decreasing on $(-\infty, 1)$

When $1 < x < 2$, $\frac{dy}{dx} = 4(+)\,(-)\,(-) = +\text{ve}$

$\Rightarrow$ y is strictly increasing on $(1, 2)$

When $2 < x < 3$, $\frac{dy}{dx} = 4\,(+)\,(+)\,(-) = -\text{ve}$

$\Rightarrow$ y is strictly decreasing on $(2, 3)$

When $x > 3$, $\frac{dy}{dx} = 4\,(+)\,(+)\,(+) = +\text{ve}$

$\Rightarrow$ y is strictly increasing on $(3, \infty)$

5. *Extreme values* :

$$\therefore \qquad \frac{d^2y}{dx^2} = 12x^2 - 48x + 44$$

$$\therefore \quad \left.\frac{d^2y}{dx^2}\right]_{x=1} = 12(1)^2 - 48(1) + 44 = 12 - 48 + 44 = 8 > 0$$

$\therefore$ y is a minimum, when $x = 1$

$\therefore$ Local minimum value

$$y = (1)^4 - 8(1)^3 + 22(1)^2 - 24(1) + 4 = 1 - 8 + 22 - 24 + 4 = -5$$

$$\left.\frac{d^2y}{dx^2}\right]_{x=2} = 12(2)^2 - 48(2) + 44 = 12(4) - 96 + 44$$

$$= 48 - 96 + 44 = -4 < 0$$

$\therefore$ y is a maximum, when $x = 2$

$\therefore$ Local maximum value

$$y = (2)^4 - 8(2)^3 + 22(2)^2 - 24(2) + 4 = 16 - 64 + 88 - 48 + 4 = -4$$

$$\left.\frac{d^2y}{dx^2}\right]_{x=3} = 12(3)^2 - 48(3) + 44 = 12(9) - 144 + 44$$

$$= 108 - 144 + 44 = 8 > 0$$

$\therefore$ y is a minimum, when $x = 3$

$\therefore$ Local minimum value

$$y = (3)^4 - 8(2)^3 + 22(3)^2 - 24(3) + 4$$

$$= 81 - 216 + 198 - 72 + 4 = -5.$$

6. *Table of values* :

x	0	1	2	3	4
y	4	-5	-4	-5	4

$\therefore$ The graph of the function is as shown in the figure.

Example 9. *Sketch the curve* $y = \sqrt{4 - x^2}$

Solution. We have

$$y = \sqrt{4 - x^2} \qquad \text{...(1)}$$

1. *Symmetry*. By changing x to $-x$, the equation of the curve is not changed, so the curve is symmetrical about y-axis.

2. *Origin* : The curve does not passes through the origin.

3. *Points* of intersection with axes :

Put $x = 0$ in (1), we get

$$y = \sqrt{4 - 0} = 2$$

$\therefore$ The curve intersects y-axis at (0, 2).

Put $y = 0$ in (1), we get

$$0 = \sqrt{4 - x^2} \Rightarrow 4 - x^2 = 0 \Rightarrow x^2 = 4 \Rightarrow x = \pm 2$$

$\therefore$ The curve intersects x-axis at (+ 2, 0) and (– 2, 0).

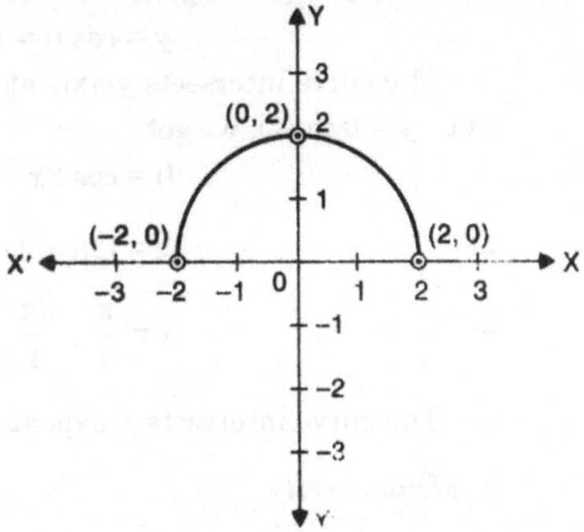

4. *Monotonocity* :

Differentiating (1) w.r.t. x, we get

$$\frac{dy}{dx} = \frac{1}{2\sqrt{4-x^2}}(-2x)$$

$$\Rightarrow \quad \frac{dy}{dx} = \frac{-x}{\sqrt{4-x^2}} \quad \therefore \quad \frac{dy}{dx} = 0$$

$$\Rightarrow \quad \frac{-x}{\sqrt{4-x^2}} = 0 \Rightarrow x = 0.$$

When $x < 0$, $\frac{dy}{dx} = \frac{(+)}{(+)} = +\text{ve}$

$\Rightarrow$ y is strictly increasing on $(-2, 0)$

When $x > 0$, $\frac{dy}{dx} = \frac{(-)}{(+)} = -\text{ve}$

$\Rightarrow$ y is strictly decreasing on $(0, 2)$

5. *Extreme values* :

Since, $\frac{dy}{dx} = 0$ at $x = 0$.

The monotonocity of the curve shows that $x = 0$ is or maximum.

$\therefore$ Local maximum value

$$y = \sqrt{4-0} = 2.$$

$\therefore$ (0, 2) is a point of maximum.

6. *Table of values* :

x	– 2	– 1	0	1	2
y	0	1.732	2	1.732	0

The graph of the function is as shown in the figure.

Example 10. *Sketch the curve* $y = \sqrt{9-x^2}$

Solution. Please try yourself.

[**Hint.** Proceed as in Example 11.]

Example 11. *Sketch the curve* $y = \cos 2x$

Solution. We have

$$y = \cos 2x \qquad ...(1)$$

1. *Symmetry* : Changing x to $-x$, we get

$$y = \cos 2(-x) = \cos(-2x) = \cos 2x$$

$\Rightarrow$ $y = \cos 2x$, which is same as (1)

$\therefore$ The curve is symmetrical about y-axis.

2. *Origin* : The curve does not passes through the origin

3. *Points of intersection with axes* :

Put $x = 0$ in (1) we get

$$y = \cos 0 = 1$$

$\therefore$ The curve intersects y-axis at (0, 1).

Put $y = 0$ in (1), we get

$$0 = \cos 2x$$

$$\Rightarrow \quad 2x = (2n+1)\frac{\pi}{2} \quad \Rightarrow \quad x = (2n+1)\frac{\pi}{4}$$

$$\Rightarrow \quad x = \frac{\pi}{4}, \frac{3\pi}{4}, \frac{5\pi}{4} \text{ ....}$$

$\therefore$ The curve intersects x-axis at $\left(\frac{\pi}{4}, 0\right), \left(\frac{3\pi}{4}, 0\right), \left(\frac{5\pi}{4}, 0\right)$ ....

4. *Monotonocity* :

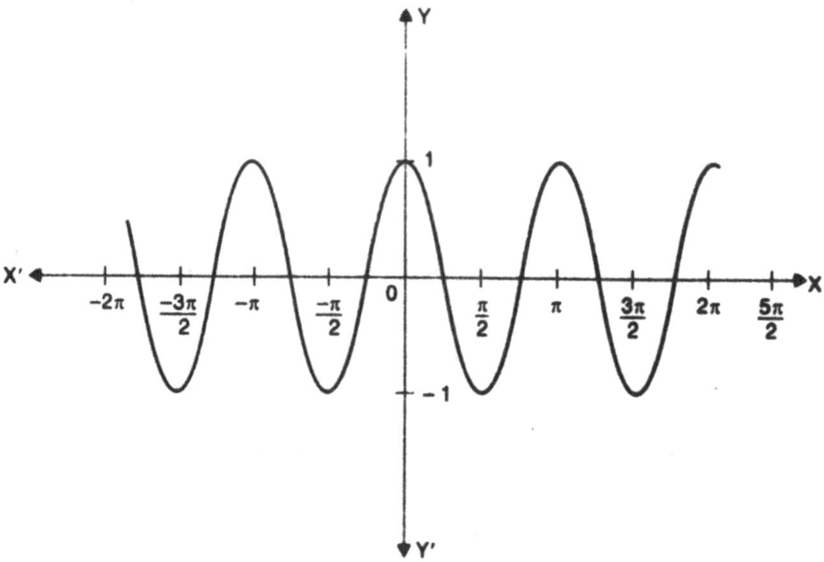

Differentiating (1) w.r.t. x, we get

$$\frac{dy}{dx} = -2\sin 2x$$

$$\therefore \quad \frac{dy}{dx} = 0 \quad \Rightarrow \quad -2\sin 2x = 0 \quad \Rightarrow \quad \sin 2x = 0$$

$$\Rightarrow \quad 2x = n\pi$$

$$\Rightarrow \quad x = \frac{n\pi}{2}$$

$$\Rightarrow \quad x = 0, \frac{\pi}{2}, \pi, \frac{3\pi}{2} \text{ ......}$$

Also,

$$\frac{d^2y}{dx^2} = -4\cos 2x$$

$$\therefore \quad \frac{d^2y}{dx^2} = (-\text{ve}), \text{ when } x = 0, \pi, 2\pi \text{ ......}$$

$\Rightarrow$ y is maximum, when $x = 0, \pi, 2\pi$ ......

$$\frac{d^2y}{dx^2} = (+\text{ve}), \text{ when } x = \frac{\pi}{2}, \frac{3\pi}{2}, \ldots\ldots$$

$\Rightarrow$ y is minimum, when $x = \frac{\pi}{2}, \frac{3\pi}{2}$; ....

5. *Extreme values* : Monotonocity shows that the function has minimum value when

$$x = \frac{\pi}{2}.$$

$\therefore$ Local minimum value

$$y = \cos 2\left(\frac{\pi}{2}\right) = \cos \pi = -1.$$

When $x = (2n + 1)\frac{\pi}{4}$ *i.e.,* $x = \frac{\pi}{4}, \frac{3\pi}{4}, \frac{5\pi}{4}$ ......

$\Rightarrow$ $y = \cos 2x = 0.$

When $x = n\pi$ *i.e.,* $x = 0, \pi, 2\pi$ ....

$\Rightarrow$ $y = \cos 2x = 1.$

When $x = (2n + 1)\frac{\pi}{2}$ *i.e.,* $x = \frac{\pi}{2}, \frac{3\pi}{2}, \frac{5\pi}{2}$, ...

$\Rightarrow$ $y = \cos 2x = -1.$

The graph of the function is as shown in the figure.

Example 12. *Sketch the curve $y = -\sin 2x$*

Solution. We have

$$y = -\sin 2x \qquad \ldots(1)$$

1. *Symmetry* : Put $x = -x$ and $y = -y$ in (1), we get

$\Rightarrow$ $-y = -\sin 2(-x) \Rightarrow -y = -\sin(-2x)$

$\Rightarrow$ $-y = \sin 2x \Rightarrow y = -\sin 2x$, which is same as (1)

$\therefore$ The given curve is symmetric in opposite quadrants.

2. *Origin* : The curve passes through the origin.

3. *Points of intersection with axes* :

Put $x = 0$, in (1), we get

$$y = -\sin 0 = 0$$

$\therefore$ The curve intersects y-axis at (0, 0) *i.e.,* origin.

Put $y = 0$ in (1), we get

$$0 = -\sin 2x \Rightarrow \sin 2x = 0$$

$\Rightarrow$ $2x = n\pi$

$\Rightarrow$ $x = \frac{n\pi}{2}$

$\Rightarrow$ $x = \frac{\pi}{2}, \frac{3\pi}{2}, \frac{5\pi}{2}$, ...

$\therefore$ The curve intersects x-axis at $\left(\frac{\pi}{2}, 0\right), \left(\frac{3\pi}{2}, 0\right), \left(\frac{5\pi}{2}, 0\right)$ ......

4. *Monotonocity* :

Differentiating (1) w.r.t. x, we get

$$\frac{dy}{dx} = -2\cos 2x$$

$$\therefore \quad \frac{dy}{dx} = 0 \Rightarrow -2\cos 2x = 0 \Rightarrow \cos 2x = 0$$

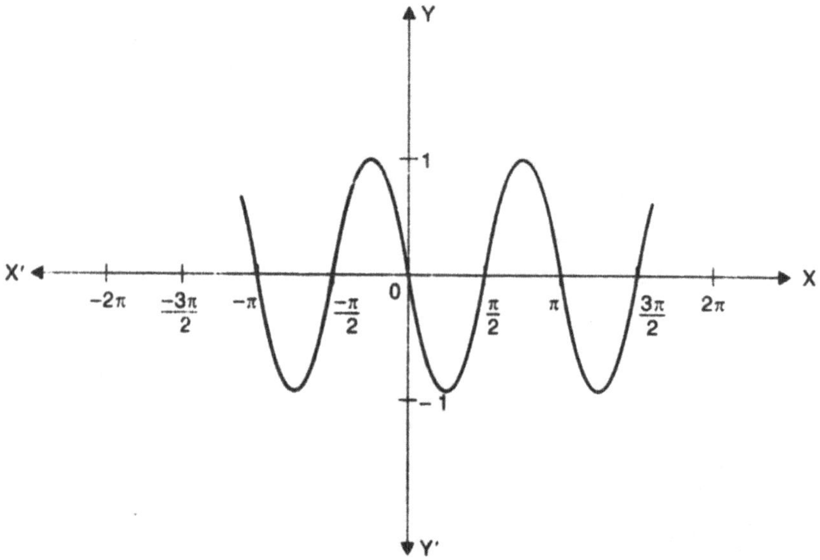

$$\Rightarrow \quad 2x = (2n+1)\frac{\pi}{2}$$

$$\Rightarrow \quad x = (2n+1)\frac{\pi}{4}$$

$$\Rightarrow \quad x = \frac{\pi}{4}, \frac{3\pi}{4}, \frac{5\pi}{4}, \ldots\ldots$$

Also, $$\frac{d^2y}{dx^2} = 4\sin 2x$$

$$\therefore \quad \frac{d^2y}{dx^2} = +\text{ve, when } x = \frac{\pi}{4}, \frac{5\pi}{4}, \frac{9\pi}{4}, \ldots\ldots$$

$\Rightarrow$ y is minimum, when

$$x = \frac{\pi}{4}, \frac{5\pi}{4}, \frac{9\pi}{4}, \ldots\ldots$$

$$\frac{d^2y}{dx^2} = -\text{ve, when } x = \frac{-\pi}{4}, \frac{3\pi}{4}, \frac{7\pi}{4}, \ldots\ldots$$

$\Rightarrow$ y is maximum, when

$$x = \frac{-\pi}{4}, \frac{3\pi}{4}, \frac{7\pi}{4}, \ldots\ldots$$

5. *Extreme values*. Monotonocity shows that the function has minimum value as − 1 and maximum value as 1.

The graph of the function is as shown in figure.

Example 13. *Sketch the curve $y = \sin^2 x$.*

Solution. We have

$$y = \sin^2 x \qquad ...(1)$$

1. *Symmetry* : Put $x = -x$ in (1), we get

$$y = \sin^2(-x) = (\sin(-x))^2 = (-\sin x)^2 = \sin^2 x$$

which is same as (1)

$\therefore$ The curve is symmetrical about y-axis.

2. *Origin* : The curve passes through the origin.

3. *Monotonocity* :

Differentiating (1) w.r.t. x, we get

$$\Rightarrow \quad \frac{dy}{dx} = 2 \sin x \cos x \qquad [\because \ 2 \sin A \cos A = \sin 2A]$$

$$\Rightarrow \quad \frac{dy}{dx} = \sin 2x$$

$$\therefore \quad \frac{dy}{dx} = 0 \ \Rightarrow \ \sin 2x = 0$$

$$\Rightarrow \quad 2x = n\pi \ \Rightarrow \ x = \frac{n\pi}{2}.$$

$$\Rightarrow \quad x = 0, \pm\frac{\pi}{2}, \pm\pi, \pm\frac{3\pi}{2}, \pm 2\pi \ ....$$

Also.

$$\frac{d^2y}{dx^2} = 2 \cos 2x$$

$$\therefore \quad \frac{d^2y}{dx^2} = +\text{ve}, \quad \text{when } x = \pm\pi, \pm 2\pi, ...$$

$\therefore$ y is minimum, when $x = \pm\pi, \pm 2\pi, ...$

$$\frac{d^2y}{dx^2} = -\text{ ve, when } x = \pm\frac{\pi}{2}, \pm\frac{3\pi}{2}, \pm\frac{5\pi}{2}, ...$$

$\therefore$ y is maximum, when

$$x = \pm\frac{\pi}{2}, \pm\frac{3\pi}{2}, \pm\frac{5\pi}{2}, ...$$

4. *Extreme values.* Monotonocity shows that the function has maximum value as 1 and minimum value as 0.

The graph of the curve is as shown in the figure.

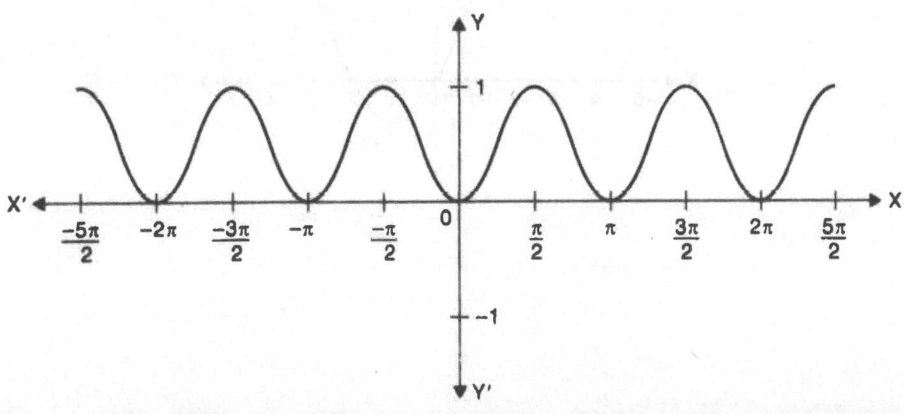

EXERCISE FOR PRACTICE

Sketch the graph of the following functions :

1. $y = x^3 + 1$
2. $y = x^2 - 2x$
3. $y = \sqrt{9 - x^2}$
4. $y = x^4 - 1$
5. $y = 2 \cos x$
6. $y = \tan x$

Answers

1.

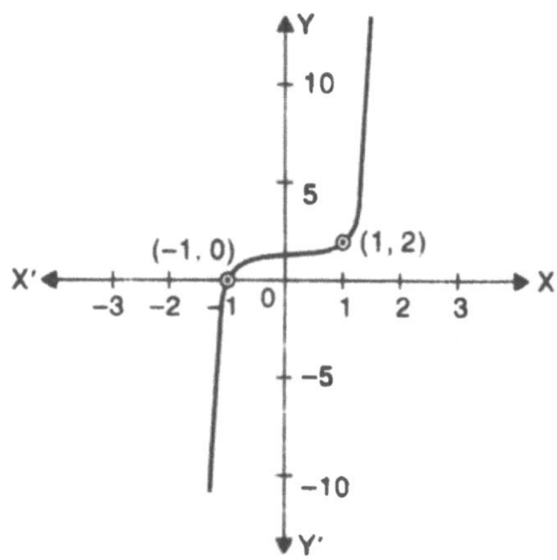

2.

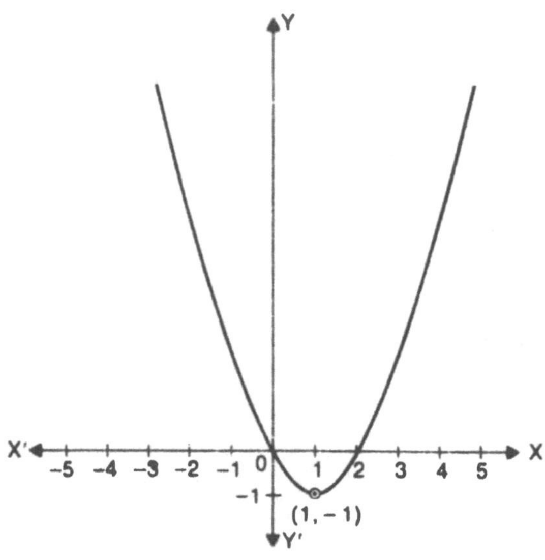

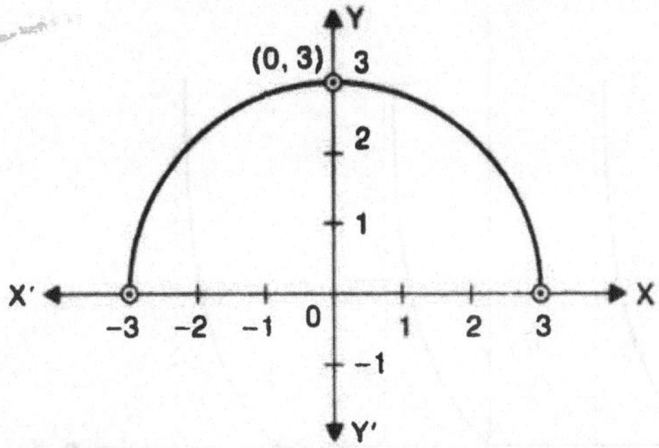
Y
(0, 3)
3
2
1
X′
X
−3
−2
−1
0
1
2
3
−1
Y′

4.

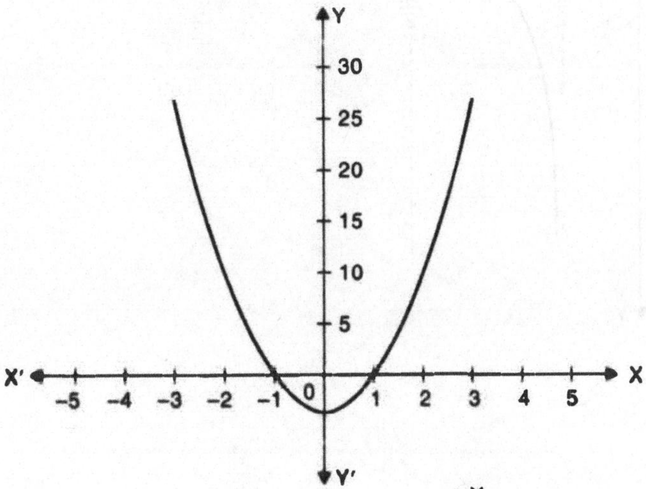
Y
30
25
20
15
10
5
X′
X
−5
−4
−3
−2
−1
0
1
2
3
4
5
Y′

5.

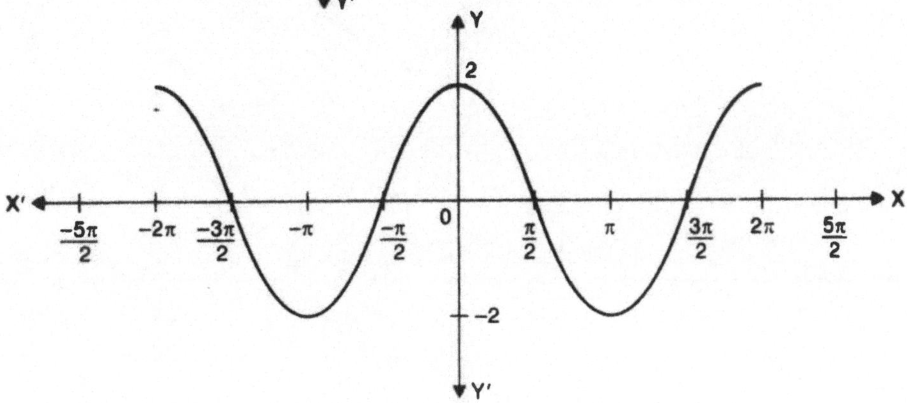
Y
2
X′
X
0
−5π/2
−2π
−3π/2
−π
−π/2
π/2
π
3π/2
2π
5π/2
−2
Y′

6.

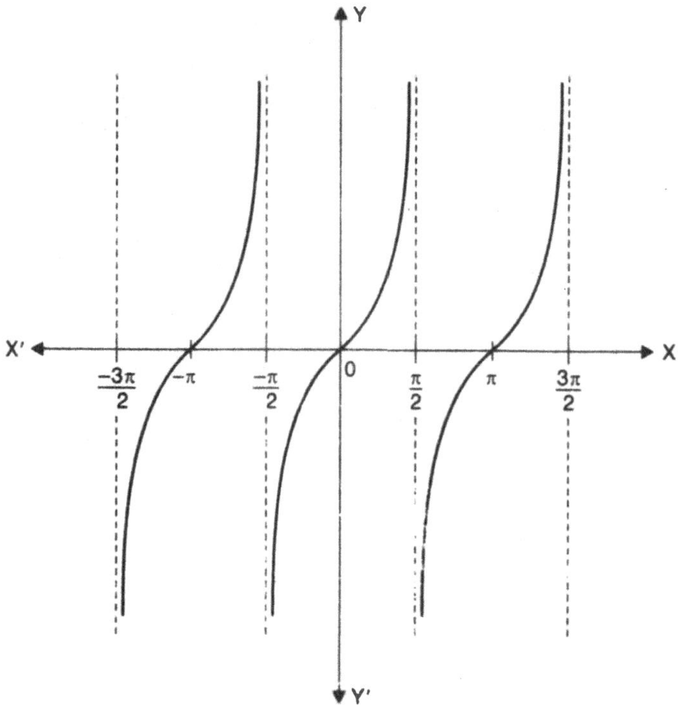